THE CLEVER HANS PHENOMENON:
COMMUNICATION WITH HORSES, WHALES, APES, AND PEOPLE

Left to right: Wilhelm von Osten, Clever Hans, and Oskar Pfungst. (From KRALL, K. 1912. Denkende Tiere. 2nd edit. Friedrich Engelmann, Leipzig. Courtesy of the General Research Division, The New York Public Library, Astor, Lenox, and Tilden Foundations.)

ANNALS OF THE NEW YORK ACADEMY OF SCIENCES

Volume 364

THE CLEVER HANS PHENOMENON: COMMUNICATION WITH HORSES, WHALES, APES, AND PEOPLE

Edited by Thomas A. Sebeok and Robert Rosenthal

The New York Academy of Sciences
New York, New York
1981

Library of Congress Cataloging in Publication Data
Main entry under title:
The Clever Hans phenomenon.
    (Annals of the New York Academy of Sciences; v. 364)
    Papers resulting from a conference entitled Conference on the Clever Hans Phenomenon: Communication with Horses, Whales, Apes, and People held by the New York Academy of Sciences, May 6-7, 1980.
    Includes index.
    1. Human-animal communication—Congresses. 2. Human-animal communica-tion—Research—Congresses. 3. Interpersonal communication—Congresses. 4. In-terpersonal communication—Research—Congresses. 5. Errors, Scientific—Con-gresses. I. Sebeok, Thomas Albert, 1920-   . II. Rosenthal, Robert, 1933-   . III. Con-ference on the Clever Hans Phenomenon: Communication with Horses, Whales, Apes, and People (1980: New York, N.Y.) Q11.N5 vol. 364 [QL776]   500s   81-2806   [001.4'01'9]
AACR2

CCP
Printed in the United States of America
ISBN 0-89766-113-3 (cloth)
ISBN 0-89766-114-1 (paper)

ANNALS OF THE NEW YORK ACADEMY OF SCIENCES

VOLUME 364

June 12, 1981

THE CLEVER HANS PHENOMENON: COMMUNICATION WITH HORSES, WHALES, APES, AND PEOPLE*

Editors and Conference Chairmen
THOMAS A. SEBEOK AND ROBERT ROSENTHAL

———————

CONTENTS

* This series of papers is the result of a conference entitled *Conference on the Clever Hans Phenomenon: Communication with Horses, Whales, Apes, and People*, held by The New York Academy of Sciences on May 6–7, 1980.

Financial assistance was received from:

- THE HARRY FRANK GUGGENHEIM FOUNDATION
- THE WENNER-GREN FOUNDATION FOR ANTHROPOLOGICAL RESEARCH

Introduction

EVER SINCE THE BYZANTINE EMPIRE was ruled by Justinian (A.D. 483–565) there have been reports of clever animals. But no animal intelligence so captured the imagination of laymen and scholar alike as that attributed to Clever Hans, the horse of Mr. von Osten. Hans gave every evidence of being able to add and subtract, multiply and divide, read and spell; and he could solve problems of musical harmony. Hans communicated with his questioners by converting all answers into a number and tapping out that number with his foot.

Clever Hans constitutes a famous "case" in the history of psychology and it has offered many lessons to those who have examined its details. But as its lessons are many, so are the views and perspectives of those who have studied these lessons and written about them. As we shall see in the volume that lies ahead, the major lesson of Clever Hans is something of an inkblot.

For some, the major lesson of Clever Hans is the subtlety of processes of communication, witting and unwitting, between organisms of the same or different species. For others, the major lesson is how easy it is for us to be taken in by deceit we visit upon ourselves, or by deceit others may wish to visit upon us. For still others, the major lesson is that people and other organisms may be susceptible to the operation of self-fulfilling prophecies. For yet others, the major lesson is the set of methodological precautions that must be taken to guard against our obtaining data in our research that fits too well with the data we want or expect to obtain—fits too well not because we have been so adroit in divining Nature's ways but because our wishes and our expectations have led us to affect our research subjects to respond in accordance with our expectations, or because we interpret their behavior as consistent with our expectation when in fact it is not.

The conference at which these papers were presented was exciting. Deep scholarship was leavened with strong affect as first-rate scientists tried to convince one another that their view of the lessons of Clever Hans was the right view, and that other views were wrong. But aren't scientists above that sort of thing? Certainly not, and it's a good thing said William James:*

* JAMES, W. 1948. Essays in Pragmatism. p. 102. Hasner, New York, N.Y.

... science would be far less advanced than she is if the passionate desires of individuals to get their own faiths confirmed had been kept out of the game. ... If you want an absolute duffer in an investigation, you must, after all, take the man who has no interest whatever in its results: he is the warranted incapable, the positive fool. The most useful investigator, because the most sensitive observer, is always he whose eager interest in one side of the question is balanced by an equally keen nervousness lest he become deceived.

THOMAS A. SEBEOK   ROBERT ROSENTHAL
National Humanities Center   *Harvard University*
and
Indiana University

The Clever Hans Phenomenon from an Animal Psychologist's Point of View

HEINI K.P. HEDIGER*
Department of Zoology
University of Zürich
Zürich, Switzerland

IT IS CERTAINLY a unique and highly unusual event for The New York Academy of Sciences to dedicate a conference to a horse named Clever Hans, which lived in Germany, in Berlin, about 75 years ago.

If Clever Hans were still alive, according to his admirers' conviction, he would undoubtedly feel very honored, for they believed firmly "that an animal can think in human way and can express human ideas in human language." This is literally the declaration made in 1914 by one of the authoritative experts, the psychiatrist Gustav Wolff from Basel.[1] The amazing fact is that today again there are numerous people who, based on other observations, would sign this same declaration.

Not only did uninformed people have this opinion on Clever Hans, but also prominent scholars, scientists, psychologists, psychiatrists, medical doctors, and many others.[2,4]

What had happened? In the year 1904 it was reported that in Berlin the retired schoolteacher Wilhelm von Osten had succeeded in producing evidence that animals—for the time being, horses—could think, talk, and calculate if instructed by the right method. The method invented by Herr von Osten consisted of giving a number to each letter of the alphabet. This association between letter and number the horse had to learn by means of a blackboard.

By tapping the right number with a front hoof on a board mounted in front of him the horse could combine letters into words, words into sentences and so express his thoughts. Through this tapping method the front leg of the horse became a kind of a speaking organ. For each correct answer the horse was awarded with a delicacy. Thomas A. Sebeok, on whose initiative this conference was organized, has described this method and its history very extensively;[4] so I do not need to enter into this in detail.

Clever Hans, who at the time drew worldwide attention, was the first

---

* Current address: Gfennstrasse 29, CH-8603 Schwerzenbach, Switzerland.

0077-8923/81/0364-0001 $01.75/0 © 1981, NYAS

and most famous of the thinking animals. Soon he was to be followed by others, not only horses, but also dogs and cats. By 1937 there were already over 70 of these so-called thinking animals.[5] Since then their number has continued to rise, and allegedly dialogues have been set up with many more representatives of the animal world, in America with dolphins and apes especially.

I would like to confine myself to try to present the Clever Hans phenomenon in a larger sense from the standpoint of an animal psychologist. Above all there are three basic biological phenomena which in this connection I feel are important to consider. They are:

(1) The age-old burning desire of mankind to take up language contact with animals, that is, to understand the language of animals.

(2) The assimilation tendency, that is, the deep-rooted tendency present in all higher living beings — man included — to see in creatures of a different species, with whom there exists a certain familiarity, creatures of their own kind and to treat them accordingly.

(3) The catalytic effect from human being to animal, which can often be extremely strong.

The history of man–animal relationships has been almost completely neglected by modern behavioral research, but I think it might be useful to consider some experiences of practical men who live in direct contact with a variety of animals not only in scientific laboratories but also in zoos and circuses.

If we want to understand higher animals we remain widely dependent on their specific ways of communication and on the interpretation of their expression. To my knowledge up to this day an actual conversation on the basis of a real language has not yet been established.

To begin with, it is important that we recognize objectively the before-mentioned general, intensive, and ancient desire, for it represents a very essential source of mistakes not only in the matter of animal language, but also in animal experiments generally. In behavior research the well-known proverb, "He who searches will find," plays an important and often crucial role.

I see a certain relationship between the placebo effect in Medicine and what Robert Rosenthal[6] in such an impressive way described as the Experimenter's Expectancy in behavior studies. Just as a patient feels a strong effect from nothing, an experimenter may read a positive result or induce such in the behavior of an animal. Experiments certainly do not always take an objective course in the sense that one stimulus of a known quality and quantity releases a corresponding objectively measurable reaction.

In other words, it depends on what sender and receiver make of a stimulus or a sign; sometimes as much is made of it as the satisfaction of the wishes requires. In this respect, greatest significance should be given to semiotics, especially zoosemiotics, for it endeavors to examine all possible channels of communication as they have been enumerated by Sebeok in 1976.[7]

Communication, as it preoccupies us here in connection with the Clever Hans phenomenon, is not only a series of flow-processes between a chemical substance and a subject; it is a much more complex connection between the experimenter and the animal.

In many animal experiments, whether psychological or physiological, the effect of the design of the apparatus, of the laboratory, the atmosphere, in short, the whole envelope in which the signs (stimuli) work, is often neglected. One believes in dealing with only one relationship, one channel of communication between the animal and the experimenter, that is to say, the official one. In reality there are complex bundles of relationships, of channels. An animal experiment does not begin only when the animal starts in the training apparatus. In a certain sense it starts by planning, with the invention of the design of the experiment. It will be cut out for the animal in a specifically human way of thinking, based on the posing of the question; and, in addition, the animals are selected in this sense.

As often with the study of man–animal relationships, the conditions of the circus can be helpful since we have to do with big animals, with wide swinging movements, and acoustic effects, which means with conspicuous signs in a typical surrounding. Despite their importance, the micro-signs manifesting themselves there too will be ignored for the time.

When a group of lions on pedestals are to "sit up," the trainer in front of them, in a bent or kneeling position, will give the command and get up with corresponding stimulating calls and, with stretched out arms, will raise stick and whip way up.

If the same animal tamer with the same movements and calls puts himself in front of a group of free-living lions, which, for example, he might encounter in a national park in East Africa, his behavior will induce a general flight reaction. Why? The drastic difference is illuminating: The circus lions are trained and understand the signs given by their master. Free-living lions, however, have no such intimate relationship with man. Like practically all wild animals in nature, they see in him just an enemy, their main enemy from whom they have to run away.

On one and the same stimulus, on identical signs, the two lion groups react in a basically different way according to their individual experience, depending on what man has invested or not invested in them. A free-living untamed and untrained animal reacts completely differently from one well-adapted to the experimental situation, regardless of whether the animal is lion, ape, horse, parrot, rat, or even planaria. This is due not only to the training invested, but also even more to the increased comprehension possibilities and better developed communication channels reached during the close contact with man.

Clever Hans represents an excellent example of these facts: A wild horse or even an ordinary domestic horse does not pay attention to human movements that barely consist of one millimeter, as Oskar Pfungst has shown.[8] It is only on the basis of extraordinary familiarity between Clever Hans and his master, gained during the course of teaching, that the horse became able to interpret as decisive signs movements of the head of his master of even one-fifth of a millimeter deflection.

I would like to emphasize again the importance of familiarity between man and animal. It may fluctuate between 0% and 100%. The free-living animal knows only the flight reaction to an approaching man. The other extreme is absolute tameness, the lack of any flight-tendency that results from a generation-long domestication or what has lately been labeled "laboratorization," or from individual taming and training.

We humans are still much biased by the idea that all animals of a given species are equal, in a way, stereotype figures right out of a textbook of Zoology, fixed genetically to the last detail. In my opinion this is a naive and simplified concept to which we should no longer adhere. Through the catalytic effect of man, surprising changes and an increase in behavior may be caused which surpass by far the genetically given ethogram of a species.

I would like to remind us of the famous wolf "Poldi," which as a young animal was trapped in the wilds and trained by an Austrian police officer, Rudolf Knapp, to become a police dog.[9]

The Swiss National Circus shows a white rhino (*Ceratotherium simum*) that was caught in the South African bush as a young animal and which after a few months showed its astonishing acts in front of thousands of people and this in the glare of spotlights and to the sound of blaring music. It also may be freely guided through dense city traffic.

In its original African homeland this rhino, "Ceyla," knew only one reaction towards man and this was the flight reaction. Other relations to man were nonexistent, but these relations changed to the maximum degree, that is, to perfect tameness and trainedness. Today man is its intimate friend, its socially superior companion to whose tiny signs the rhino reacts with great attention and reliability. In a certain sense Ceyla is no longer simply a zoological rhino. Enormous human influence has been invested in it.

The famous chimpanzee, "Washoe," was also born in the African jungle and within a short time reached the highest level of animal–man relationship, a level at which it is believed that even a certain conversation becomes possible. Through the catalytic effects of man extraordinary latencies have been activated in Washoe similar as in "Sarah," "Lana," "Koko," and others. All these surprising activities would never have manifested themselves in the wild.

By visualizing how tremendous the difference of habitat is in which Washoe was born in Africa and of the one in which she later received her language training in America, it will become obvious how very much the animal, its surroundings, the stimuli to which it is exposed have been changed. The whole communication system has been altered. The answers we receive from such an animal must be quite different from those of a free-living chimpanzee.

The biologist and biophilosopher Bernard Rensch in 1973 states:

> While the imporant experiments of the Gardeners and the Premacks show that Chimpanzees to a certain degree may replace words by the use of ASL or plastic figures as symbols these are in fact results which could only be reached through man capable of speaking. In nature nothing similar of human language has ever evolved in Chimpanzees, Gorillas and Orang-Utans.[10]

In other words, with all animals with which we try to enter into conversation we do not deal with primary animals but with anthropogenous animals, so-to-speak with artifacts, and we do not know how much of their behavior may still be labeled as animal behavior and how much, through the catalytic effect of man, has been manipulated into the animal. This is just what we would like to know. Within this lie the alpha and omega of practically all such animal experiments since Clever Hans.

I have the impression that we biologists find ourselves in a situation corresponding to the atomic physicists from whom we have to learn so much lately. It has been discovered that as soon as single atoms are observed their reactions are influenced. Of course it is a long way from the atom to a living being, a horse or a chimpanzee. However, if the simplest thing, the atom, is influenced in its behavior through the human observer, how much more must we suppose then that two living subjects influence each other through observation.

Maybe it is a general law that through observation the observed will be altered. However, we do not want to indulge in speculations, but it may be certain that every animal experiment — not only the ones concerning language — represents a relation between two living subjects and not simply a relation between a subject and an object whose behavior could be registered objectively. This means that in an animal experiment we have to work not with pure unaltered animal behavior, but always with the behavior of the animal plus the influence of the human observer.

Now, does that mean that we should resign because it is not possible to find out pure animal behavior anyway? Certainly not. We are given the urgent duty to understand the nature of the manifold signs and their mutual alteration flowing between the sender and the receiver. We can do this best by using the methods of zoosemiotics established by Sebeok in 1963 (Reference 7, page 57). To this complex belong, for example, the problems of intensive wishing, the mechanism of looking, and so forth.

The closer—in the spatial and psychological sense—the experimenter and the animal find themselves the greater is the danger of mutual influence. Oskar Pfungst (Reference 8, page 49) has found a close relationship between the performances of Clever Hans and the distance of his master. Therefore, Otto Koehler, in 1937 already the senior of most prominent of European animal psychologists has demanded with emphasis:

> During the experiment every contact between experimenter and animal has to be strictly avoided. The animal has to decide absolutely free. Experimenters who are not willing or who are not able to fullfill these conditions have to be eliminated harshly. [Reference 5, page 24.]

In his classical investigations on the counting ability of different birds —which reaches the number seven—Koehler himself had set a good example. During the critical experiments the animals were observed from an insulated side-room through the view-finder of a hidden camera. During the experiments with his famous raven, "Jakob", Koehler disappeared from the grounds altogether and read a book or held his lecture in order to exclude even a telepathic influence.[11]

Further—repulsed by the nonsense with Clever Hans—Koehler demanded that the experimenter should not wish a definite result. Literally he explained: "Who believes in advance what in reality he could only read out of experimental results, he who does not criticize himself and can not take criticism of others does not do research but he deceives himself". (Reference 5, page 21). This was exactly the case with Clever Hans.

I do not want to enter into the other precautions demanded by Koehler. But I doubt whether the active scientists who do research on animal language do adhere to the two above-mentioned basic demands of Koehler.

I would like to quote here the Oxford zoologist, Sir Alister Hardy: "Again and again faulty conclusions have been reached because the worker has been unwittingly influenced by an ardent desire for particular results."[12]

It is not easy to explain such an effect. However, with regard to big animals an approach may be found. To this I would like to mention a kind of reverse situation of the zoo where the lack of an intense desire led to failure. By chance I became a witness of an interesting dialogue between an experienced elephant-keeper and a beginner who complained very seriously that the elephants would not obey his orders although he gave them exactly as he was instructed. The old keeper's typical explanation was:

> You should not simply stand there, give orders and move accordingly, but you must participate with your whole inner strength. With all your energy you must want what the elephants should do. You must have the strong will that they do what you order them to do.

For every circus animal tamer this is a clear fact.

From this simple zoo or circus experience I believe we can draw two conclusions: First, through the strong inner effort the human expression and accompanying movements, which have been presented and analyzed so brilliantly by Paul Ekman,[13] Erving Goffman,[14] and others, are enormously intensified. Second, we have to take into consideration that many animals—elephants also—are excellent observers, often extremely skillful interpreters of the human expression.

To this observation with elephants I would like to add a short remark on a series of preliminary experiments which I made many years ago at the Zoo in Basel. It deals with the chimpanzee, "Max," and his keeper Carl Stemmler who at the time were very popular personalities in Basel. Stemmler had taught his adolescent chimpanzee a few simple tricks, so-called obedience exercises, which mainly served to establish and to secure for the future as long as possible the social superiority of man. One goal of these rather primitive experiments was to test whether any evidence of verbal understanding could be established by giving Max the orders through a loudspeaker while his master was invisible.

The chimp showed that he understood the orders but he would not execute them when his master was not to be seen. When he was present, mute, and made the normal accompanying movements while the orders came from the loudspeaker, Max would obey as if Mr. Stemmler would give the orders personally.

Certainly we do not want to attach an exaggerated significance to these old and crude pilot experiments. However, this much we may conclude from them: They demonstrate the overall significance of the gesture, the expression, that is, of the nonverbal communication during close contact between animal and man. They also show that it is a "multichannel" communication, that each channel has its own specific properties, and that their interaction creates possibiities, the number of which is greater than the sum of the possibilities of each channel taken separately as Sebeok puts it (Reference 7, page 78) in connection with the work of Bar-Hillel. In view of the Clever Hans phenomenon, this evidence may not be emphasized enough.

Last year in a Swiss circus I talked with a Polish animal tamer, Mrs. Krystina Terlikowska, who worked with four female tigers and three female lions. Mrs. Terlikowska, an analphabetist who has practiced this profession since she was eight years old, says that the most important part of her present activity was to talk with the animals. This indeed is a very old and proven recipe for all who have to work with big animals.

By speaking, our expression is activated and enriched. It is of course not the specific language nor the single words that matter but the facial expression, the intonation, the sound intensity, the posture, the movements, and maybe the color of our face, the thermal pattern of our skin, even the smell which changes from moment to moment, and many other factors that we do not know yet, channels not yet investigated.

As is well known, we humans lack, for example, the ability to perceive ultrasonic sound and ultraviolet light. We lack the sense of magnetism, a very fine sense of vibration and temperature, a sense for polarized light, and so on, and so forth. Concerning the sensory organization, we humans are very often strongly underprivileged as compared with many animals, which also enjoy a much shorter reaction time. We have nothing by far to compare with the whiskers and the ear-pouches of cats and so on.

In order to emphasize the significance of our nonverbal communication with animals I would like to mention an observation which in 1979 was made in a well-known private zoo in the Netherlands, in Burger's Animal Park in Arnheim. Dr. Frans de Waal has given a detailed report,[16] of which the author has confirmed every word personally.

According to this observation it was possible to teach a chimpanzee mother, "Kuif," who never was able to raise her own young, to give a milk bottle to a foreign young chimpanzee called "Roosje" and to raise it successfully. The whole instruction of the mother was given through the bars of her cage by showing her the necessary actions and by continuous repetition of six words or orders like: raise the bottle, give it to Roosje, hold the bottle tight, hold it a bit higher, stop (take the bottle out of the mouth), and return the bottle. After only three weeks of instruction through the bars of the cage one dared to trust the stepmother with the baby who, up to now, had been raised artifically by human caretakers. The raising by the chimp succeeded perfectly. Later on, the stepmother was able to provide her adopted child partly with her own milk.

This is the first time that such an astonishing experience with bottle feeding through an animal was realized. It shows how effective com-

munication between man and animal can be. This marvelous performance of the stepmother was reached within the short period of 3
weeks and without an artificial language — only through the demonstration and through the spontaneous manifestations of her attendants.

This amazing act of training causes one to ponder the manifold efforts of several researchers to enter into language contact, into a
dialogue with apes. The chimpanzee Kuif did not answer with any language signs but by correctly tending the bottle to Roosje. With this
doubtlessly she confirms that she understood the human instructions in
every detail.

In principle, I recognize here performances similar to those in some
language experiments. In each case the chimpanzees were demonstrated the desired actions with the hope that they would react in a certain way. With Kuif it was the correct handling of the bottle; with
Washoe, Sarah, Lana, and so forth, it is the production of certain signs
in which we would like to see a language. But how can we prove that
such answers are to be understood as elements of a language, and that
they are not only reactions to certain orders and expression, in other
words simply performances of training?

When an outside observer follows and tries to understand the signalling of a chimpanzee trained in American Sign Language (ASL), he has
great difficulty in distinguishing preparative from conclusive movements, especially since these movements succeed each other very rapidly. The noninitiated person has a hard time deciding where the preparing movement stops and where the real signal starts and ends, for
each signal is embedded seamlessly into another movement (e.g.,
flower, bug, dirty).

I do not doubt that Washoe and other chimps have learned a number
of signs in the sense of ASL. But it seems to me that a better clarification could be reached mainly through the introduction of the orders "repeat" and "hold it." By this the chimpanzee could show that he really
understands the single elements and does not execute fast, sweeping
movements into which one possibly could read such elements.

With trained animals and animals in the process of being trained —
which is quite a different matter — there exists a general tendency not to
react in single, self-contained elements but in sequences. In the circus
many animal tamers have gotten into embarrassment or even danger of
life through this. I am referring to the phenomenon of anticipation.
This fact is also known in connection with dog training. Within a tiger
group, for example, it may happen that a certain tiger starts the next
trick (e.g., jumping through a hoop) before the animal tamer has been
able to put into safety the other tigers, that is, to bring them back to
their seats. Because of this, dangerous contacts and aggressive conflicts

may occur in the manege, which not infrequently lead to catastrophes. With dogs this undesired anticipation of orders is mostly more harmless. It is caused by an extremely fine interpretation of the expression of the dog's master. The dog then executes an order before it has been given officially to him. Are we sure that the possibility of anticipation can be absolutely excluded in language experiments with trained apes? In this situation it may easily make the impression of a spontaneous utterance. There are so many and different pitfalls when you try to interpret animal behavior.

Thomas A. Sebeok[15] recently has reminded us of the memorable story of the famous Russian physiologist Ivan Pavlov and his assistant Studentsov. By posing the same problem to five consecutive generations of mice, the two believed they had proved correct the Lamarckian theory of inheritance of learned behavior. As they expected, each generation needed fewer sessions for the mice to master their task. The original number decreased from 300 to 100, 30, 10, and 5.

It was discovered later, however, that the improvement of performance was apparent only and was by no means based on inheritance. What had improved was the treatment of the mice by the experimenter. This is by no means the only experiment by which Lamarck's theory was to be evidenced.

Let me briefly mention the well-known experiment of William McDougall who, similar to Pavlov, also tried to prove the inheritance of effects of training. This time no less than 23 generations of rats were used. Interestingly enough, McDougall was absolutely aware of his possible wishful thinking as Sir Alister Hardy[12] points out by literally quoting McDougall's apprehension in the following way:

> In this connection it is necessary to avow that, during the course of the experiment, there grew up in all of us a keen interest in, I think I must in fairness say, a strong desire for, positive results. From the first it was obvious that a positive result would be more striking, would excite more interest in the biological world, than a negative one. And when indications of a positive result began to appear, it was but human nature to desire that this result should appear as clearly cut and positive as possible. Further, on my own part, there was a feeling that a clear-cut positive result would go far to render tenable a theory of organic evolution, while a negative result would leave us in the Cimmerian darkness in which Neo-Darwinism finds itself.
>
> I was conscious, therefore, of a strong bias in favor of a positive result, and throughout I was consciously struggling against the temptation to condone or pass over any detail of procedure that might unduly favour a positive result. Such details are encountered at every point, more especially in the breeding of the animals. To have disguised from onself this bias, to have pretended that we were superior to such human weakness, would have been dangerous in the extreme. The only safeguards against its influence were the frank

avowal of it and unremitting watchfulness against it. I can conceive of no task that could make greater demands upon the scientific honesty of the worker, and it is in part this demand for unremitting watchfulness that renders the work peculiarly exhausting. I can only say that I believe we have succeeded in standing upright, and in fact, for myself, I am disposed to believe that I have leaned over backwards, as we say in America. Whether we have really succeeded in this, the most difficult part of our task, can only be proved when other workers shall have undertaken similar experiments. If our results are not valid the flaw, which has escaped our penetration hitherto, must, I think, be due to some subtle influence of this bias.[12]

In fact: is there a more conscientious and a more sincere attitude of an experimenter? It is almost identical with the severe commands of Otto Koehler. And despite all of that McDougall stumbled. Nobody could repeat his important experiment with the same results. Very grave errors had occurred as David Katz demonstrated in a very interesting chapter with the title "The Clever Hans Error and Similar Mistakes of Modern Animal Psychology."[17]

The essence of McDougall's experiment consisted — to be brief — in setting the rats at the starting point of the apparatus in a water channel. From there they had the possibility to swim towards the left or the right into a dim or glaringly illuminated channel in order to reach dry grounds. If they chose the glaringly illuminated channel they received a strong electric shock. If they chose the dimly lit channel they could reach dry grounds without punishment.[18]

The rats' performances improved — with some irregularities — fairly continuously. The total time of the rats spent in the water dropped from 2320 seconds to less than half, that is, 1020 seconds in some of the last generations. Astonishingly, the very last generation was not the best. Its total time in the water had risen again to 1620 seconds. This "anomalous" record for the last generation is according to McDougall, as quoted by Munn,[18] probably due to the death of an assistant who had been handling the rats. So there was a change of experimenter.

Of course in such an experiment it depends greatly upon how the rats have been handled and put into the water. The pretreatment and the starting phase may be decisive. Today we find it normal that the experimenter who puts the animals in the water at least does not know if he is working with trained rats of the generation to be tested or with naive control animals. However, David Katz[17] to his great astonishment discovered that the experimenter always knew about the origin of the rats. This is just as big a methodological mistake as is direct contact with the experimental animal. But there were even more and quite different sources of mistake involved, as Munn[18] points out, e.g., in the selection of the animals.

Today by means of numerous experiments it has been proved by

Rosenthal,[6] Timaeus,[19] and others that the way and means how an animal is brought into the apparatus may greatly influence the result of the experiment, even the animal's past, the intrauterine as well as the postnatal one. Morton in his fascinating paper[20] on handling and gentling laboratory animals has, for example, found that subjects shocked in infancy took longer to respond in an avoidance learning situation.

In other words: What also matters in evaluating the performances of experimental behavior is the animal's past, even in the case of rats. The human influence on the behavior of experimental animals may even start before its birth.

More and more I come to the conviction that the results of animal experiments do not so much depend on the exact sequence and make-up of the experiment itself, but to a great deal also on the past of the individual animal and on the personal attitude of the experimenter. These two important facts are very often what conventionally is neglected most.

Clever Hans is only one example of this but one that should cause us to rethink the whole experimental situation.

With this I come to Factor 2 of my introduction, to the assimilation tendency, that is, the anthropomorphizing of the animal through man and the zoomorphizing of man through the animal. This is not at all a theoretical speculation but a proved practical experience gained through a huge pile of facts.[2,3,21,22,23]

If, for example, we ask the gorilla Koko: "Are you an animal or a person?" as it happened according to Francine Patterson's report,[24] Koko's answer could never be authentic if it is, "Fine animal gorilla." Koko has studied neither zoology nor anthropology. She cannot distinguish between man and animal as two different categories of the zoological system. As with each animal she lacks the notion of species. Therefore she could not know that she belongs to the species *Gorilla gorilla* and the human beings surrounding her to *Homo sapiens*.

According to all we know today about big animals living in close contact with man we have to assume that Koko, based on her assimilation tendency, sees in her mistress (for the time being) a superior specimen of her own species, that is another gorilla. The same stands for the attendants of chimpanzees who at present apply themselves with language studies. There is no reason to believe that, in this regard, they behave differently than apes in a zoo where they are in close contact with their keepers.

As long as we deal with very young animals everybody probably agrees that, to the bottle baby, the caring human being is nothing else but its mother, a mother of the same species. The assimilation tendency to one's own species here is apparent and uncontested.

At the zoo the caretaker of adolescent apes as a rule goes through a significant change of meaning (Bedeutungswandel in the sense of Uex-küll) leading to a dangerous situation: he becomes to be a social rival. This may lead to dangerous conflicts and to direct attacks on the keeper. Many — probably most — of the zoo accidents have this motivation not only with apes but also with deer, big cats, antelopes, elephants, and so forth.

In scientific research institutions, the decades-long experiences of zoological gardens should not be ignored completely. Not all zoos are mere showbusinesses. There have always existed a number of zoos — and the Bronx Zoo is one of the leading — in which scientific research has been done, in spite of the often chronic lack of money and personnel, which in all departments necessitates very strict economical operation.

This gives rise to the following purely practical consideration. If apes really dispose of the great intelligence and the highly developed communication ability that one has attributed to them lately — why in no case in the zoos of the world, where thousands of apes live and reproduce, has it been possible to get one to clean his own cage and to prepare his own food?

As far as I know in all zoological gardens of the world as well as in all Primate stations people still have to be employed to do these simple jobs of daily life. Paradoxically in zoos as well as in laboratories *Homo sapiens* still is the servant of the ape.

One may argue that apes have no interest in such work, that they have no perseverance and that they simply do not want to. By all means I do agree with these arguments for I have always emphasized that one may never conclude incapability when an animal does not obey an order. There is always the possibility that it has understood the order but does not want to carry it out. Apes have no notion of work. We might perhaps teach an ape a sign for work but he will never grasp the human conception of work.

The world-wide evidence of unwillingness of apes to carry out the simplest household activities, that obvious disinterest, almost necessarily leads to the question as to what interests apes at all. With this we are confronted with the problem as to how a research program of future efforts should be designed in order to possibly enter into a dialogue with animals, to lead a conversation with them.

So we try to arrive at a program that makes sense. On what subject may we best converse with an animal? Unfortunately it is not much: Apes, dolphins, or horses have no interest at all in things that are of general interest to humans. By this I mean all that is written in books and the media:

> Culture in the broadest sense
> Politics, Sports
> Business, Finance, Traffic, Technique
> Work, Research, and so forth.

What else then remains? Very little it seems to me. Especially when we think that the animal has no access to the future. It lives entirely in the present time.[25] Also the past is mostly out of its reach, except for very recent experiences. Therefore there remain the essential daily needs, above all metabolism, food and drink, social and sexual contact, rest and activity, play and comfort, conditions of environment in connection with the sensations of pleasure and dislike, some objects, and possibly a few more things. This is indeed rather modest.

Experimenters who try to enter into a conversation with animals should keep this in mind. Clever Hans has taught us clearly what happens when we force this simple repertoire with an exaggerated program. It again would lead us to believe that "the animal can think in a human way and express human thoughts in human language," as I have quoted in the beginning.

The work of Oskar Pfungst[8] on Clever Hans, going back to the year 1907, has not really given us a satisfactory solution of involuntary signaling. According to my opinion this problem is far from being solved. Concerning the work of Pfungst I would like to say the following:

(1) It has remained widely unknown.
(2) It has been almost forgotten.
(3) It is not complete.
(4) It gives rise to important criticism.
(5) It has never been repeated and confirmed.

To the Clever Hans critics, Oskar Pfungst's work was so welcome and appeared to be so perfect that they believed it to be 100% right and that they could accept it forever as such. Here again we have to deal with the basic phenomenon, with the idea of wishful thinking, with the experimenter's expectancy, which is central to every animal experiment and which is so difficult to eliminate.

Pfungst also was an experimenter. He too was obsessed by the wish to reach positive results, to prove that a horse or any other animal could not think in a human way and not express human ideas in human language.

He apparently thought he had found evidence that Clever Hans was doing nothing more than watching that famous relaxation jerk of his master when the correct number of knocks was reached, that involuntary almost microscopic head movement, which could measure 1/5 mm or even less (Reference 8, page 120). I do not doubt this result but I would be much more convinced had it been confirmed at least once.

The work of Pfungst does in fact contain all kinds of curiosities. The greatest one for me is what he mentions in connection with the so-called laboratory experiments. These are the basis of most of his work and have been carried out in the Psychological Institute of the University of Berlin. By contrast the other experiments and observations took place in the open, in the courtyard where Clever Hans normally would perform, or in a tent which occasionally was set up in that courtyard, or in the stable.

In laboratory experiments one generally expects greater precision and more exact experimental conditions than in a stable or courtyard. In reality, during the so-called laboratory experiments a horse was not even present. The role of the horse to be tested was taken over by Oskar Pfungst himself (Reference 8, page 77). You will understand now why I permit myself to speak of curiosities of the Pfungst expertise. This, however, is not the only one.

On page 123 of his work Pfungst clearly mentions:

> With the standing of our knowledge of today all attempted explanations have a more or less hypothetical character (Bedeutung). Should as a result of further research the herewith attempted explanation prove to be untenable one would have to take under consideration abilities yet unknown in the horse's eye or search for the reason in the brain. Experiments with other specimens of this genus will have to teach us if all other horses have the same ability or if some individuals are privileged.

This shows that Pfungst felt insecure and that he demanded a repetition of his investigations. However, up to now — more than 70 years later – this has never happened. One did not wish to recognize as questionable or even wrong the apparently so clear and convincing explanation given by Pfungst, which he himself considered to be hypothetical. Again and again desire proves to be the most powerful factor in an animal experiment. This fact seems to be unchanged even today.

It is not possible now to consider critically every point of Pfungst's work. To me personally the following statement matters most of all. On page 125, Pfungst confirms that Clever Hans, compared with all his critics, was a superior observer.

Basically this fact puts the human observer into a very difficult position. As I have mentioned before, we have to resign ourselves to the fact that many animals are far superior to man in terms of their sensory organization. This is true not only for horses but also for primates, rats, and even many invertebrates, maybe even for planaria and protozoa like paramecium.

This is of primary importance for assumedly physiological experiments and also, for example, for the experiments started by James V. McConnell[26] on the transfer of learned information. From this general situation I can only draw one logical and biological conclusion: training is always part of all the aforementioned experiments. Training is a most intense relation between animal and man, between animal and experimenter, and this within an artificial surrounding which has been imposed upon the animal.

In an experiment we mostly observe results, which we may regard as proof of a hypothesis. However, we still know much too little about how these results have been arrived at, about what has taken place between man and animal during preparations, and during the critical experiments, in other words, all the different communication channels.

Since Clever Hans—more than 70 years ago—I fear we have not made much progress in this regard. Otto Koehler has shown us in an exemplary way how we may practically exclude man as a source of mistake, but he has never demonstrated precisely what we want to eliminate thereby, what human influences.

It is my firm conviction that there is only one way to clear up this complex of problems. We have to grasp the phenomenon at its roots; that is, we have to repeat the Clever Hans experiment from the very beginning under exact laboratory conditions with the tools of zoosemiotics.

Since here we are primarily interested in direct communication with the animal, in animal language, I would like to conclude by quoting the British biologist J.B.S. Haldane.[27] In Paris, 1954, during a lecture at the Sorbonne University he made the following statement:*

> When a child says to his mother, "I am hungry" or "I want to sleep," he is still animal. When he says, "This I have done this morning," he begins to be man.

To my knowledge, up to now, no animal, not even an ape, has ever been able to talk about a past or a future event.

REFERENCES

1. WOLFF, G. 1914. Die denkenden Tiere von Elberfeld und Mannheim. Süddeutsche Monatshefte. pp. 456–467.
2. HEDIGER, H. 1967. Verstehens- und Verständigungsmöglichkeiten zwischen Mensch und Tier. Schweiz. Zs. Psychol. und ihre Answendungen **26**: 234–255.
3. HEDIGER, H. 1974. Communication between man and animal. Image Roche **62**: 27–40.
4. SEBEOK, T.A. 1979. The Sign & Its Masters.University of Texas Press, Austin, Texas.

---

* Quand un enfant dit à sa mère: "J'ai faim", ou "Je voeux dormir", il est encore animal. Quand il dit: "Voici ce que j'ai fait ce matin", il commence à être homme.

5. KOEHLER, O. 1937. Die "zählenden" Tauben und die "zahlsprechenden" Hunde. Der Biologe 6: 13–24.
6. ROSENTHAL, R. 1966. Experimenter Effects in Behavioral Research. Appleton, New York, N.Y.
7. SEBEOK, T.A. 1976. Contributions to the Doctrine of Signs. Indiana University, Bloomington, Ind., and Peter de Ridder Press, Lisse.
8. PFUNGST, O. 1907. Das Pferd des Herrn von Osten (Der kluge Hans). Joh. Ambrosius Barth, Leipzig.
9. KNAPP, R. 1936. Poldi, die bosnische Wölfin. Pirngruber Linz a. Donau.
10. RENSCH, B. 1973. Gedächtnis, Begriffsbildung und Planhandlungen bei Tieren. p. 241. Paul Paray, Berlin.
11. KOEHLER, O. 1943. "Zähl"-Versuche an einem Kolkraben und Vergleichsversuche an Menschen. Z. Tierpsychol. 5: 575–712.
12. HARDY, A. 1965. The Living Stream. A Restatement of Evolution Theory and its Relation to the Spirit of Man. p. 156–8. Collins, London.
13. EKMAN, P., W.V. FRIESEN & P. ELLSWORTH. 1972. Emotion in the Human Face. Pergamon Press, New York, N.Y.
14. GOFFMAN, E. 1971. Relations in Public. Allen Lane, The Penguin Press, New York, N.Y.
15. SEBEOK, T.A. 1979. Reference 4: Chap. 5. Looking in the Destination for what should have been sought in the Source.
16. WAAL, F.B.M. DE. 1980. Schimpansin zieht Stiefkind mit der Flasche auf. Tier 20: 28–31.
17. KATZ, D. 1937. Animals and Men. Studies in Comparative Psychology, p. 7 Longmans, Green, New York, N.Y.
18. MUNN, N.L. 1933. An Introduction to Animal Psychology. The Behavior of the Rat. p. 39, 40 Houghton Mifflin Company, Riverside Press, Cambridge.
19. TIMAEUS, E. 1974. Experiment und Psychologie. Verlag Hochgrefe Göttingen.
20. MORTON, J.R.C. 1968. Effects of Early Experience. "Handling" and "Gentling" in Laboratory Animals. In Abnormal Behavior in Animals. M.W. Fox, Ed. p. 269. W.B. Saunders Company, Philadelphia.
21. HEDIGER, H. 1940. Ueber die Angleichungstendenz bei Tier und Mensch. Die Naturwiss. 28: 313–315.
22. HEDIGER, H. 1951. Grundsätzliches zum tierpsychologischen Test. Ciba Z. 11: 4630–4636.
23. HEDIGER, H. 1965. Man as a Social Partner of Animals and vice-versa. Symp. Zool. Soc. London 14: 291–300.
24. PATTERSON, F. 1978. Conversations with a Gorilla. Natl. Geogr. 154: 438–465.
25. HEDIGER, H. 1973. Tiere sorgen vor. Manesse Verlag, Conzett & Huber, Zürich, Switzerland.
26. McCONNELL, J.V. 1968. Biochemie des Gedächtnisses. Med. Prisma 3: 3–21.
27. HALDANE, J.B.S. 1954. La signalisation animale. Ann. Biol. 30: 89–98.

Behavior in Context: In What Sense Is a Circus Animal Performing?

PAUL BOUISSAC

Victoria College
University of Toronto
Toronto, Ontario M5S 1K7

IN A CELEBRATED ESSAY Roman Jakobson[1] says that a Russian actor with whom he was acquainted could utter the phrase "this evening" in fifty different ways with respect to intonation, i.e., intensity, pitch, rhythm, and juncture. He was thus able to create fifty different messages that evoked in the mind of the listeners fifty different situations. The emotive cues combined with the semantic, morphologic, and syntactic components of the linguistic message could easily be related by native speakers to culturally congruent social settings and psychological moods. This was later experimentally verified under the auspices of the Rockefeller Foundation.

Advances in the pragmatics of human communication (e.g., Argyle,[2] Hinde,[3] Watson[4]) have shown that social interaction in all its forms can be construed as a model whose parameters can be experimentally controlled. Such an analytical approach enables the observer to shift the focus from the message to its context and its situation and to study the covariations that may occur. "Context" should be understood here as not only whatever immediately precedes and follows the linguistic message itself but also whatever is emitted simultaneously through all the other available channels; "situation" refers to the type of social interaction that is identified by the observer within a specified cultural domain, such as conversation between peers, confrontation between rivals, ritualized testing of the affective bond between lovers, and so forth. The spontaneous identification of the various categories of situations is an on-going business in everyday life and an important aspect of the cultural competence of any individual. Obviously these various situations are construed from some minimal cues that can be manipulated once they have been isolated (e.g., respective class of age, sex, natural posture, role expectation, distance, place, and degree of normality or abnormality of the interaction.) It would be possible to design an experiment inspired by the one recounted by Roman Jakobson, in which the utterance of the linguistic message would remain invariant and the context or the situation would vary, the hypothesis being that a

0077-8923/81/0364-0018 $01.75/0 © 1981, NYAS

given sign would convey various meanings depending on the manipulation of the context and situation of this sign. As a matter of fact, such experiments are performed daily in one of the most popular institutions of our culture, i.e., the circus. Performing animal acts are indeed patterned events that are two-sided. On the one hand, the trainers interact with their charges on the basis of their socio-biological competence, on the other hand they frame these interactions in particular situations relevant to the system of social interactions shared by the public for which they perform. Therefore all animal acts are the combination of a biologically patterned behavioral sequence and a constructed social situation. Once an animal is trained, the behavioral sequence can be considered as the equivalent of the invariant aspect of a message whose meaning can be modified, within certain limits, by the variations of the situational construct. This does not mean, naturally, that the behavioral sequence is produced independently from any situation but that the situation through which the behavioral sequence is elicited simply overlaps, sometimes minimally, with the situation outwardly constructed for the audience. For instance, in A. Zoppe's Rhesus Monkey act, one of the two trainers dressed as a clown starts a mild argument with one of the monkeys who, at a given moment, slaps him twice. This monkey wears a masculine garment. If it were dressed as a woman and if the clown had whistled at "her," this patterned segment that is triggered by the trainer, i.e., slapping the man on the face, would have communicated a quite different meaning, shifting from confrontation between males of equal status to inappropriate behavior towards a proper lady by an ill-educated male. Such changes of frame for a given trick are common practice in the circus. By thus manipulating both the animal's behavior and the context of this behavior the trainer utilizes, at the same time, two different semiotic systems. As a result, such manipulation generates for the public, and to a lesser extent for the trainer, the illusion that the relevant context is the one they perceive and that the animals share this perception of the situation that is constructed in the ring. This paper will attempt to show the generality of this illusion in the art of training and to point out the nature of the performing animal's performance.

Let us take a few simple cases. In the 1980 program of the blue unit of Ringling Brothers, Barnum and Bailey Circus, a bear trainer, Ursula Bottcher, concludes her act by performing what is sometimes called "the kiss of death" with one of her charges. This is a well-known circus trick often done with bears as well as big cats although the technique is different with the latter. One animal leaves its stool, rises on its hind legs and walks across the ring until it stops in front of the woman whose head is at the level of the animal's chest. The bear then lowers its head toward the woman's face until its mouth apparently touches her lips.

Even if you are seated on the first row in a one-ring circus, you may not notice that the trainer has put in her mouth a piece of carrot, that sticks out between her teeth because your attention is being irresistibly drawn to the huge white animal that is menacingly proceeding toward the frail trainer standing helplessly, her hands behind her back. Once the bear has caught the carrot, or whatever convenient bait is used, it is led back to its place in the cage. The illusion is completed by the fact that in the printed program a beautiful close-up photograph of this sequence does not show the bait, either because it has been artificially removed from the photograph or because the bear has already caught it with its tongue or its lips. In any case it is very clear in the photo that the trainer has a pouch full of "goodies" attached to her belt. It is obvious from this elementary example that the situation created to elicit a certain behavior from the bear, i.e., he approaches the trainer in order to obtain a delicacy, is different from the one created by the various cues provided to the public by the announcement, the music, and the posture of the woman, which bring into focus the complex and powerful themes of bestiality, love, and death.

Incidentally, this example shows why television more often than not destroys circus acts from the point of view of their semiotic efficacy, by constantly varying the distance and the focus. This is done with the best of intentions by cameramen in search of interesting details, but the viewer loses the frame within which the situation is constructed and is made to perceive elements of the acts that are not congruent with this situation, such as the actual direction of the animal's glance or visual expressions that stand in contrast with its assumed mood and goal. The unfortunate result of this visual dismembering of the circus act is the destruction of the semiotic illusion that was constructed with great care by the trainer. This is also true of many acrobatic acts that make sense only if they are perceived in relation to the global situation created by the totality of the apparatus.

My second example is a childhood memory. At the end of a lion act, after the trainer had sent all the animals back to their individual cages, a lioness remained on her stool, stubbornly refusing to leave the ring in spite of the shouting and whip cracking of the man. Then at the suggestion of the ring master, the trainer put his whip and stick aside and made a very polite and formal gesture to the lioness, inviting her to retire. She immediately did so with the utmost grace. There was no doubt in my ten-year old mind that this event was still another proof of the deep understanding between the wonderful trainers and their wonderful lions. Since then I have seen the same trick performed several hundred times by dozens of trainers. It is an excellent example of the minimal overlapping of two situations, the one, a constructed frame perceived by the public at large and the other perceived and negotiated by the trainer and the lioness. There are several procedures to obtain this effect; I will describe three of them: as a rule, circus wild animals have an urge to return to their territory, i.e., their own individual cage. It is usually sufficient to open the door connecting the central ring to their cage for them to rush back there. As long as the door remains closed they will not dash for the exit. Therefore, the only valid auditory and visual cue for the animal is the opening of the door by the cage attendant, regardless of whatever else can be hinted at by the trainer's behavior. Another possibility is for the trainer to stand in the way of the exit and to crack his whip while shouting at the animal to leave. All that the animal perceives is the man as an obstacle, regardless of what he says. A few moments later, it will be enough for the trainer to step aside while making his emphatically polite gesture to trigger the rush to the exit. For the animal, the relevant modification of the situation is that the path is now free of obstacles. A third procedure consists of manipulating the flight distance, a phenomenon that has been thoroughly described and evidenced by Hediger.[5] Any animal in a given environment tends to flee from man—this is a complex system with many variables that I do not want to deal with in detail here. Suffice it to say that in the cage one of the most relevant dimensions of the man–animal interaction is the distance that is maintained between them. This is an individual factor; some tame animals accept and enjoy a close contact. These are the ones who are man-imprinted; i.e., they consider their trainer as a conspecific rather than an alien. This phenomenon has vast consequences for the training process and beyond. More on this later. Others have responses appropriate to the variation of physical distance. Until the critical threshold is passed, they do not have any reason for moving. If this threshold is transgressed, then they flee in the opposite direction. If they cannot flee—i.e., if they are cornered—then they usually attack. To return to the case of the well-mannered lioness's, the trainer will not cause her to move as long as he stands outside of the lioness's critical distance and cracks his whip sideways. But if he steps toward her while making the polite gesture, this reduction of the distance will trigger the flight reaction especially if it is combined with the opening of the exit door.

These examples show clearly the two-level manipulation in which a circus trainer is engaged, and how he carefully constructs situations that are socially relevant for the spectators but that include as building blocks, so to speak, segments of interactions based on totally different biological systems.

A first conclusion that could be drawn at this point is that circus trainers are professional deceivers. This is true at a certain level, but their imperfect knowledge of the ethology of their charges also leads them to what amounts to self-deceptions, a phenomenon that accounts for the accidents, often tragic, occurring from time to time in the circus ring.

From an investigation conducted in 1971 under the auspices of the Research Center for Language and Semiotic Studies of Indiana University and supported by the Wenner Gren Foundation, it appears that many trainers tend to see themselves as teachers and may overlook the socio-biological significance of some aspects of the interactions with their animals. I am not questioning here the practical knowledge of many experienced trainers who reliably manage to elicit complex behaviors on the part of their charges, but the situation they perceive in the very process of training may be at times quite different from what is known to be the biological meaning of the animal's behavior. For instance, there seems to exist a permanent competition among tiger trainers regarding the number of tigers they can line up and make do a trick known as the "roll over" at the same time. This consists in having the tigers lie down in front of the trainer then rolling over laterally several times. I have attended training sessions of this trick and have seen that some tigers seem prone to perform this trick whereas others are very reluctant to do so. I have seen a stubborn trainer spend many hours of exhausting effort in trying to "make the tiger understand what he wanted it to do and to persuade it that it had better do so." The result was eventually a painful poking game during which the tiger was enticed to roll over as a necessary segment of a defensive behavior aimed at catching the stick that the trainer poked painfully in its side. The contrast with the relative ease with which other tigers could "learn" this trick puzzled me until I read Charles McDougall's book, *The Face of the Tiger*,[6] which is devoted entirely to the ethology of this beautiful animal. It is clear from the many segments of natural behavior in context that are described in this book that "roll overs" belong to the tiger's repertory of social behavior. It indicates submission and it is performed both by the mature tigress in the mating ritual and by cubs in front of adult animals. It has also been observed during chance encounters between subadults or transient males and the resident male tiger of a given territory. Its purpose is to prevent a fight. Having discussed this with some trainers whom I have known personally for years, I found out that they were not aware of the fact and simply thought that their more successful colleagues had discovered a better pedagogical trick than theirs or had been luckier with their animals, i.e., they had come across some more obedient and intelligent animals.

True enough, most trainers are remarkably sensitive to the biological constraints of the situations they negotiate in their daily confrontations with those large predators, but their perception of these situations nevertheless is more often than not at odds with the animal's own perception of these situations, at least to the best current ethological knowledge. What makes training possible is only a slight behavioral overlapping between the two species, in this case man and tiger. However, this very overlapping often causes misreadings and misgivings that entail fatal consequences. The chronicle of the circus is replete with accidents that bear witness to this fact. As a result, even the wisest of wild animal trainers seems to oscillate between a narcissistic and a paranoid attitude towards his animals. Whoever has once in his/her life enjoyed the playful company of friendly tiger cubs or has seen at close distance the frightful charge of an adult male lion can easily understand this double attitude. It is not uncommon to hear trainers calling their "pupils": "good boy!" "good girl!," with a paternal or maternal soft inflection of the voice, when the animals behave according to their plans. On the other hand, in their dealings with some of their most impressive and difficult animals, the same trainers may use the most abject insults. Both types of interactions are obviously experienced by these men and women as heavily anthropomorphized and strictly within the domain of the repertory of human values and emotions.

These last remarks open up our perception of a circus act to the third dimension of the performance: the animal's point of view. In addition to the situation constructed for the audience and the one perceived and manipulated by the trainer interacting with the animals, there exists also, necessarily, a situation that is experienced and negotiated by the animals within their own semiotic system, i.e., the system provided by the structure and programs of their brain. Although their adaptability enables them to survive in an artificial environment and to adjust accordingly their ethogram, such modification nevertheless remain within the boundaries of a strict socio-biological code that forms an absolute obstacle to all attempts at cross-specific communication. Naturally, it would be ludicrous to deny that some form of interaction takes place first in the training process, then in the performance: attention is obtained, orders are obeyed, behavior is shaped by appropriate conditioning, cues are recognized, moods are sensed, and so forth. Hediger has brilliantly discussed the flow of information that circulates between the participants during the performance of an animal act.[7] However, if one considers communication as the vicarious sharing of experience, a term other than "communication" should be used in order to describe adequately the sort of interaction that models an animal act. Crucial features of the training process such as the structure of attention, the establishment of a social hierarchy, and even the learning ability of the young belong to the species-specific competence of the animals that are used in circus acts. Tigers, for instance, are genetically programmed to learn patterned behavior relevant to their particular environment from their mothers during their first years. This is a condition for their survival. The fact that a trainer undertakes to model some aspects of their behavior when they are about a year old, imposes himself or herself as a leader and becomes the focus of their attention, makes sense "tigerwise." It is also in the order of things that later on, when the male tigers mature and reach full adulthood, the situation evolves from deference to confrontation; although the trainer may then manage to maintain his/her leadership — and this is not always the case — the same patterned behavior has a different meaning with respect to the socio-biological chart of the animals. In this sense, a circus animal performs, i.e., negotiates social situations by relying on the repertory of ritualized behavior that characterizes its species. The ritualization may have been slightly modified by the human input during its learning stage, but far less than is usually thought. It seems obvious that the biological score according to which they perform is not the same as the one that the trainer has conceived and keeps perceiving. The trainer can elicit at will some segments of behavior and frame them in a situation of his/her choice, but the animal's behavior is never performed out of its own socio-biological context, which transcends the trainer's understanding of the animal's performance. In the Rhesus Monkey act mentioned earlier in this paper, the animal performers display most of their specific social repertory in the category of agonistic pattern and related movements, such as "submission and withdrawal," "threat, attack, and approach," and "affinitive movements," as they are listed and described in Altman.[8] The trainers have built situations easily identifiable by their audience, that are mostly irrelevant to the monkey's concern. However, these animals find themselves in a series of situations that they negotiate according to their own terms and through which they may be led to actual attacks if the trainers are not fully aware of the animal's point of view, as is often the case.

The psychologists who have attempted during the last decade to educate some apes in order to test their nonverbal linguistic abilities have found themselves in the same situations as circus trainers. There are indeed indications that accidents are not infrequent, although they have never been publicized; the recent attack of the celebrated "Washoe" on Karl Pribram, in which the eminent psychologist lost a finger (personal communication, June 13, 1980) was undoubtedly triggered by a situation that was not perceived in the same manner by the chimpanzee and her human keepers and mentors.

Trainers are not always aware of the discrepancies existing between their context of action and that of their charges, and situational overlapping often creates the illusion that a given context is fully shared. The fact that accidents occur unexpectedly has generated an abundant circus lore regarding the nature of wild animals, in which moral or psychological interpretations distort the ethological significance not only of the segment of behavior that is incriminated but also the totality of the animals' comportment in their daily routines. "Never trust a lion," some trainers will say, "because a lion always goes through an outburst of madness in its life." The concepts of "trust" or "madness" are totally irrelevant in this context, but it is through such concepts that trainers can cope with the socio-biological otherness with which they are confronted daily, an otherness that poets have sometimes sensed and expressed in felicitous forms such as William Blake's "fearful symmetry" or Victor Hugo's lines on "la nuit qu'un lion a pour âme" (the dark night of a lion's soul), an expression that, incidentally, has found its way into some trainers' jargon.[9]

REFERENCES

1. JAKOBSON, R. 1960. Linguistics and Poetics. *In* Style in Language. T.A. Sebeok, Ed. pp. 350–377. Indiana University Press, Bloomington, Ind.
2. ARGYLE, M. 1969. Social Interaction. Atherton Press, New York, N.Y.
3. HINDE, R. A. 1974. Biological Bases of Human Social Behavior. McGraw-Hill Book Company, New York, N.Y.
4. WATSON, O.M. 1970. Proxemic Behavior. Mouton, The Hague.
5. HEDIGER, H. 1968. The Psychology and Behavior of Animals in Zoos and Circuses. Dover, New York, N.Y.
6. McDOUGAL, C. 1977. The Face of the Tiger. Rivington Books and André Deutsch, London.
7. HEDIGER, H. 1974. Possibilités de communication entre l'homme et l'animal. Image Roche **62**: 27–40.
8. ALTMAN, S.A. 1965. Sociobiology of Rhesus monkeys. Ill. Stochastics of social communication. J. Theoret. Biol. 8: 490–552.
9. THÉTARD, H. 1928. Les Dompteurs. Gallimard, Paris.

Who Feeds Clever Hans?*

DUANE M. RUMBAUGH
Department of Psychology
Georgia State University
Atlanta, Georgia 30303

THIS PAPER CONSIDERS research methods and problems entailed in an area that has come to be known as ape-language research. This area has attracted a great deal of interest for a variety of reasons, but primarily because of wide-spread interest in the relationship between man and animal. In what ways is man distinct? What are his unique characteristics? Although I and my colleagues of the Language Formation Studies Program in Atlanta do not deny interest in questions of that type, these questions are not the ones that compel us to research the area. Our main interest reflects the fact that the ape is a valuable animal model for research into questions that will help us understand the maturational processes of the human child. At present, our focus is upon the maturation of communication skills.

CAN ANIMAL MODEL BEHAVIOR RESEARCH HELP THE INTELLECTUALLY IMPAIRED?

Our original question as we launched this Study program in 1971 was, "Can research into the parameters of the language-like behaviors of the age be used to facilitate the development of language research and teaching programs with retarded persons who frequently have pervasive deficits in language production and comprehension?" The answer to this question is very clearly *yes!* The computer-based keyboard language training situation that has been developed by Harold Warner, Victor Speck, and S. Tom Smith,[1] of the Yerkes Regional Primate Research Center of Emory University, has been applied with great success to the research and teaching of language skills with the severely and profoundly retarded persons of the Georgia Retardation Center, also in Atlanta. That particular project has been developed by Dorothy Parkel, Royce White, and Caren Millen of the Research Department of that Center.

* This research was supported by a grant from the National Institute of Child Health and Human Development (HD-06016) and from the Division of Research Resources (RR-00165) of the National Institutes of Health.

0077–8923/81/0364–0026 $01.75/0 © 1981, NYAS

In brief, we have worked with nine children, whose mental ages, formally assessed, do not exceed that of the average three-year-old child. Characteristically, the children of this study had essentially no ability to speak intelligibly. Only those very familiar with a given child and the context within which he or she made an utterance could approximate anything like a reliable decoding of what was being said.

Through use of training procedures developed with chimpanzee subjects at the Yerkes Primate Center, this deficit was ameliorated to a very significant degree in five of those individuals. Not only have they learned the meanings of arbitrary symbols, which they use referentially and communicatively, they use those symbols as words to engage in novel conversation with their mentors. The consequence of this learning has been remarkable. The children's general social skills have been profoundly advanced; there has been a significant attenuation of their behavioral management problems, even in the cottages in which they live; and there has been an enrichment of the social commerce between the children and between them and the variety of persons who work in that particular project.

Initially there were real questions, quite properly posed, as to whether the research of our first project, with the chimpanzee "Lana," would have anything to offer by way of remedy to the communication deficits of these children. There was also proper questioning as to whether or not the research to be continued, not only with Lana, but with other chimpanzees as well (notably "Sherman" and "Austin"), could serve to define the training methods that would work better with these children than would methods devised through research only with those children. Over the course of the past five years, not only is the answer to us and to those who work on the project at the Georgia Retardation Center clearly *yes*, the answer is also *yes* in the perspective of large numbers of staff and professionals of the Georgia Retardation Center who have no direct affiliation with the project.

What does this mean? It means to us very clearly that we have made significant inroads into the definition of the parameters of initial language acquisition through exploitation of the ape as a behavioral animal model, paralleling the exploitation of animals as models for study by researchers in the field of biomedicine. As we do so, however, we do not view the chimpanzee as an identity with the human child any more so than the biomedical researcher equates the physiology and disease processes of an animal with a sick human being. Rather, from a comparative psychological perspective we search for insights and methods that will enable us to make the gains which we cherish in our understanding of the dynamics of communications. If our observations within the ape research effort had not been valid, how could the training tactics uniquely formulated therefrom have had the observed advantage when applied to the children?

What Is A Word?

Communication and language, along with all other terms in the behavioral sciences, are very difficult to define. Definitions of terms constitute major stumbling blocks to communication among scientists as to what has been done and observed and, consequently, what is to be properly concluded. None should deny that, as a direct consequence of the ape-language efforts, much has been learned about what communication and language are and are not. For instance, the research has made it abundantly clear that the essence of what is called a *word* is at the heart of language. Through words we generally believe that we can accurately transmit information amongst members of our own species. At times we might doubt that, however. It is clear, even in this meeting of the Academy, that there is reason to question whether scientists can communicate information they would like to communicate with precision. Furthermore, the use of words to transmit information is frequently confused and confounded by the introduction of extraneous factors, reflecting motivations other than those that would be constructive and stimulating of efforts to accrue knowledge.

What attributes does a word have that responses in general do *not* have? What attributes must a response accrue for it to be properly termed a word? Is a response even to a conditioned stimulus properly viewed as a word? We would say no. On the other hand, a response of any type can accrue the qualities of word*ness* to the degree that the response clearly serves to refer to, to represent, something not necessarily present in space and time in a way that facilitates communication between two animates.

Research from certain aspects of the ape-language effort, and particularly the earliest of the efforts, have served to make the majority of us in the field sensitive to the conclusion that the simple execution of a response in the presence of a stimulus, regardless of its high reliability, does *not* mean that that response necessarily has the attributes that words have as symbolic referents that have meaning and that facilitate communication. Even the recent study by Epstein, Lanza, and Skinner[2] serves to make this point very clear, though it was conducted for quite another purpose. Simply because pigeons manifest certain surface behaviors through the highly ritualized, highly predictable pressing of keys, there is no justification for concluding that wordness, semanticity, is entailed. Simply because an automatic banking machine, of the type now so common in this country, responds predictably, faithfully, and validly to the instructions which I convey to it as I put in my plastic card and press the various lighted buttons with their surfaces etched with various words and comments, does not compel the conclusion that there has been a linguistic exchange or communication between it and me. It is simply the case that the machine has been designed and is operating so as to bring out a desired effect. The machine has no concept, no meaning whatsoever as its button is pushed that instructs it to vend for me ten dollars in cash from my checking account. Likewise, the pecking by Jack and Jill in the Epstein *et al.* study, tells us only that through basic conditioning procedures, known by even the average student of undergraduate experimental psychology, that pigeons can develop surface behaviors (superficial behaviors) which emulate the production of language as generated by man through use of highly complex cognitive processes.

As accurately quoted by Umiker-Sebeok and Seboek,[3] we have said in an unpublished manuscript, "Frankly, we are not interested in whether or not langauge is the exclusive domain of man. That question leads all who address it to a quagmire of confusion, despair, and impatience. We want none of that!" That fact notwithstanding, we stand firm in our conclusion that ape-language research is a very productive area of study for the development of knowledge regarding the origin of initial communication and language skills which can be used to advance our understanding of the impediments to the acquisition of communication and language skills by the mentally retarded child. The transition from nonverbal behavioral communication to language is a gradation, a continuum, and animal research plays a critical role to the end of our understanding the rich dynamics of a word, of language.

THE PROPER ROLE OF THE SKEPTIC

A very proper facet of the scientific endeavor is for the scientist and layman alike to be skeptical, to ask for clarification regarding the data base for conclusions drawn as a result of research. In our own research effort, we engage in careful review and criticism of that which we do and that which we plan to do. We are not at all reluctant to initiate discussion of our methods and the bases for our conclusions with any skeptic. How *long* we will sustain that discussion, however, is determined by the following: (i) Has the skeptic studied the research reported? (ii) Does the skeptic bring a sense of balance to the debate? By this we mean, does the skeptic give us reason to believe that he or she is educable, or is this skeptic so persuaded by his or her own rhetoric as to be deaf to the dialogue? (iii) Has the skeptic, particularly if he or she purports to be a scholar and/or scientist, taken the opportunity to gain in-

formation through such things as invitations extended to visit a labora-
ry, to participate in prior discussions, and to use letters and phone calls
to seek clarification? To the degree that such opportunities are side-
stepped by the skeptic who purports to be a scientist, for whatever
reasons of time or convenience, there is reason to question the motivations behind the skepticism, regardless of how scholarly its garb might
be.

In brief, the skeptic plays a very proper role in science, and each
researcher is well advised to be the most dedicated skeptic of his or her
own work. On the other hand, the contributions of the career skeptic,
who draws conclusions prior to debate, even with this forum which is at
hand, contributes only noise, confusion, and negativism to the evaluation of research offered by well-trained researchers whose experience
base may be far greater than that of the skeptic and whose honesties
may not be properly impugned.

It is to be emphasized that the field of ape-language research is very
young and that, contrary to the beliefs of some, it has not been an area
into which vast financial resources have been invested. By far and away,
the greatest resource invested has been the lives and time of those who
have ventured their careers into this area. The financial investment is
palled by comparison. The history of science will surely show that, at
this point in time, we have at best a modicum of data relative to that
generated in other areas of science and that it is anticipatory and
counterproductive for anyone to presume that "the data are in" for a
valid assessment as to what is to be offered from this area of research.

I venture these positions with the best of intent and sobriety. I have
risked venturing them because I fear that the Academy, unless alert,
might have its attention distracted from the most constructive of the
processes vital to the development of healthy science produced only by
interactions of open minds in open discussion and honest exchange,
unencumbered by unyielding prejudice from any quarter.

For all of its faults, faults inherent in all human endeavors including
those of science, ape-language research has contributed significantly
even in this early day to our understanding of communication, language, the ape, and man himself. The insemination of pervasive fear
regarding the Clever Hans phenomenon, with no apparent regard for
balance, serves only to remind us of the old Flemish prayer, "From
goulies and ghosties and things that go bump in the night, Dear Lord,
deliver us." Broad, blatant injection of Clever Hans into every study of
the ape-language area smacks of a witch hunt. Every scientist, behavioral or otherwise, has heard Clever Hans neighing in the stable. Every
scientist of repute will do all that is possible to keep Clever Hans where
he belongs—in the stable. Contrary to the allegation made by Sebeok,

the *conclusion* really is not that Clever Hans has confounded each and every contribution to the data base. Rather, the *question* should be, has Clever Hans been properly held at bay when the researcher goes about obtaining the critical data in test situations that follow designated training regimens, which might be informal.

How Competent Is Clever Hans?

We are all familiar with Clever Hans and I need not do more than to remind the reader that the term is now used to refer to inadvertent cuing of behaviors. Whenever there are Clever Hans effects, it is clearly unfortunate because they preclude the goal of science — to define the parameters of that which is observed. As we consider the potency of Clever Hans, we should bear in mind the following points: (1) To allege Clever Hans is not to prove Clever Hans. (2) That inadvertent cuing might be potentially available does not mean that it in fact is used by the research subject. (3) For the subject to employ Clever Hans cues is not necessarily easier for him than it is for him to employ past learning. (4) That not all sounds, odors, and so forth can become sources of Clever Hans.

It is an irrational fear that is manifest when the skeptic makes the broad generalization of the type that, because other people were possibly present in adjoining rooms and doing their chores irrelevant to the research, there was a real chance for Clever Hans to confound the validity of the data collection. Inadvertent cues, regardless of their source, are of potential worry in the unhaltering of Clever Hans *only* to the degree that they are *reliably* associated with the presentation of opportunities for the research subject to respond.

The perfect, totally controlled experiment has never been conducted and it will never be conducted. Even the best will always allow for the committed skeptic to say, "But what about this, what about that, what about all of these things for which no control was made?"

Graduate students in psychology programs at most universities are taught to be critical of that which they read. In particular, they are taught to be critical with regard to controls and with regard to whether the data base as operationally defined by methods justifies the conclusions offered by the authors. This generally is a healthy part of their education. On the other hand, it can become a way of life in which the student comes to enjoy the game, "Let's pick," which invites those present to tear apart a study that is drawn even at random. Any researcher can take a paper and pick at it until a thread is unraveled, which then, of course, can feed efforts to the end of destroying and rejecting that particular research effort. As noted above, within limits it is a healthy thing to become critical and to become a skeptic, particularly of one's own work. But, to be a skeptic with no sense of balance, with no sense of regard for the credentials of those who do the work, is, at best, a dubious contribution to the goal of any scientific forum.

MISREPRESENTATIONS OF REVIEWS

One of many reasons to question the credentials of some skeptics is the degree to which they draw upon what is argued in secondary references, authored in the main by nonscientists, and the degree to which they selectively report only negative conclusions of research efforts. For example, Umiker -Sebeok and Sebeok[3] have alleged that a recent "rigorous analysis of the manipulation of Yerkish computerized symbols by the chimpanzee Lana . . . " by Thompson and Church[4] has resulted in their concluding that "the animal's behavior can be attributed to two basic processes — namely, (1) conditional discrimination learning where situational cues determine the selection of one sentence from a small set of six stock sentences, and (2) paired-associate learning, or the coupling of a lexigram with an object, person, or event" (Reference 3, page 41a). In point of truth, we very happily cooperated with Thompson and Church in their study. We extended the data base available to them for their building of a computer program that would account in so far as possible for Lana's language-like productions, and we included them as prime participants in a recent symposium at the Southeastern Psychological Association, to which Sebeok was also invited. Thompson and Church did not conclude that the animal's behavior can be attributed to two basic processes. Rather, they concluded that the majority, but not all, of Lana's behavior could be attributed to two basic processes. There is a real differences between "some," even "a majority," and "all." Even if Thompson and Church had been able to account for all of Lana's behavior through the development of a computer program, it would not serve to discredit Lana's productions. Thompson and Church themselves disavow that their effort serves to discredit either Lana's performance or the research program that generated it. Rather, and very appropriately, they asked the empirical question, "Can we write a computer program that will account for Lana's novel productions?" And the answer was in the majority, but not in toto, *yes*. Their very correct argument was that to the degree that one can account for behavior through the writing of a computer program, one has defined what it is that one has.

Lana's productions by any yardstick were of interest and a challenge to explain, but Lana was not the end-all of ape-language research. Rather, we always declared that she was a pilot animal in our first effort to determine whether or not we could develop a computer-based language training situation that would lead to new systems and research methods that would facilitate research into the parameters of initial language skills of the mentally retarded person. Thompson and Church are scholars, and they may expect to have our full cooperation at every turn of the road. We are gratified when their research serves to hone our mutual understanding as to what probable truth is in the situation.

What Is a Symbol?

Umiker-Sebeok and Sebeok, in passing, allege that "examples of true symbols can be found throughout the animal world, including insects" (Reference 3, page 96). One of them (Umiker-Sebeok) commented in 1976, "that what has *not* been proven by the ape 'language' research is that the symbols used by the animals are any more propositional than the circus tricks taught apes in circuses."[6] Now, I do not deny that they may conclude that "true symbols can be found throughout the animal world." I would question the validity and the usefulness of such a conclusion, however, for the use of symbols is very much at the heart of ape-language research. A symbol must be more than just a stimulus. A symbol must be demonstrated to have referential, representational, and communicative potential and value. It is something that is learned and it entails very complex psychological processes, not generally found in the animal world. I submit that probably a more precise conclusion by the Sebeoks would be that animals, even insects, can learn to respond to stimuli as though it has some meaning to them. But that does not mean that the stimuli serve as symbols.

Of no small interest on this point is that had Sebeok attended the entirety of the Southeastern Psychological Association's convention in reference, a symposium in which he was a participant in transit, he would have learned of data that gives us good cause to conclude that, indeed, for our apes the symbols are referential, representational, and communicative in value.[5] Data obtained and reported by Savage-Rumbaugh at that convention made it clear that the chimpanzees Sherman and Austin categorize learned symbols as foods and tools (nonedibles) just as they categorize the physical referents themselves. These data were obtained from tightly controlled test situations in which the animals had no human present in the room at the keyboard to influence their choice of keys for purposes of categorizing. Furthermore, in tightly controlled situations even where no humans are present (the exchanges by Sherman and Austin are observed by remote TV monitors), Sherman and Austin used their symbols and also used untaught gestures to moderate their linguistic exchanges, their requests and the giving of foods and tools to one another. Tightly controlled as the situation was for the collection of data, we hasten to add for the benefit of the skeptic that we did not control for such things as the following: (1) barometric pressure, (2) alternations in the amperage of the electrical lines that feed the laboratory, and (3) people in adjacent rooms doing their chores, along with a host of other factors, which collectively can be declared *irrelevant* to the situation and Clever Hans effects.

SUMMARY

I recommend that everyone look very carefully at the credentials and motivations of the radical skeptics to determine whether or not they are truly participants of the scientific enterprise to advance knowledge. For the skeptic to be taken seriously, it should be clear that he or she (i) has appropriately equipped himself with knowledge, (ii) has taken every reasonable opportunity to clarify questions prior to conferences and symposia, (iii) has shown a sense of balance and perspective, and (iv) has the ability to benefit from exchanges and data as availed. To fail to examine thus the credentials of the skeptic will set back the advance of research and knowledge, a consequence very foreign to the interests of scientists everywhere.

REFERENCES

1. RUMBAUGH, D. M., Ed. 1977. Language Learning by a Chimpanzee: The LANA Project. Academic Press, New York, 1977
2. EPSTEIN, R., R. P. LANZA & B. F. SKINNER. 1980. Symbolic communication between two pigeons (*Columba livia domestica*). Science 207: 543–545.
3. UMIKER-SEBEOK, J. & T. A. SEBEOK. 1980. Clever Hans and smart simians. Paper distributed at the convention of the Southeastern Psychological Association, Washington, D.C., March, 1980.
4. THOMPSON, C. R. & R. M. CHURCH. 1980. An explanation of the language of a chimpanzee. Science (April 18): 208.
5. SAVAGE-RUMBAUGH, E. S. 1980. The importance of communication to ape language learning. Paper presented at the convention of the Southeastern Psychological Association, Washington, D.C., March, 1980.
6. UMIKER-SEBEOK, J. 1976. Comments on "Language, Communication, Chimpanzees," by G. Mounin. Curr. Anthropol. 17(1): 17–18.

Can Apes Use Symbols to Represent Their World?*

E. SUE SAVAGE-RUMBAUGH

Yerkes Regional Primate Research Center
Emory University
Atlanta, Georgia 30322

WHEN CHIMPANZEES USE symbols or signs which they have been taught by human beings—do they know what those symbols reference? Do they know that symbols can reference objects, people, places, events, action, states, relationships, and so forth? Do they use symbols intentionally, i.e., in order to transmit information that is not apparent, given the context? A large body of research based on affirmative assumptions regarding these questions has been published.[1-4] There also has arisen a recent awareness that the initial conclusions regarding these linguistic capacities of apes were reached too rapidly.[5,6]

The issue of referential capacity is absolutely fundamental to any concept of language. It is more basic than the issue of modality and more important than the issue of syntax. If the symbols of a language are not referential, if they cannot be used to represent things, people, places, events, states, and relationships that are absent in space and time, then they have no more information value than emotive facial or gestural expressions. To put it bluntly, communication that is not referential is not language, regardless of what other qualities it may possess.

The purpose of the series of studies reported below was to determine whether or not the abstract symbols used by three chimpanzees, "Sherman," "Austin," and "Lana," were functioning at a referential level. In order to answer this question in a concise fashion, we devised a paradigm that required the chimpanzees to learn to make a categorical decision about two classes of objects within their environment: edibles and inedibles. We trained the chimpanzees to label the *names* of three inedibles (stick, key, and money) as tools, and the *names* of three edibles (beancake, orange, and bread) as foods. The "names" in this case were the arbitrary "Yerkish" geometric symbols with which these tools and foods had become associated through prior training.[7,8] The labels "food"

* This work was supported by Grants NICHD-06016 and RR-00165 from the National Institutes of Health.

and "tool" were also Yerkish symbols. We then presented the chimpanzees with the *names* of 17 other foods and tools and asked them to categorize these additional names as foods or tools.

Two chimpanzees were able to do this on the first trial, without specific training while the experimenter was absent from the room. In order for them to make a categorical judgement of this sort on the first trial, it was necessary for the chimpanzees to recall some sort of image of the actual object, since the specific names of these foods and tools had never been paired with these categorical labels. Simply stated, after being taught to call a stick, a piece of money, and a key "tools," and to call an orange, a piece of bread, and a piece of beancake "foods," the chimpanzees were asked what they would call sponges, levers, straws, magnets, candy (M & Ms), potatoes, bananas, and so forth, when presented only with the *names* of these other objects. Only by knowing what these names represented, could they reliably answer such a question.

WORDS AND THEIR MEANINGS

There exists no empirical evidence, collected under conditions that preclude human cuing, that demonstrates that chimpanzees can use symbols in a referential manner.[5,6] Several reports, from both the Lana and the Washoe projects, indicate that chimpanzees can use symbols, in combination or singly, to obtain a wide variety of things from their human experimenters.[10,11] However, when the chimpanzees use these symbols, it is not clear that they comprehend their referential function and that they use them to convey information. Terrace[5] has recently shown that the sentences of apes are not like the sentences of children in many important regards. We suggest that this is because most, if not all, of the vocabulary signs or symbols learned by the apes Nim, Washoe, Sarah, Koko, and Lana did not have referents. These apes knew when to say many words, but they did not know that for their human listeners, these words represented things.

It is not, we contend, the presence or absence of a systematic syntactical language system that separates man from ape. Rather, it is the emergence within the human species of a simultaneous array of referential and representational skills, *none* of which occurs in the ape in its natural state. About the time human children begin to talk, they also begin to use objects upon other objects, to draw, to engage in symbolic play, and to order and categorize groups of objects in play.[9]

THE BEGINNINGS OF REFERENCE

Prior to the onset of the study reported in this paper, each of the three chimpanzee subjects, Lana (8 years), Sherman (5 years), and Austin (4 years), had received extensive experience with the Yerkes computer-based language training system. Details of their previous training and accomplishments can be found elsewhere.[5-8, 11-15]

The Yerkes computer-based system is simple in concept and in design. Nine geometric symbols are combined in arbitrary ways to form the equivalents of written English words.[†] These symbols are displayed on vertically mounted boards throughout the laboratory. Under each symbol is a touch-sensitive plate. Whenever a human, or a chimpanzee, makes contact with that plate, the symbol brightens, is displayed upon a row of projectors above the board, and is entered in the permanent computer record. When a symbol is lighted, it is reacted to by the experimenters as a spoken word. When the experimenters wish to communicate with a chimpanzee, they engage the animal's attention and then touch various symbols. The symbols need not be mounted on these boards, nor is it necessary that they brighten in order to be responded to by the chimpanzees.[‡]

The words used in the present study and their Yerkish equivalents are shown in TABLE 3. The original training of these items was similar for Sherman and Austin, but differed for Lana.

From the initial phases of training with Sherman and Austin, we hoped to achieve interanimal communication. Therefore, as soon as they were able to use food lexigrams to request foods, the experimenters began to use these lexigrams to transmit information about hidden foods. We then required that they use these lexigrams to transmit similar information between themselves. Next, they learned to regulate the exchange of a variety of foods between themselves by use of these symbols.[7,8] To accomplish these tasks, it was necessary for them to be able to use these food symbols for three distinct semantic functions: (a) the *request function* — to use specific symbols to request specific foods when a variety are available; (b) the *labeling function* — to use

---

† Lana originally learned a syntactical system of symbol ordering that was referred to as "Yerkish." The syntax of "Yerkish" was not equivalent to the syntax of English. Sherman and Austin have not been taught "Yerkish" syntax. With Sherman and Austin, the experimenters have employed typical English word order, although they regularly omit conjunctions, articles, etc. For Lana, the background color of the symbol originally indicated word class. However, all of the words used in the present study were assigned a single background color to prevent confounding. For Sherman and Austin, background color has not indicated word class.

‡ Symbols may be combined by simply touching them in sequence. Each symbol production is evaluated by an experimenter immediately after it is completed. This is done by means of a small hand coding device located near each room. The evaluation of the symbol production encodes (1) the name of the individual who lighted the symbols, (2) whether or not the symbol production was correct, given the context, and (3) the semantic function of the symbol production (statement, request, question, etc.).

symbols to label or indicate specific foods when a variety are available;§ and (c) the *receptive function*—to respond to the symbolic requests of others by selecting and giving or showing the item that is symbolically referenced.

Once Sherman and Austin were able to communicate with one another about specific foods, we attempted to increase their vocabularies by adding object names. They experienced difficulty learning to label reliably nonfood objects such as bowl, cup, blanket, and so forth.[7,8,12] We therefore introduced objects that had a specific function. More precisely, each object was used as a tool to procure food. Every tool was used in a specific manner, and for any given problem, only one tool would suffice. The chimpanzee's task was to decide which tool was needed and then to use the keyboard to request that tool from the experimenter. As we had done with the food symbols, once Sherman and Austin could use symbols to request specific tools of the experimenter, we placed them in a situation that required them to use these symbols to request tools of one another. Again, in order to be able to accomplish their task, Sherman and Austin needed to be able to use these tool symbols for three distinct semantic functions: (a) the *request function*—to decide, based upon a visual survey of their environment, which type of tool was needed to procure the embedded food and to select the proper symbol to request that tool; (b) the *labeling function*—to name a tool properly, regardless of whether or not they had an immediate use for the tool, and to name one tool while using another;* and (c) the *receptive function*—to select and give or show a tool from the set of all tools in response to the use of symbol by others [the chimpanzees must be able to respond to the requests of others for tools even when they see no need for tools themselves (i.e., when they cannot see the hidden food or when no food has been hidden)].

Thus, the experience of Sherman and Austin with food and tool symbols, prior to the onset of the present study, involved a variety of specific skills and was organized around the exchange of symbolically encoded information about foods, tools, or both. They shared every reward obtained by their mutual communication and they accompanied their mutual symbolic communications with a constantly increasing repertoire of facial expressions and gestures. With both food and tool symbols,

§ *Labeling* differs from *requesting* in that when the symbol is functioning as an indicator or label, the use of the symbol is not followed by consumption of the food. In order to show that a chimpanzee can actually *label* foods, it is necessary for him to reliably and repeatedly name one food while receiving another food as a reward for the correct label.

* This skill is necessary in order to insure that the chimpanzee has separated the name of the tool from the use of the tool itself. Initially, Sherman and Austin could not label tools accurately without also using them.

they had learned: first, to identify and request; second, to label but not consume or use; and third, to select and give to another individual in response to a symbolically encoded request.

Throughout this training, emphasis was placed upon communicative referential specificity within a class. Global or ill-defined requests for objects in the possession of others were unsatisfactory. The informative value of these interchimpanzee communications lay in their ability to identify specific referents, the nature of which could not be precisely determined from context alone.

Lana's training differed from that received by Sherman and Austin in a variety of ways. Because she was a pilot animal, our initial strategy had been to determine whether or not she could recognize and respond to the symbols. Also of interest was whether or not she could sequence them and what sort of ordering rules she might acquire. Lana initially learned to request foods that she saw loaded into vending devices out-side her room, and later to name these foods without eating them, but was not taught to select and given these foods in response to the re-quests of others. She also learned to label various objects that were held up outside her room and to give the color of these objects. Although she was allowed to play with these objects occasionally (box, shoes, bowl, can, etc.), the objects served no *specific* and immediate function for her, and it was not possible to distinguish between an instance of "request" and one of "label" because the physical environment did not dictate for Lana a need for a specific object. Lana initially learned to *label* nonfood objects because of the food reward she received for a correct response. Sherman and Austin initially learned to *request* specific nonfood objects by name because they needed these objects to obtain embedded foods. These and other important differences in training strategies between Lana and Sherman and Austin are summarized in TABLE 1.

Prior to the onset of the present study, we attempted to eliminate some of the skill deficiencies listed in TABLE 1. At this time, Lana weighed over 70 pounds, and while she enjoyed playing with Sherman and Aus-tin, she could not be induced to work cooperatively with them. We began by attempting to improve upon Lana's receptive competencies, but she did not want to hand the experimenter portions of food displayed in front of her, and, therefore, we used only photographs. We intro-duced the eight tools used by Austin and Sherman to Lana, and our manner of instruction with Lana followed exactly the procedures that had been used with Austin and Sherman. Lana learned to request tools rapidly. Her rate of acquisition in this task was comparable to those of Austin and Sherman, with one important difference—Lana was able to *label* all the tools correctly long before she could *request* them correctly. By contrast, even when Sherman and Austin could request all the tools

TABLE 1

SYMBOL-USE SKILLS ACQUIRED BY THE CHIMPANZEES
SHERMAN, AUSTIN, AND LANA

Sherman and Austin	Lana
(a) Not taught.	(a) Use multiple symbols and sequence all symbol productions.
(b) Request specific foods to eat.	(b) Request specific foods to eat.
(c) Label foods without eating them	(c) Label foods without eating them
(d) Respond to information about hidden foods provided by symbolically encoded statements of others.	(d) Not taught.
(e) Give foods in response to symbolically encoded request of another.	(e) Not taught.
(f) Cooperatively divide and share food with another chimpanzee by means of symbols.	(f) Not taught.
(g) Request objects for a specific use— tools used to procure foods.	(g) Not taught until after skills (h)–(j) were acquired.
(h) Label objects without using them.	(h) Label objects without using them.
(i) Not taught.	(i) Label colors of objects.
(j) Not taught.	(j) Give either color or name of object as requested.
(k) Give objects in response to symbolically encoded requests of others.	(k) Not taught.
(l) Cooperatively request of and give tools to another chimpanzee by means of symbols.	(l) Not taught.

correctly, they still experienced considerable difficulty labeling them and often it was necessary to demonstrate the function before they could recall the label. Accurate labeling, divorced from actual usage, required several months of practice by Austin and Sherman. Lana, however, continued throughout her training to experience difficulty requesting tools which she could easily label. This difficulty suggests that the ability to produce the label for a specific tool, such as a key, when a key *is needed* to open a lock, may not necessarily be related to the ability to label the key when it is displayed. In the case of Sherman and Austin, the demonstration of object function helped them to recall its name, suggesting that the label of the object and the use of the object were internally linked to a greater degree for Sherman and Austin than for Lana. In Lana's case, the two skills appeared to be separate, as if she had learned two distinct tasks with the same symbol, but had formed no conceptual connection between these skills. Once Lana had learned to request and label tools correctly, we were able to get her to cooperate, attend to, or work with Austin and Sherman. Her age and her differential rearing apparently worked against any ability she might have had to learn to use her symbols to transmit information to other chimpanzees.

SORTING AND LABELING ACCORDING TO FUNCTION

At the beginning of the present study, Lana's vocabulary consisted of 108 symbols, and that of Sherman and Austin, 72 symbols. As we have repeatedly noted, statements regarding vocabulary size merely reflect the number of symbols available to the chimpanzees for use. The competency with these symbols varies considerably, across the symbols themselves and across animals.

Prior to any training with the categorical classifications of "food" and "tool," we conducted blind tests of the skills of all three animals on the specific food and tool names that were to be used in the present study. Test results of request skills, labeling skills, and receptive skills are given in FIGURE 1. In each case, the tests were conducted with the ex-

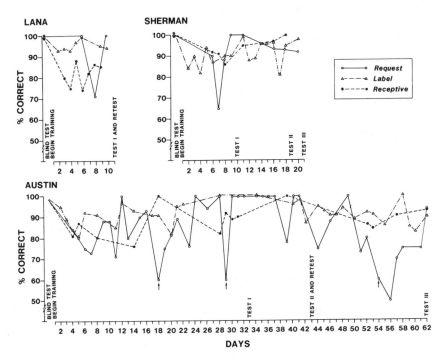

FIGURE 1. This figure shows scores on blind tests prior to categorization training and scores throughout training. The scores during training drop from 100% because additional items not as well learned as those used in this study are also presented during these regular review sessions. The very low scores for Austin (indicated by arrows) and the one low score for Sherman reflect a decrement in performance that is due to (or at least coincides with) either the introduction of a new experimenter or a new tool name. These scores reflect performance on the 19 items used in the blind tests and on 10–15 additional items.

perimenter out of the room. The design of the Yerkes computer-based system permits the experimenter to evaluate and respond to the chimpanzee's symbol productions even though the chimpanzee is out of sight. FIGURE 2 depicts the blind test conditions under which the data presented in FIGURE 1 were collected. During these blind tests, all vocabulary symbols on the animals' keyboards were lighted.

Although the blind tests demonstrate that all three chimpanzees could select the appropriate objects and symbols without experimenter cuing, they do not reveal how well these skills were maintained across days as the new symbols "food" and "tool" were acquired. This information is also given in FIGURE 1, which depicts the daily performance on the request, label, and receptive tasks throughout the duration of the recent study. These data were collected under essentially the same blind conditions, with the exception that the experimenters did go into the chimpanzees' room and help if the chimpanzees experienced repeated confusion on an item.†

Categorical sorting of foods and tools were begun by requiring the animals to sort three foods (orange, bread, and beancake) into one bin, and three tools (key, money, and stick) into another. Identical bins were placed directly in front of the animal, with a piece of food in one bin and a tool in the other. (The left–right locations of these bins differed among the animals, but not for a given animal.) The initial piece of food or tool was placed in the bin to serve as a marker, and each training item was used equally often as a bin marker. The chimpanzee was then handed either a food or a tool to be sorted. None of the foods or tools resembled each other physically, thereby precluding a match-to-sample response. The dimension for sorting was a functional one, i.e., the foods could be eaten and the tools could not. Initially, the chimpanzees evidenced difficulty with this functional sorting task.‡ Their responses were hesitant, they looked to the experimenter for cues, and they tried to eat the food. We therefore attempted to increase the saliency of the sorting dimension by allowing them a small bite of each piece of food that was to be sorted and by encouraging them to demonstrate manually the use of each tool. Their sorting abilities improved and we were able to eliminate these pre-sorting behaviors. The animals were reinforced with food for each correct response. The type of food reinforce-

---

†These reviews included all of the tools and foods used in the present study plus a variety of other items. Difficulties experienced during these reviews were not with items used in the present study, but with newer vocabulary items not yet adequately mastered. Virtually all review was done under blind conditions to continually discourage any situations that might lend themselves to cuing problems. In new tasks, however, the experimenter purposefully stays with and helps the chimpanzee.

‡ Chimpanzees have been reported to sort objects, but these sorting responses have been based on physical similarities, not functional ones (Reference 16).

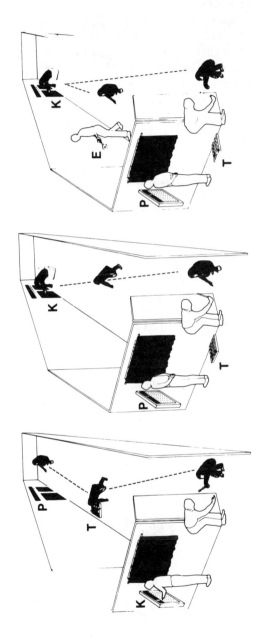

FIGURE 2. Blind tests of receptive skills are shown in the left panel. The experimenter (standing outside the room) uses his keyboard (K) to request an item while the chimpanzee (seated in the room) closely watches the projectors (P) above his keyboard to see which item is being requested. When the item is flashed on his projectors, the chimpanzee goes to the bin of objects, photographs, and tools (T), where he selects the correct item, and takes it to the experimenter.

Blind tests of labeling skills are depicted in the center panel. The experimenter selects an item from the bin of objects, food, and tools (T), and shows it to the chimpanzee. The chimpanzee looks at the item, then walks to his keyboard (K) to select the proper symbol. The experimenter turns to observe on his projectors (P) the symbol that is selected.

Blind tests of tool request skills are shown in the right panel. The experimenter enters the room and places food in one of many embedded food sites (E). The chimpanzee watches the food placement with his keyboard (K) disabled. The experimenter then leaves the room and activates the chimpanzee's keyboard. The chimpanzee then selects the symbol that represents the tool which the chimpanzee needs to extract the embedded food. The experimenter watches his projector (P) to determine which tool the chimpanzee requests. As the chimpanzee approaches the door, the experimenter gives him the requested tool.

ment varied from day to day, since the animals were allowed to select the foods which they wished to eat.§

The reader will realize at this point that it would be possible for the animals to learn the above sorting task in either of two ways: (a) by forming a specific association between each item and the relative location (left vs right) of the appropriate bin for that item; or (b) by formulating a classification rule that says to sort all inedibles into one bin and all edibles into the other. We did not know which solution the animals had adopted, or even if they had all adopted similar solutions. However, the fact that training was facilitated in all three cases by the emphasis of the *functional* distinction suggested that the animals might be classifying these items along the edible–inedible dimension. The performance of the animals during the acquisition of sorting skills is shown in Figure 3. It is clear that Lana learned this task far more easily than Austin or Sherman.

When the animals reached a sorting criterion of 90% or better across 60 trials, we introduced the lexigrams for *food* and *tool*. These lexigrams were initially located on the animals' keyboard in the same relative positions as the sorting bins, and the bins were placed directly under the correct lexigram. The specific names of the training foods and tools (orange, beancake, bread, money, stick, and key) were rendered inactive (by turning off the dim light behind the key that signals the active state of the key), but all other keys remained active. The chimpanzee's task was to sort a food or a tool into the proper bin and then to select the lexigram representing either food or tool.¶

Once the chimpanzees reliably selected a key after sorting each object, the bins were removed. The chimpanzees's task then was to label each of the six training objects as they were held up by the experimenter.* Training in this phase continued until the animals met all of the following criteria: (a) ability to label all training items correctly without eating the food or using the tool; (b) ability to label all training correctly *on trial 1*, after food and tool lexigrams were relocated on the keyboard; (c) ability to label all training items correctly with the experimenter out of the room; and (d) ability to label all training items correctly under the conditions listed in items (a)–(c) above for two consecutive sessions of more than 25 trials at 90% or above correct.

§ They did this either by requesting specific foods at the keyboard or by going to the refrigerator or the cabinet and pointing to the food they wanted. Often they chose to mix together a variety of foods.

¶ Many keys were actually located above each bin, but the keys were familiar to the subjects and were not selected because they did not label any of the specific training items. The only new keys above the bins were the keys "food" and "tool."

* If the animals began to make frequent errors, they were again given small bits of the training food and asked to manually demonstrate the use of the tools prior to labeling any item as either a "food" or a "tool." The lexigrams for food and tool were then moved randomly about the keyboard.

The number of training trials needed to acquire these skills varied widely across the three chimpanzees, as can be seen in TABLE 3. Although we cannot be certain, we believe that Austin's learning difficulties stemmed from a number of unpleasant experiences and illnesses during this period.† It will be recalled that during the sorting phase, we did not know whether the animals had learned to place specific objects in specific bins, or whether they had abstracted a more general concept, i.e., "This bin is for items that I eat and the other bin is for items that I do not eat." Likewise, in this phase, we did not know whether the keys "food" and "tool" had become secondarily associated with specific objects so that the illumination of those keys was simply a conditioned response to object presentation, or whether a more general functional concept had emerged and become linked to those keys. If such a concept or rule had emerged and had become symbolically encoded by the symbols "food" and "tool," then we would expect that the chimpanzees could use these generic symbols to categorize other items with no additional training.

We tested the generalizability of this skill by presenting five additional foods and five additional tools. These were foods and tools with which the animals were highly familiar and for which they had learned specific lexigram symbols. These items were presented once each, in a random order, interspersed with trials of training items. Prior to the presentation of the first novel item, the animals had to respond correctly to the training items for 20 consecutive trials. During the entire test, the experimenter remained outside the room, as shown in FIGURE 2. The chimpanzee approached the door, looked at the object displayed by the experimenter, then re-entered the room and labeled the item on the keyboard. The experimenter could see neither the keyboard nor the chimpanzee once the chimpanzee left the doorway. The specific names

---

† He contracted chicken pox, which interrupted the training sessions for one week. When he began working again after his illness, he seemed to have forgotten the task entirely and it was necessary to resume sorting training. When we again placed the sorting bins under the keyboard, Austin could not label properly until we placed a piece of tape on the keyboad leading from the correct key to the sorting bin below. At this point, a new experimenter was hired and began working with Austin. This individual found chimpanzees difficult to get along with, and Austin did not work well with her. Every time she attempted to relocate the positions of the keys, Austin would stop attending, begin to alternate, then rock, and refuse to work. When this individual left the project, Austin began working well with others more familiar to him. However, he had now developed a fear that he would be left alone (we had left him with the new person when he did not want us to) and he would perform well only when we were in the room. When we went out of the room in preparation for a blind test, he would again start to alternate between keys, rock, whimper, and eventually refuse to use the keyboard. It was necessary to slowly decondition this fear by moving toward the door only a few inches each day. Thus, in Austin's case, it is necessary to take into account a variety of psychosocial factors when viewing his performance. Neither Lana nor Sherman encountered any such problems and, in their case, the differences reflect performance factors more directly.

TRAIN   **TEST**

SORTING OBJECTS

	Total Trials to Criterion	Total Errors During Training
Lana	160	19
Sherman	1115	200
Austin	1210	252

	Correct/Incorrect
Lana	10/10
Sherman	Not given
Austin	Not given

LABELLING OBJECTS

	Total Trials	Total Errors
Lana	1493	199
Sherman	852	68
Austin	3239	429

	Test	Retest
Lana	3/10	1/10
Sherman	9/10	
Austin	10/10	

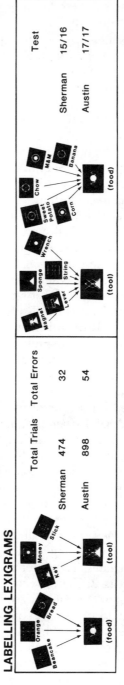

FIGURE 3. This figure serves as a flow chart of the present study. The animals learned the items on the left and were tested, in a blind setting, with the items on the right. The numbers of trials and total errors are given for training and the number of correct trial 1 selections is given for testing.

of the 10 novel exemplars were deactivated during this test, but all other keys remained on (FIGURE 3).

Austin correctly categorized each of the ten novel items on trial 1. Sherman correctly categorized each of the ten novel items except sponge, the tool which he occasionally eats portions of as he uses it (Austin does not eat sponges). Lana correctly identified only three items (chance would be five). Thus, it appeared that Sherman and Austin had acquired a concept of "food" and "tool" that was functionally based, generalizable, and symbolically encoded. Lana had not; she had learned specific associative responses. This raised three questions: Had the differences in the training that Sherman and Austin had received versus that which Lana had received enabled Sherman and Austin to: (a) conceptualize functional relationships that Lana did not? (b) symbolically encode referential relationships that Lana perceived but did not encode at a symbolic level? or (c) recall specific food–food or tool–tool associations that Lana had not mastered during her training?

In order to answer these questions, three additional tests were given in this phase. First, Lana was simply retested with the novel items to determine whether or not the first test might have been in error. On this second blind test, Lana correctly identified only one novel item. We then turned off Lana's keyboard and returned to the initial sorting procedure. Following the same pre-test criterion (20 correct consecutive trials), we began to randomly intersperse the ten novel items used in the two tests described above. This test was also given blind; Lana simply came to the door, took the item from the experimenter, then went into her room and sorted it into one of the bins. Lana sorted all ten novel items correctly on trial 1, thereby indicating that her failure on the earlier tests had not been due to an inability to conceptualize the functional relationships between the foods and the tools. Rather, it had been due to an inability to encode symbolically this perceived relationship.

To determine whether or not learned second-order food–food associations and tool–tool associations could account for the abilities of Sherman and Austin, we presented them with 28 items (14 foods and 14 tools) with which they were generally familiar, but which had not been used in any specific training paradigm and were, therefore, not associated with lexigrams. These items were common household items and are listed in TABLE 2. Sherman correctly categorized 24 of these 28 items and Austin correctly categorized 25 of 28 items. All but one of their errors were errors of a single type: tools were termed foods. Most of these tools were those which are used in food preparation. Since Sherman and Austin often help in food preparation and afterward, they are allowed to lick these tools, the salient "function" of such tools for Austin and Sherman is to apparently eat them. These "edible tools"

TABLE 2

UNNAMED ITEMS PRESENTED FOR CATEGORIZATION TESTING*

Tools	Foods
Scrub brush (A)	Ice Cube
Shovel	Peanut
Screw driver	Celery
Juice squeezer (S)	Can opener
Steel ball bearing	Peanut butter
Cage locking pin	Jelly
Spoon	Raisins
Cooking pan (A)	Cabbage
Hammer	Grapefruit
Sink stopper	Cucumbers
Knife (S)	Chimp crackers
Scissors	Turnip
Cutting board (S)	Potato (white)
Can opener (A & S)	Lemon

* Errors made by Sherman are marked "S," those by Austin, "A."

were: can opener, cooking pan, knife, juice squeezer, and cutting board. (The remaining error was scrub brush.) It is clear that the ability of Austin and Sherman to categorize objects as foods or tools goes far beyond any specific training that they have received with named foods and tools.‡

CATEGORIZING PHOTOGRAPHS

The purpose of Phase III was to broaden the categorization abilities of these chimpanzees in order to permit them to categorize not only real and present objects and foods, but also nonedible representations of objects and foods, in this case, photographs. Only Sherman and Austin were continued in this later phase, since Lana's inability to encode referential relationships symbolically implied that in her case it would be fruitless to move from real objects to photos.

During Phase III training, we used the same foods and objects that were used in Phase II training: stick, key, money, orange, bread, and beancake. We began by taping photographs of these objects to the objects themselves, and we then held up the training object and its photograph and asked the chimpanzees to label it as a food or an object. The photographs were encased in plastic (Lexan) to prevent the chim-

‡Prior to the presentation of these items, Sherman and Austin again had to reach a criterion of 20 consecutive correct responses on the training items. These tests were also conducted in the blind setting described above.

panzees from bending and tearing them. The same training photographs were used throughout the training in this phase: however, Sherman and Austin regularly identified novel slides of foods and tools projected above their keyboard during their regular review work, which continued during this study.§

Sherman readily transferred the categorization skills acquired with real objects to photographs, and, by the end of the second training session, he accurately labeled all the photographs without being shown any of the actual items. Austin, however, experienced difficulty: when the photographs were no longer accompanied by the actual objects, he tended to call all of them tools.

All of the photographs were, themselves, inedible, and we wondered if this caused Austin to discard the basic dimension of distinction, i.e., items that could be eaten versus those that could not. It was necessary to re-pair the actual objects and foods with their photographs for many trials, in Austin's case. Training on photographs continued until the chimpanzees reached the same criterion with photographs that they had achieved with actual objects. Novel photographs were then presented. The order of presentation was random, test items were interspersed with training items, and the experimenter was out of the room. The novel photographs were of the following items: wrench, magnet, straw, sponge, candy (M&Ms), banana, sweet potato, chow, and corn.¶

Sherman correctly labeled all nine novel photographs (100%), while Austin labeled only five novel photographs correctly (55%) (FIGURE 3). This suggested either that Austin had simply learned specific responses to specific photographs, or that when presented with novel photographs, he had, for some reason, not treated these photographs as representations of real objects.

In order to determine whether the training with photographs had somehow altered Austin's ability to categorize these items as foods or tools, we repeated the test with the real items instead of the photographs. He labeled each of these novel items without error or hesitation, just as he had done when initially tested on these items. This indicated

---

§These photographs were not changed simply because it was expensive and time-consuming to cut and seal the plastic (Lexan) casings, and no experimental need for different photographs existed. The critical event was to have Sherman and Austin respond categorically to the pictures as they did to real objects.

¶ The lever, which was included in Phase II, was not presented during the Phase III test because we had no photographs of this item that the animals could reliably identify. This was due to our inability to take a good picture of this tool, not to any particular difficulty on the part of the chimpanzees. Professional photographs of this tool (made later) were easily identified.

that the training with photographs had not interfered with his ability to categorize real objects.

We then readministered the blind novel photographs test, believing that it was possible that Austin had treated all the novel photos as pieces of plastic and had guessed on these trials instead of looking closely at the object depicted in the photo in order to make a correct categorical response. The plastic casing surrounding the photographs frequently reflected back enough light to render the enclosed picture invisible at certain angles. Sherman accommodated for this by moving his head to change his line of regard. Austin at times also accommodated his line of regard, but at other times he did not. Consequently, he often made naming errors with plastic-encased photographs but rarely made errors with slides of objects. During this test, as Austin came to the door of the room, we encouraged him to look carefully and slowly at each picture and we rotated the angle of the photo. Under these conditions, Austin correctly identified nine of the nine novel photographs. Since there had been no interim opportunity for Austin to have learned the correct response to give to these particular photos, it is reasonable to conclude that, in this case, the trial 2 data, although not as strong as the trial 1 data, suggest that Austin, like Sherman, was able to categorize not only novel items but also novel photographs of items.*

Categorizing Symbols

This phase of the study is the most critical of all, for it alone is clearly and unequivocably a test of the referential value of the symbols. Unlike all other tests of this capability with apes,[1-4,12] this test included the following critical constraints: (a) it required a completely novel response, one never before given by the chimpanzee *or by the experimenter* when working with the chimpanzee; (b) it was administered under blind conditions; and (c) it required that the chimpanzee use one symbol to classify another, thereby forcing him to cognitively refer to the specific referent of one symbol and, based on the recalled perceptual characteristics of that referent, assign it to a class of functionally related items — he then had to recall the symbol that had been used to reference that functional relationship in the past and to employ it. The first two condi-

---

* If the hypothesis that Austin initially looked at the plastic casing, not through it, is correct, then we would predict that Austin would have less difficulty in the final phase when the lexigrams themselves were shown to him to be categorized. While the final task is more difficult conceptually, it is simpler perceptually, since the lexigrams were not enclosed in plastic because they did not taste interesting, as did the photographs, and the chimpanzees were not inclined to lick them.

tions are simply standard test conditions that should be employed (but have not been) when the purpose of the test is to determine whether or not the capability under question can be attributed to the chimpanzee being tested and not to incidental learning or to experimenter cuing.

Training in this phase began as it had in the previous ones. We returned to the three training foods and the three training tools. Initially, the lexigrams for these items were taped to photographs of the items, again to provide a bridge between the levels of stimuli presented.† We did this not because we felt that the chimpanzees could not easily move between responding to real items, photographs, and lexigrams, but because we knew that they otherwise might perceive the task as one that was entirely new, and, instead of applying a categorical assessment strategy, they would begin by guessing until they discovered the parameters of the task. Because we were interested in trial 1 data, we wanted to make certain they understood that, simply because we were using different stimuli, we did not expect them to try alternative interpretations of the task question.

As in the previous phase, we then began to remove the object or photo and asked the chimpanzee to categorize the Yerkish symbol alone as either a food or tool. Sherman demonstrated rapid transferral and, within three sessions, had met the test criteria (TABLE 3). He was, in fact, able to categorize the training lexigrams alone at 90% correct from the first presentation. Austin, however, again proved to have difficulty. Whenever he made an error, he began to refuse to look at the stimuli and simply to alternate between the keys "food" and "tool." His lengthy acquisition phase represents attempts on the part of the experimenter to make the task easy by delaying the removal of each additional real object until Austin evidenced self-confident responding and showed no tendency to lapse into an alternation strategy after each error.

Once the chimpanzees had reached the training criterion, they were presented with the lexigrams shown in TABLE 3 and FIGURE 4. This test was also administered with the experimenter out of the room, and again novel items were interspersed randomly with training items. *It is to be emphasized that prior to this test, these chimpanzees had never been asked to make a categorical assessment of these symbols.* In some cases (starred), they had not been requested to make a categorical assessment of the real objects that these symbols represented. Because the Yerkish symbols are arbitrarily assigned to all objects, it is not possible to decide, simply by looking at the symbol, whether it represents a food or a tool. Only

---

† Because of the plastic casing reflection problems that Austin had shown during the previous training, we also taped the lexigrams to real objects for a few sessions in this case.

if one can recall its referent can such a decision be made. Sherman categorized the lexigrams correctly on 15/16 novel trial 1 presentations, and Austin categorized them correctly on 17/17 novel trial 1 presentations (FIGURE 3).

The physical arbitrariness of the lexigrams used in this task is illustrated in FIGURE 4. On the left of TABLE 3 are the English glosses of the lexigrams which Sherman and Austin were asked to classify as foods or tools. The table lists the choices the chimpanzees would make it they were simply to choose either "food" or "tool" based on the physical similarity of these lexigrams relative to each specific lexigram displayed by the experimenter. It is clear that the choices made by Sherman and Austin were not controlled by the degree of physical similarity between the categorical lexigrams and the specific lexigrams. If they had been, Sherman and Austin would have been correct only half the time. Between them, only a single error was made on the entire test and that error (sponge was called a food by Sherman) was one that would not have been predicted on the basis of physical similarity. In fact, the lexigram for sponge looks more like the correct categorical lexigram for tool than does the specific lexigram representing any other tool. The only explanation for this mistake is the fact that Sherman often eats small parts of the sponge, which is filled with juice, and, thus, to him it is more appropriately classified as a food than as a tool.‡

‡ There has been a recent suggestion (Reference 17) that it is possible to unintentionally cue chimpanzees by the way in which one holds objects, touches the chimpanzees, looks at the stimuli, and so forth. Those who work closely with apes are aware that such cues are not the ones to which the chimpanzee primarily attends. The subtle cues that the ape looks for are signs that its response is correct or incorrect *as it is in the process* of making that response. Typically, a chimpanzee will start to form a sign, touch a key, or select a chip, and then look quickly at the face of the experimenter for signs of approval or disapproval. The slightest body or facial movement can cause an alteration in the response. We have eliminated this problem by requiring that the chimpanzee select a key when there is no experimenter present to whom a questioning glance can be directed. While this prevents the experimenter from cuing an in-progress response (and it also stops the chimpanzee from looking for cues), it does not preclude other forms of cuing. It would be possible, for example, that the experimenter held all symbols to be labeled "food" somewhat higher than the symbols that were to be labeled "tool." Since three different experimenters participated in all phases of testing, they would have had to be in collusion, and also extraordinarily careless, in order to allow a consistently reliable cue of this sort to determine the responses of three different chimpanzees (additionally, we would have had to be deliberately omit such a cue when testing Lana). In any event, in order to preclude such criticism, we reran the final phase of the study with Austin and Sherman. During this retest: (1) the experimenter did not know which lexigram the chimpanzee was viewing, (2) all lexigrams used in TABLE 3 were presented, and (3) lexigram presentation was completely random; *any* lexigram could be followed by itself or by any other lexigram any number of times. No constraints were placed on the number of consecutive food and tool responses. Sherman was correct on 68 of 70 trials and Austin was correct on 65 of 70 trials. All trials were given in a single session.

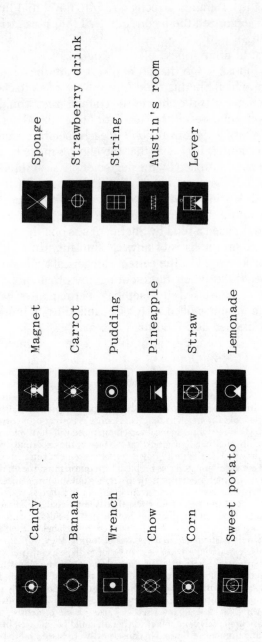

TABLE 3

PHYSICAL SIMILARITY BETWEEN CATEGORICAL LEXIGRAMS ("FOOD" AND "TOOL") AND SPECIFIC LEXIGRAMS*

Categorical Items (Tool or Food)	Elements in Common with Food Lexigram	Elements in Common with Tool Lexigram	Number of Uncommon Elements	Physical Similarity Prediction	Correctness of Prediction Based on Physical Similarity	Correctness of Response	
						Sherman	Austin
1. Candy (M&Ms)	2	1	1	Food	Correct	R	R
2. Banana	0	1	1	Tool	Wrong	R	R
3. Wrench	1	1	1	Either	Chance	R	R
4. Chow	0	2	1	Tool	Wrong	R	R
5. Corn	1	1	2	Either	Chance	R	R
6. Sweet potato	1	2	1	Either	Chance	R	R
7. Magnet	1	2	0	Tool	Correct	R	R
8. Carrot	2	2	1	Either	Chance	R	R
9. Pudding	1	0	1	Food	Correct	R	R
10. Pineapple	0	1	1	Tool	Wrong	R	R
11. Straw	0	2	2	Tool	Correct	R	R
12. Lemonade	0	1	1	Tool	Wrong	R	R
13. Sponge	0	2	0	Tool	Correct	W	R
14. Strawberry drink	0	1	1	Tool	Wrong	R	R
15. String	0	1	2	Tool	Correct	R	R
16. Austin's room	1	0	1	Food	Wrong	†	R
17. Lever	1	1	2	Either	Chance	R	R
Total Correct						15/16	17/17

* Lexigrams are shown in FIGURES 3 & 4.
† Not given.

EXPANSION OF THE BIDIMENSIONAL SYSTEM

The goal of the final phase of this study was to determine whether or not it would be possible to expand the bidimensional categorization system of "food" and "tool." As stated earlier in this paper, although we have given the categorical lexigrams the English glosses of "food" and "tool," we are aware that the lexigrams themselves probably reflect the more global division of the world into things which can be eaten and things which cannot. We began to expand the dichotomization by breaking the edible class into two divisions, "food" and "drink," and the inedible class into "location" and "tool".§ At the time of writing, only Sherman has received training in this four-way task.

Training of these additional categorical labels followed the procedures used for the training of foods and tools, except that now the chimpanzee was required to make a four-way judgement (food, tool, drink, or location), and both real objects and photographs were intermixed as exemplars from the start.¶

Following acquisition of the four-way categorical distinction (TABLE 3), Sherman was presented with photographs and/or actual instances of three novel foods (yogurt, frozen lemonade, and a lemon), five novel drinks (liquid lemonade, oil, juice, milk, and strawberry Kool-Aid), two locations (sink and evening housing area), and one tool (wheel).* The unevenness across categories reflects the fact that these were the only available well-learned novel items left in Sherman's vocabulary at this time. Sherman categorized ten of these items correctly, his only error being milk, which was termed a food. Following this test, we attempted to teach Sherman to categorize milk as a liquid. Training notwithstanding, Sherman insists on classifying milk as a food.

---

§ The additional categories of locations and drinks were chosen because they reflected the only word classes that the chimpanzees had mastered at this point in training. Other word groups are used regularly by the animals (people's names and actions), but we have no evidence to indicate that the use of these words has passed beyond the pure performative level; that is, the animals knew when to use these words in a general way in order to achieve desired actions and rewards, but they did not give clear evidence of comprehending the specific referential function of these words.

¶ It was not necessary to demonstrate the function of these additional categories, as it had been when food and tool were first introduced. Two training exemplars were presented for the location group (photos of the kitchen area and outdoors) and three exemplars were used for drinks (orange drink, coffee, and Coke). Only two training exemplars were used with the location symbol because the number of specific location terms used by Sherman at this time was only four. The remaining two were withheld for the transfer tests.

* Photographs and actual objects were used as exemplars because, in the case of locations, it was impossible to use other than photographs. Therefore, photographs of some items in other categories were also presented to preclude the possibility that Sherman could merely label all photos as "location."

We have not yet determined whether or not these four-way categorization skills extend to those lexigrams that represent specific instances of each of these categories. Adequate tests of this ability must await vocabulary expansion.†

Conclusion

Arbitrary symbols can come to function in a referential manner in apes. When they do, it is possible for the ape to categorically organize these symbols and to respond to symbolically represented instances by producing the symbolic categorical label. These abstract referential labels can be applied to new members of the categorical subset on a trial 1 basis, without specific training. The ability to do this with categorical instances—the linkage of which is based on functional, rather than physical, similarity—implies that the ape can abstract and organize sensory input in a way that takes into account the ape's own actions upon its environment. However, this ability to organize sensory input along a dimension of functional similarity *does not necessarily* give rise to a corresponding ability to organize this information similarly at a referential symbolic level, as Lana's inability to transfer the food and tool labels to novel exemplars indicates.

We suggest that Sherman and Austin were able to treat "food" and "tool" as referential labels, and to expand the use of these labels to novel exemplars because, prior to this study, they had begun to use many other symbols in a referential manner. They, in contrast to Lana, were able to do this because of training that encouraged the appearance of functional symbolic communication between chimpanzees.‡

The present study offers a paradigm that makes possible, for the first time, an unequivocal determination of the presence or absence of referential symbolic function. The author doubts that other apes[1-6,12]

---

† Individuals who are interested in seeing the work reported in this paper may purchase a color video cassette which clearly depicts the chimpanzees performing all aspects of the present study. The tape shows the skills of the animals, their manner of acquisition, and the use of blind controls during testing. The tape is not self-explanatory; however, it does complement and expand upon the descriptions presented in the article.

‡ Although the training differences emphasized between Lana and Sherman and Austin refer to specific tasks, it should also be noted that all three animals have used symbols in many other communicative settings. We work with Sherman and Austin in an uncaged social setting, all day, everyday. They are never apart from one another, and are only apart from their human companions at night. The extent of their socialization, both with chimpanzees and humans, has been far more extensive than that of other apes. Lana's only close bond was with Tim and Gill and, although she did not evidence behavioral depression at his departure, it is possible that her performance in the present study would have been different had Tim been her teacher.

have reached the level of symbolic functioning achieved by Sherman and Austin, because training in other ape language projects has emphasized only the skills of labeling and combining. In any event, the paradigm presented here could be employed by others to test this view. We submit that, in ape language research, representational symbolic function should be demonstrated before instances of spontaneous symbol combinations can be presumed to reflect attempts to convey novel meanings. Ape language studies have rushed to show that Chomsky[18] was wrong regarding the innateness of a species-specific language acquisition device which programmed syntax. Far more would now be understood had they, instead, attempted to operationalize the characteristics that separated the behaviors they classified as "word usage" from those of simple associationistic processes, as Skinner suggested.[19]

SUMMARY

Two chimpanzees were able to abstract functional definitions of edibleness and inedibleness from six training items and to attach labels to these functional distinctions. The meaning of these labels extended far beyond the training context. The chimpanzees were able to categorize objects in their world as human children do, on the basis of function. They then categorized their world of symbols in a similar manner. The paradigm used in this study provided a means of determining whether or not the symbols used by language-trained apes were functioning in a referential manner. Referential symbolic function was offered as a necessary, although not sufficient, condition for the emergence of language.

REFERENCES

1. GARDNER, B. & R. GARDNER. 1971. *In* Behavior of Nonhuman Primates. A. Schrier & F. Stollnitz, Eds. Vol. 4: 117–183. Academic Press, New York, N.Y.
2. FOUTS, R. 1973. Science 180: 978.
3. PREMACK, D. 1972. Science 172: 808.
4. RUMBAUGH, D., E. VON GLASERSFELD & T. GILL. 1973. Science 182: 731.
5. TERRACE, H. 1979. Science 206: 891.
6. SAVAGE-RUMBAUGH, E., D. RUMBAUGH & S. BOYSEN. In press. Am Sci.
7. SAVAGE-RUMBAUGH, E., D. RUMBAUGH & S. BOYSEN. 1978. Science 201: 641.
8. SAVAGE-RUMBAUGH, E., D. RUMBAUGH & S. BOYSEN. 1978. Behav. Brain Sci. 4: 539.
9. NELSON, K. In press. J. Am. Acad. Child Psychiatr.
10. GARDNER, R. & B. GARDNER. 1978. Ann. N.Y. Acad. Sci. 309: 37.

11. RUMBAUGH, D., Ed. 1977. Language Learning by a Chimpanzee: The LANA Project. Academic Press, New York, N.Y.
12. PREMACK, D. 1976. Intelligence in Ape and Man. Lawrence Erlbaum, New York, N.Y.
13. SAVAGE-RUMBAUGH, E. & D. RUMBAUGH. 1979. *In* Language Intervention for Ape to Child. R. Schiefelbusch & J. Hollis, Eds. pp. 278–294. University Park Press, Baltimore, Md.
14. SAVAGE-RUMBAUGH, E. & D. RUMBAUGH. 1978. Brain Lang. 6: 265.
15. SAVAGE-RUMBAUGH, E. & D. RUMBAUGH. In press. *In* Children's Language. K. Nelson, Ed. Vol. 2. Halsted Press, New York, N.Y.
16. HAYES, K. J. & C. H. NISSEN. 1971. *In* Behavior of Nonhuman Primates, Modern Research Trends. A. M. Schrier & F. Stollnitz, Eds. Vol. 4: 59–115. Academic Press, New York, N.Y.
17. SEBEOK, T. & J. UMIKER-SEBEOK. 1979. Psychol. Today 13(6): 78.
18. CHOMSKY, N. 1965. Aspects of a Theory of Syntax. MIT Press, Cambridge, Mass.
19. SKINNER, B. F. 1945. Psychol. Rev. 52: 270.

The Clever Hans Phenomenon, Cuing, and Ape Signing: A Piagetian Analysis of Methods for Instructing Animals

SUZANNE CHEVALIER-SKOLNIKOFF*

Medical Anthropology Program
Department of Epidemiology and International Health
University of California, San Francisco
San Francisco, California 94143

INTRODUCTION

RECENTLY, IT HAS been proposed that the use of sign language by apes can be attributed to cuing, or, more specifically, to the Clever Hans phenomenon.[1-3] However, it is not clear exactly what these phenomena are, or how they relate to one another. The Clever Hans phenomenon and cuing often are equated and the two terms used interchangeably. At times the terms are applied to teaching and eliciting behaviors that are very different from Clever Hans's behavior. Most notably, the terms may be applied to the cuing of performances that are far more complicated, like seals' ball games or chimpanzees' tea parties. This paper represents an attempt to compare systematically the Clever Hans phenomenon and cuing with ape signing.

In order to compare these behaviors, one must have a scale of measurement that is broad enough to encompass all three, and to show how they are similar and how they differ. Piaget's model of human cognitive development during the Sensorimotor Period (birth to 2 years of age) will be employed as the scale of measurement.

PIAGET'S MODEL AND ITS APPLICATION TO SYSTEMATIC COMPARISONS

Piaget's model[4-6] offers a systematic and relatively objective framework for comparison. It is made up of six qualitatively different stages that appear sequentially during human development. Each stage in the

* Address reprint requests to Dr. S. Chevalier-Skolnikoff, 205 Edgewood Ave., San Francisco, Calif. 94117.

0077-8923/81/0364-0060 $01.75/0 © 1981, NYAS

series can be defined by a specific set of behavioral parameters that are characteristic of that stage.† These parameters reflect the development of progressively complex cognitive functioning. The six stages are characterized by increasing voluntary control, elaboration of the number of motor patterns and contextual variables involved in a single act, and increased variability of motor patterns. Thus each stage in the series is characterized by a new, more advanced level of functioning as new abilities become incorporated into the behavior.

The Clever Hans phenomenon, cuing, and ape signing will be examined in terms of two of Piaget's sensorimotor series, the Sensorimotor Intelligence Series and the Imitation Series. The Sensorimotor Intelligence Series pertains to the development of general sensory and motor adaptations to the environment. Its six stages are: (1) reflex; (2) primary circular reaction—repetitive self-oriented behavior with acquired adaptations to impinging stimuli; (3) secondary circular reaction—reaching out to the environment and repeatedly eliciting reactions in objects; (4) coordinations—combining behaviors to achieve goals; (5) tertiary circular reaction—experimenting with the relationships between objects, forces, and space; (6) invention or solution of problems through mental combinations or insight (TABLE 1).

The Imitation Series relates to the unfolding of a sequence of progressively more advanced forms of imitation. Its six stages are: (1) reflexive contagious imitation; (2) sporadic imitation of behaviors already in the subject's repertoire; (3) purposeful imitation of behaviors already in the subject's repertoire; (4) attempted imitation of new motor patterns; (5) imitation of new motor patterns through matching; (6) precise imitation of new motor patterns without matching, and delayed imitation (TABLE 2).

CLEVER HANS

During the early 1900s, a German trotting horse named Clever Hans astounded audiences by his ability to solve complicated mathematical problems, to read, and to answer verbal questions. He pawed the ground with his hoof to indicate the correct answers to numerical problems and to spell; he shook his head "yes" and "no," turned it "right" and "left," and moved it "up" and "down" to answer questions; and he named colors by picking up rags of the appropriate colors with his teeth. His owner-

---

† See Chevalier-Skolnikoff[7,8] and Parker[9] for further discussions on classification in terms of behavioral parameters. The rationale for classifying behavior into stages has been discussed in Chevalier-Skolnikoff.[10]

TABLE 1

CHARACTERISTICS OF THE SENSORIMOTOR INTELLIGENCE SERIES* AS MANIFESTED BY HUMAN INFANTS

Stage	Age (months)	Description	Major Distinguishing Behavioral Parameters	Example
1. Reflex	0–1	Stereotyped responses to generalized sensory stimuli	Involuntary Stereotyped	Roots & sucks
2. Primary circular reaction	1–4	Infant's action is centered about his *own body* (thus "primary") which he learns to repeat ("circular") in order to reinstate an event. First acquired adaptations occur	Self-oriented Recognizes objects and contexts Acquired adaptations to impinging stimuli Repetitive coordinations of own body	Repeats hand-hand clasping
3. Secondary circular reaction	4–8	Repeated (circular) attempts to reproduce *environmental* ("secondary") events initially discovered by chance	Environment-oriented Semi-intentional (initial act is not intentional, but subsequent repetitions are) Establishes object/action relationships Simple orientations toward a single object or person	Swings object & attends to the swinging spectacle, or to the resulting sound; repeats
4. Coordinations	8–12	Two or more independent acts become intercoordinated, one serving another	Intentional Establishes goal from onset Establishes relationships between two objects	Sets aside an obstacle in order to obtain an object behind it

Stage	Age (months)	Description	Major Distinguishing Behavioral Parameters	Example
			Coordinates several behaviors toward an object or person Applies familiar behaviors to new situations Begins to attribute cause of environmental change to others	
5. Tertiary circular reaction (experimentation)	12–18	Child becomes curious about an object's possible *functions* & about object–object relationships ("tertiary"); he repeats his behavior (circular) with variation as he explores the potentials of objects through trial & error experimentation	Behavior becomes variable and non stereotyped as the child invents new behavior patterns Repetitive trial & error experimentation Coordinates object–object, person–object, object–space & object–force relationships Considers others entirely autonomous	Experimentally discovers that one object, such as a stick, can be used to obtain another object
6. Invention through mental combinations (insight)	18+	The solution is arrived at mentally, not through experimentation	Mentally represents objects & events not present Solves problems mentally	Mentally figures out how one object can be used to obtain another object

* Abstracted from Piaget, and adapted from Chevalier-Skolnikoff.[7,8]

TABLE 2

CHARACTERISTICS OF THE IMITATION SERIES* AS MANIFESTED BY HUMAN INFANTS

Stage	Age (months)	Description
1. Reflexive contagious imitation	0–1	Reflexive behavior (e.g., crying) is stimulated by model's behavior
2. Sporadic "self-imitation"	1–4	The model's imitation of the infant's own motor patterns (self-imitation) elicits similar behavior in the infant. It is unclear whether the infant distinguishes the model's behavior from his own
3. Purposeful "self-imitation"	4–8	Self-imitation in which matching becomes more precise
4. Attempted imitation of new motor patterns	8–12	Attempts to imitate new behavioral acts, but often fails to match precisely
5. Imitation of new behavior patterns ("true" imitation)	12–18	Precisely imitates new motor acts, through repeated attempts at matching
6. First-try imitation of new motor patterns, delayed imitation	18+	Precisely imitates new motor acts, without preliminary attempts at matching, through symbolic representation, manifests delayed imitation

* Abstracted from Piaget, and adapted from Chevalier-Skolnikoff.[7,8]

trainer, Mr. von Osten, who had been a high school teacher, maintained that Hans was so clever because he had instructed him as one teaches a high school student. He provided the horse with elaborate visual paraphernalia, such as blackboards, letter cards, word flash cards, and number boards, to aid in his education, which took four years.

Circus trainers, cavalry officers, and experts in training dressage horses tried to ascertain how the horse was able to answer such complex questions—but all remained baffled. Finally, a German experimental psychologist, Oskar Pfungst,[11] solved the mystery using the double blind experiment; if no one in the horse's presence knew the answers, the horse was unable to answer the questions. In this way, Pfungst discovered that the horse was not actually understanding the problems or questions that were posed, nor was he calculating the answers through conceptual thought. Instead, the horse was acting upon extremely subtle, unintentional cues such as unconscious body inclinations and orientations, head movements and orientations, and vocal signals. He was able to pick up these cues from his trainer, from other people who questioned him, and even from spectators. Leaning the body toward or away from the horse were the cues that prompted him to start pawing and then to stop. Up-and-down or side-to-side movements of the trainer's head cued him to move or shake his head up and down or from side to side. And identifying colors was cued by a combination of orienting the body toward the rag and vocal rejection of incorrect choices.

Since Pfungst's revelation, psychologists have attributed seemingly intelligent performances by subliminally cued animals to the Clever Hans effect. However, it has been unclear exactly what the Clever Hans phenomenon is, how the animal was trained, or upon what cognitive processes his training was based.

While no detailed record of the horse's training exists, Pfungst briefly described Mr. von Osten's methods as they were related to him. He also attempted to analyze the cognitive processes involved. However, his study was made in 1904, before *classical conditioning* (Pavlov, 1927),[12] *operant conditioning* (Skinner, 1938),[13] the Sensorimotor Development Series (developed by Piaget between around 1921 and the present), and other 20th century learning theories had been formulated.

Clever Hans was trained to do three things: (1) to paw the ground a specific number of times with his hoof, (2) to move or shake his head up and down or from side to side, and (3) to pick up a rag. In order to train the pawing behavior, Mr. von Osten reported to Pfungst that he first *guided*, or *molded* the movement of the horse's hoof with his hands in order to show him how many times to paw the ground.‡ Gradually he was able to simply touch the hoof, and soon, a mere gesture of the hand towards the hoof elicited the behavior. He rewarded the horse with bread and carrots for satisfactory responses. He then drilled the animal until he responded correctly to the mathematical questions. Evidently, the pawing behavior had been taught by *molding*, as the animal's hoof was physically guided through the desired motor pattern. As the animal learned the behavior pattern—that is, to paw—he gradually learned to perform it in response to the hand movement that was an abbreviated version of the original *guidance*. The most protracted portion of the training evidently was the development of the subtle (and unconscious) cues—leaning towards the horse and then straightening up—that indicated when the horse was to start pawing and when he was to stop.

---

‡ This is not the usual method used for training horses to paw. A horse is generally trained to paw by pinching its ankle. The horse stomps or paws with its hoof to get rid of the unpleasant stimulus, and is rewarded for this behavior with food. Gradually, the trainer reduces the pinch to a touch and touches the horse higher and higher up the leg until, finally, a touch on the shoulder elicits the behavior. Other cues, such as sounds, can also be trained to elicit the behavior (personal observations). This method of training is classical conditioning.

These cues probably were developed simultaneously by the man and the horse, and Mr. von Osten probably learned them from the horse as much as the horse learned them from him. Even though he was generally unaware that he was cuing the horse, he commented that at one point he discovered if he suddenly straightened up his body as the horse was pawing out an answer, the horse stopped counting. So he learned never to make a sudden straightening-up movement when the horse was counting. While the cues the horse learned were very subtle and required fine sensory discrimination, they evidently were not complicated.

Mr. von Osten trained the horse to make his different head movements in a similar way.§ The behavior first was *molded*, as the horses' head was pulled to the desired position with the bridle. Presumably, desired responses were again rewarded with food. Again, the most protracted part of the training period evidently was the simultaneous development by the man, and learning by the horse, of the man's subtle head cues that finally elicited the shakes and nods. Unfortunately, Pfungst gives little information as to how the horse was trained in the third task, to pick up rags.¶

The training methods employed by Mr. von Osten to teach the horse to paw and to turn his head appear to be classifiable in terms of the Piagetian Sensorimotor Intelligence Series (TABLE 1). The process of *molding* is a motor adaptation to stimuli that impinge upon the subject (the behavior is not spontaneous, and it does not involve environmental orientation). These are among the major distinguishing characteristics of the second stage of this series (FIGURE 1).*

Molding, or guidance, is one of the most common methods used for training circus animals. Hediger[14] has emphasized that the behaviors performed by circus animals usually are familiar motor patterns that the

§ Nodding and head shaking are also more commonly trained by classical conditioning. The horse is pinched under or on the side of the neck. When he nods or shakes his head to get rid of the annoying stimulus, these head movements are rewarded. Gradually, more subtle touches on the neck, or hand movements toward the neck, cue the behavior.

¶ This behavior is most commonly taught through the method of *successive approximation* (K. Pryor, personal communication) (see below, *Operant (Instrumental) Conditioning*).

* In the case of the head shakes and nods, which were ultimately cued by similar movements of the trainer's head, it is possible that imitation also was involved in cuing the behavior (but not in training it). Imitation of behavior already in the species repertoire, as head shakes and nods, would be classified as stage 2 or 3 imitation, depending on whether the horse was purposefully (i.e., voluntarily) or unwittingly imitating his trainer (TABLE 2). Unfortunately, there is not sufficient information available to show whether imitation was actually involved in cuing the horse's head movements. I think it is unlikely that imitation was involved, since arbitrary cues, such as sounds, are easily taught to cue these kinds of performances.

SENSORIMOTOR STAGE	LEARNING MECHANISM & LEVEL OF SENSORIMOTOR BEHAVIOR DISPLAYED
1 REFLEX	
2 PRIMARY CIRCULAR REACTION	MOLDING
3 SECONDARY CIRCULAR REACTION	
4 COORDINATIONS	
5 TERTIARY CIRCULAR REACTION (EXPERIMENTATION)	
6 INVENTION THROUGH MENTAL COMBINATIONS (INSIGHT)	

  LEVEL OF SENSORIMOTOR BEHAVIOR DISPLAYED

FIGURE 1. Sensorimotor analysis of the Clever Hans phenomenon.

animals must learn to perform on cue under unfamiliar conditions. The animal is first "put through the action," that is, physically put into the desired position, or made to perform the desired behavior. For example, an elephant will at first be forced to sit on a pedestal with ropes and pulleys. Eventually, an abbreviated gesture representing the method of "putting through" becomes the cue to elicit the performance. While molding is most commonly used to train and cue the performance of motor patterns already in the species repertoire, it can also be used (though with more difficulty) to train new motor patterns. By physically lifting a bear's rear legs, and forcing him to walk, he can be trained to walk on his front paws.[14]

CUING

Sensorimotor Training and Cuing

While the term, Clever Hans phenomenon, pertains to a single type of cognitive process at the stage 2 sensorimotor level, cuing is a generic term that refers to several kinds of cognitive processes, including the Clever Hans phenomenon. Cuing can be used to train behaviors of different levels of cognitive complexity that are classifiable within stages 2 to 5 of Piaget's Sensorimotor Intelligence Series.

Classical (Pavlovian) Conditioning, as well as molding, can be used to train and cue behavioral performances. During classical conditioning, a stimulus (e.g., a piece of food) that automatically elicits a response, characteristically a reflex (e.g., salivation), becomes associated with a new stimulus (e.g., a bell) that is presented at the same time. The animal is passive, and the stimulus impinges upon him. Classical conditioning involves the recognition of specific objects or contexts and adaptations to these stimuli as they impinge upon the subject. Like molding, classical conditioning can be classified as a stage 2 sensorimotor phenomenon.

Classical conditioning also is used to train circus animals. The large cats often are trained by classical conditioning. Their training capitalizes on their aggressive inclinations,[14] and the animals' fight and flight reactions are classically conditioned and eventually elicited by subtle cues. Thus the lion first roars and paws at the trainer as he approaches the cat and enters his critical flight distance, eliciting the cat's fight reactions. Eventually the cat becomes trained to perform the behaviors, cued by more subtle movements towards him (usually movements of the trainer's whip).

Operant (Instrumental) Conditioning also can be used to train and cue behavior. During operant conditioning the active animal reaches out into its environment and spontaneously performs a behavior (e.g., pushing a bar or lifting its paw) and is rewarded by a reinforcing stimulus (e.g., a piece of food). Operant conditioning can also be combined with *selective reinforcement*, and the animal can be taught to perform the conditioned behavior (pushing the bar) under one specific condition (e.g., if a particular noise is made), by rewarding the performance only when it is made under this condition.

Operant conditioning is environmentally oriented, is spontaneously initiated, involves the establishment of object–action relationships, and is semi-intentional (that is, the animal performs the behavior unintentionally the first time, but learns to perform it intentionally). These are the behavioral parameters that are characteristic of stage 3 of the Sensorimotor Intelligence Series. In fact, operant conditioning is essentially the same thing as the "secondary circular reaction," the most characteristic phenomenon of the third stage.

Circus animals may be trained by operant conditioning as well as by molding and classical conditioning. For example, sea lions can be enticed with food to perform certain desired behaviors (like climbing onto a pedestal) spontaneously.[14] This spontaneous behavior is then rewarded. Either abbreviations of the original enticement, or selectively reinforced cues of a different nature can later be used to cue the behavior.

Clever Hans's behavior, as well as traditional classically and operantly conditioned behaviors, represents the conditioning of single behaviors that occur naturally in the species repertoires. Circus animals often perform behaviors not normally found in their species repertoires. These behaviors can be trained, and learned, in different ways. The effectiveness of different training methods is dependent upon the cognitive capacities of the species being trained.

New motor behaviors can be taught to animals with relatively low (stage 3) cognitive capacities by operant conditioning. This is called the method of *successive approximations*, or *shaping*. The trainer takes advantage of variations in the operant responses, and rewards only those responses that approximate the desired results. In this way, a chicken or a pigeon can gradually be conditioned to hold her head abnormally high as she walks around, first by reinforcing her for holding her head up normally high, and then only when she holds her head up slightly higher than normally, and, finally, only when she stretches her neck as high as possible.[15] Once the new behavior is learned, the trainer can use selective reinforcement to train the animal to respond to the desired cue for its performance. Most of the "dressage" training of horses occurs in this manner (personal observations).

More complex behaviors that involve combinations of several motor patterns can most easily be taught to animals that have at least stage 4 sensorimotor capacities. A chimpanzee will pick up a hat and put it on his head (a combination of two motor behaviors, picking up the hat, and putting it on the head, as well as a conceptual coordination of an object, the hat, and a body part, the head). Once the animal voluntarily performs the behavior, it can be rewarded with food or with social rewards, and cues for eliciting the behavior can be taught through selective reinforcement. While the method of training and cuing such a coordination is operant conditioning, a stage 3 mechanism, the behavior that is entrained is of a higher (stage 4) cognitive level.

New and variable behavior patterns involving an understanding of object–object, object–space, and object–force interactions can be taught most easily to species that naturally possess more advanced (stage 5) sensorimotor abilities.† Only highly intelligent species will manifest these interactions naturally. They can be trained to perform them on cue by operantly conditioning the desired acts and selectively reinforcing the cues. Thus, while these behaviors, like the coordinations, are stage 4 and 5, the training methods and cuing are stage 3 mechanisms. In this way, highly intelligent species, such as seals, dolphins, and chimpanzees, can be taught ball games (that require an understanding of object–space and object–force relationships). Intelligent species that have prehensile extremities, such as chimpanzees, can be trained to set the table (requiring comprehension of object–space and object–force—i.e., gravity—relationships), and to eat with fork and spoon (requiring object–object coordinations).

TABLE 3

SPECULATIVE SENSORIMOTOR ABILITIES OF ANIMALS

Highly Intelligent (Attain Stages 5-6)	Moderately Intelligent (Attain Stage 4)	Less Intelligent (Attain Stages 2-3)
Great apes	Gibbons	Horses
Capuchin (Cebus) monkeys	Most monkeys other than	Cows
Dolphins	Cebus	Sheep
Seals	Parrots	Deer
Dogs		Pigs
Cats		Camels
Bears		Rabbits
Elephants		Rats
		Sloths
		Aardvarks
		Chickens
		Turkeys

The sensorimotor capacities of only a few animal species have been studied systematically from a Piagetian perspective. Several species of monkey (stumptail macaques,‡ *Macaca arctoides*;[9] spider monkeys, *Ateles* sp.; and howler monkeys, *Alouatta palliata*)[16] achieve stage 4 in the Sensorimotor Intelligence Series.§ Parrots attain stage 4, and dogs achieve at least stage 5 (personal observations). During unsystematic observations, I have observed that seals and dolphins¶ also achieve at least stage 5. Cebus monkeys (*Cebus capucinus*),[16] gorillas,[7,8,18] chimpanzees,[8] and orangutans[19] all achieve stage 6 in this series.

Chimpanzees, seals, and dolphins are among the more intelligent animals, and I would speculate that many other animals, such as horses,* sheep, and rabbits, are not as smart (TABLE 3). It is no accident that these smarter animals are the circus stars.

† Some kinds of seemingly advanced and complex behaviors can occasionally be taught to less intelligent animals using the method of approximations.[15]

‡ Parker observed stage 4, but not stage 3 secondary circular reactions in the stumptail monkey she studied.

§ Some animal species may possess different levels of ability in different sensory modalities, or in different body parts. For example, stumptail monkeys possess advanced stage 5 and 6 abilities in their gross body tactile/kinesthetic modality, but do not possess advanced manual/gestural or vocal abilities. Apes possess advanced body tactile/kinesthetic and manual/gestural abilities, but not vocal abilities.[7] Thus a stumptail monkey probably could be taught to swing on a swing, but not to set the table.

¶ Tayler and Saayman also report behaviors in dolphins—e.g., tossing and catching objects—that can be interpreted as stage 5.[17]

* The reason that horses are so easily trained probably is not because they are particularly intelligent, but because they are domesticated animals that have been selected for high trainability.

Readers may have noted that up to this point I generally have been using the term "training"; according to Webster, "to train" is "to instruct by discipline or drill." "Training" in this sense does not occur at the stage 6 sensorimotor level, at which solutions are arrived at mentally, through insight. Rather, instruction on this level is best called "teaching" (Webster: "to guide; to impart knowledge, to instruct by example, and to cause to think for themselves"). Cuing of truly insightful, stage 6 sensorimotor behavior probably cannot occur, for if the behavior is cued, it would not be a true manifestation of insight. However, behavior that *appears* to be insightful may actually be cued behavior of a lower cognitive level.

Thus, cuing can occur on two sensorimotor levels, stage 2 or stage 3. Three kinds of cuing and training can occur: molding (stage 2), classical conditioning (stage 2), and operant conditioning (stage 3). The Clever Hans phenomenon refers to only one kind of training and cuing, i.e., molding, which occurs at the stage 2 level. In addition, cuing can be used to elicit behaviors from stages 2 to 5 in the Sensorimotor Intelligence Series, but the Clever Hans phenomenon involved the cuing of only stage 2 behaviors.

Imitative Learning and Cuing

Animals also can be taught and cued imitatively. As can be seen in

SENSORIMOTOR STAGE	LEARNING MECHANISM & LEVELS OF SENSORIMOTOR BEHAVIOR CUED
1 REFLEX	
2 PRIMARY CIRCULAR REACTION	MOLDING CLASSICAL CONDITIONING
3 SECONDARY CIRCULAR REACTION	OPERANT CONDITIONING SUCCESSIVE APPROXIMATIONS
4 COORDINATIONS	
5 TERTIARY CIRCULAR REACTION (EXPERIMENTATION)	
6 INVENTION THROUGH MENTAL COMBINATIONS (INSIGHT)	

   LEVELS OF SENSORIMOTOR BEHAVIOR CUED

FIGURE 2. Sensorimotor analysis of cuing.

TABLE 2, imitation, like sensorimotor intelligence, can occur at different levels of cognitive functioning. As with sensorimotor training, the levels of imitative training that can be used are dependent upon the cognitive capacities of the species.

Imitative abilities have been studied systematically from a Piagetian perspective in only a few animal species. Stumptail macaques achieve at least stage 3; spider monkeys achieve stage 4; cebus monkeys (personal observation) and the great apes (chimpanzees and gorillas and orangutans) achieve stage 6 in the Imitation Series.[7,8,18,19] Parrots and dogs have at least stage 5 abilities (personal observations), and dolphins have stage 6 abilities.[†][17,20] As is the case with Sensorimotor Intelligence (TABLE 3), probably only a few animal species achieve stages 5 or 6 in the Imitation Series.

Among the nonhuman primates that have been studied, the first two stages of imitation, reflexive imitation, and sporadic self-imitation, pertain only to infants, and both occur only sporadically (TABLE 2). Consequently, it is unlikely that imitation on these levels is very important for training or cuing primates. There are no data on these stages in other animals.

The ability to imitate familiar motor patterns already in the species repertoire (self-imitation, stage 3) is prominent among the primates, and this ability is probably present in most other mammals, especially in the more social species. Animals possessing stage 3 imitative abilities can be imitatively cued to perform specific behaviors already in their repertoires—such as raising a paw, or picking up an object. The imitated behaviors can subsequently be reinforced through food or social rewards (that is, operantly conditioned) as the animals are trained to perform them reliably.

Animals possessing stage 4 imitative abilities will attempt to imitate new motor patterns (though they often fail to do so precisely). They will also imitate coordinations. In this manner, imitation can be used for cuing and teaching a chimpanzee to wave goodbye (a simple new motor pattern) or to put a hat on its head (a coordination of an object with a body part). Again, appropriate imitations can be reinforced as the animals are trained to perform them reliably.

Animals possessing stage 5 imitative abilities will accurately imitate new motor patterns after repeated attempts at matching. In addition, they can be taught to perform advanced, stage 5, sensorimotor behaviors involving object–object, object–space, or object–force relation-

† Cebus monkeys and great apes achieve these advanced imitative abilities only in the body and manual modes; parrots manifest advanced imitative abilities in the vocal mode.

ships through imitation. Through stage 5 imitation a chimpanzee can be cued or taught to eat with a spoon, or a dog can be taught to catch a frisbee.

Animals that attain stage 6 in the Imitation Series can be cued to imitate and learn new motor patterns on the first try, as they match their behavior to that of the model through mental representation. These imitations can then be reinforced.

Deferred or delayed imitation is also characteristic of stage 6 imitation. While the ability to learn through delayed imitation can be used in teaching highly intelligent animals, it cannot be used to cue performances, since by definition the behavior of model and subject are separate in time.

Instruction through imitation is instruction by example. It is most accurately called "teaching," rather than "training," for the animal cannot be drilled or forced to imitate through disciplinary actions. Once the animal voluntarily imitates, his repeated performance can be encouraged through positive reinforcement.

Imitative cuing can occur on different levels of cognitive complexity, to elicit behaviors characteristic of stages 3 through 6 in the Imitation Series. It can be used to elicit familiar motor patterns (stage 3), to semireliably elicit new motor patterns (stage 4), to elicit new motor patterns through matching (stage 5), and to evoke new motor patterns on the first try, through mental representation (stage 6). However, cuing is not involved in stage 6 deferred imitation.

COGNITIVE ABILITY AND SIGNING BY APES

Sensorimotor Abilities of Apes

While the Clever Hans phenomenon appears to be based on stage 2 sensorimotor abilities, and cuing on stage 2 and 3 sensorimotor abilities, as well as stage 3 through 6 imitative abilities, ape signing is evidently based on stage 2 through 6 sensorimotor abilities and stage 3 through 6 imitative abilities.

Systematic studies on sensorimotor development in apes[7,8,19] have demonstrated that these primates go through all six stages of both the Sensorimotor Intelligence and the Imitation Series. They progress through the first four stages slightly more rapidly than human infants, but the fifth and sixth stages are greatly protracted. The fifth stage is completed at about 4 years in the apes (as compared to 18 months in human infants), and the sixth stage (which is completed by 2 years in humans) culminates even later—probably at 7 years.

During stage 6, nonsigning apes manifest the sensorimotor ability to mentally conceptualize object–object, object–space, and object–force interactions, as, for example, they invent and use tools appropriately on first try. They also manifest deception, lying, and joking. For example, a young 18-month-old gorilla repeatedly approached me with hugs and kisses as he surreptitiously attempted to obtain my watch.[7] Fossey has observed similar deceptive tactics in low-ranking wild gorillas as they attempted to monopolize highly desired foods.[8] "Koko," the gorilla who is being taught to sign by Francine (Penny) Patterson, demonstrated a striking example of nonlinguistic joking to me. When she was 4 years 6 months, she was a cross-sectional subject for my Piagetian study. While testing Koko's imitative abilities, I asked Patterson to model the behaviors of pointing to different parts of her face, her eyes, nose, mouth, ears, forehead, and chin. Koko imitated each of these behaviors perfectly on the first try and I recorded these imitations on Super 8 film and on still photographs.[7,21] A couple of weeks later I returned with a photographer to attempt to capture some of these behaviors on 16 mm film. Again, Patterson modeled the behaviors. She pointed to her eye and Koko put her face close up to Patterson's, as if trying to examine as well as possible what Patterson was doing, and then she clearly pointed to her ear. Patterson modeled pointing to her nose, and again Koko closely examined the modeled behavior, and then pointed to her chin. Koko continued to point to inappropriate parts of her face for about 5 minutes. Finally Patterson became exasperated and scolded her and signed to her that she was a "bad gorilla," whereupon Koko signed that she was a "funny gorilla," and laughed. Using the same criteria that we use for humans to label a joke, Koko's behavior can be interpreted as a joke. Furthermore, her nonlinguistic joking intentions were confirmed by her signing "funny gorilla." (See footnote below on *Interpreting Behavior*.)

Nonsigning apes also demonstrate nonlinguistic, stage 6 imitative abilities. They will imitate new motor patterns involving object–object interactions, like tool use, correctly on the first try, and they will also manifest deferred imitation. For example, sometime after she was 3 years old, "Vicki," the chimpanzee reared by the Hayses, showed delayed imitation, sharpening pencils, using a screwdriver to pry up the lid of a paint can, and cleaning windows using a solution from a spray bottle.[22]

No systematic studies have been done on the acquisition of signing in apes from a Piagetian perspective, but data available in the literature suggest that ape signing is learned through a series of steps that correspond to stages 2 through 6 of the Sensorimotor Intelligence Series, and stages 3 through 6 of the Imitation Series.

The Signing Apes

In June 1966, at the University of Nevada in Reno, Beatrice and Allen Gardner acquired an 8- to 14-month-old female chimpanzee, whom they named "Washoe." Under their guidance and with the help of several graduate students, including Roger Fouts, Washoe was intensively schooled in American Sign Language.[23-25] By the time the first phase of the experiment was terminated, when Washoe was 4 years old, she had learned 85 signs. In addition, Washoe combined her signs and used them in context-appropriate situations. The Gardners interpreted these combinations as syntactical phrases, similar to the 2- and 3-word phrases used by 2- to 3-year-old humans.[23]

From this time, Washoe's schooling has been continued by Roger Fouts at the University of Oklahoma. Her vocabulary has continued to grow. By 6 years it included more than 160 signs,[25] and it now includes over 200 signs.[26] In addition, Fouts has replicated the Washoe experiment, teaching signs to about six other 2½- to 4-year-old chimpanzees, as well as an infant chimpanzee, "Salome."[25] In 1972 the Gardners initiated a replication of Project Washoe, teaching signs to several infant chimpanzees from birth. These new infant subjects were "Moja," "Pili," "Tatu," and "Dar."[27,28] In March 1979, Fouts initiated a new experiment, designed to examine the cultural transmission of signs among apes. Washoe was presented with a male infant, "Loulis," to adopt. The early results of this experiment will be presented below.

Meanwhile in July 1972, Patterson at Stanford University initiated a project of teaching American Sign Language to the 1-year-old female gorilla, Koko.[29] By 5½ years Koko had acquired a working vocabulary of 345 signs, and she had a cumulative vocabulary of 645 signs. Like Washoe, Koko used combinations of signs in appropriate contexts. This project, now in its eighth year, is still in progress.

Most scientists hailed these experiments as the first proof of linguistic capacities in nonhuman species. However, there were some skeptics.

In December 1973, Herbert Terrace and colleagues at Columbia University initiated another replication study of sign acquisition on the newborn male chimpanzee, "Nim." Terrace's data consist of a corpus of 19,203 multisign combinations, recorded without contextual notes, and 3½ hours of videotape. By 3 years 10 months, when the project was terminated, Nim had acquired 125 signs. Nim also used signs in combina-

SENSORIMOTOR STAGE	BEHAVIORAL PARAMETERS	SIGNING ABILITY
1 REFLEX		
2 PRIMARY CIRCULAR REACTION	SELF-ORIENTED ACQUIRED ADAPTATION TO IMPINGING STIMULI	SIGNS LEARNED BY MOLDING SIGNS LEARNED BY CLASSICAL CONDITIONING
3 SECONDARY CIRCULAR REACTION	ENVIRONMENTAL ORIENTATION UNDERSTANDS RELEATIONS BETWEEN ACTIONS & EFFECTS SEMI-INTENTIONAL	SIGNS LEARNED BY OPERANT CONDITIONING & SUCCESSIVE APPROXIMATIONS
4 COORDINATIONS	INTENTIONAL COMBINES BEHAVIORS GOAL ESTABLISHED FROM THE OUTSET FAMILIAR BEHAVIORS USED IN NEW CONTEXTS	2-SIGNS COMBINATIONS OF FAMILIAR MOTOR PATTERNS USES SIGNS IN NEW CONTEXTS
5 TERTIARY CIRCULAR REACTION (EXPERIMENTATION)	BEHAVIOR BECOMES VARIABLE EXPERIMENTS WITH & COORDI-NATES RELATIONSHIPS BE-TWEEN PERSONS, OBJECTS & SIGNS TO SOLVE PROBLEMS CONSIDERS OTHERS AUTONOMOUS	INITIATES & USES SIGNS INVOLVING NEW MOTOR PATTERNS "AS TOOLS" TO EFFECT PERSON-PERSON, -OBJECT & -ACTION CHANGES
6 INVENTION THROUGH MENTAL COMBINATIONS (EXPERIMENTATION)	INVENTION THROUGH MENTAL REPRESENTATION	MENTALLY FIGURES OUT NEW ICONIC SIGNS TO SOLVE PROBLEMS MENTALLY FIGURES OUT NEW 2-SIGN COMBINATIONS & USES THEM IN NEW CONTEXTS USES SIGNS TO DECEIVE, LIE, JOKE, ARGUE, CORRECT, & TO EXPRESS DISPLACEMENT

FIGURE 3a. Sensorimotor determinants of ape signing.

tions. However, in 1979, in publications based on the analysis of Nim's multisign combinations, Terrace and colleagues concluded that despite syntactical regularity in many of the sign sequences, they could not be considered phrases since the analysis of the 3½ hours of videotaped samples showed that many of Nim's signs were prompted or cued by his teachers.[2,3] From these data, they deduce that previous rich interpretations of ape multisign combinations as phrases are incorrect, and apes cannot master syntactic organization, and cannot acquire a language.

This discussion will not address the question of whether or not apes have language. Instead, from a Piagetian perspective, it will examine the following two questions: At what levels of complexity does ape signing function? And to what extent can ape signing be attributed to the Clever Hans phenomenon, to imitation, and to cuing in general?

SENSORIMOTOR STAGE	LEARNING MECHANISM & LEVEL OF SENSORIMOTOR BEHAVIOR DISPLAYED
1 REFLEX	
2 PRIMARY CIRCULAR REACTION	MOLDING CLASSICAL CONDITIONING
3 SECONDARY CIRCULAR REACTION	OPERANT CONDITIONING SUCCESSIVE APPROXIMATIONS
4 COORDINATIONS	BEHAVIORAL COMBINATIONS
5 TERTIARY CIRCULAR REACTIONS	INITIATES & USES NEW BEHAVIORS TO SOLVE PROBLEMS COORDINATES PERSON-PERSON & PERSON-OBJECT RELATIONSHIPS
6 INVENTION THROUGH MENTAL COMBINATIONS (INSIGHT)	MENTALLY FIGURES OUT BEHAVIORS TO SOLVE PROBLEMS

    LEVELS OF SENSORIMOTOR BEHAVIOR DISPLAYED

FIGURE 3b. Sensorimotor analysis of ape signing.

Sensorimotor Development and Ape Signing

Fouts, the Gardners, and Terrace have found that chimpanzees who have been taught to sign from birth have learned their first signs during stage 2, which occurs from about 2 to 4 months in chimpanzees (personal observations). Moja and Pili learned their first signs at 3 months.[27] Furthermore, Fouts[26] and Terrace[2] report that the earliest signs learned by infant chimpanzees were learned through molding, or guidance, a stage 2 sensorimotor process. Though Washoe's sign learning began later, at about 1 year of age, one can interpret, from descriptions of how some of her earliest signs (e.g., "more") were learned, that they were acquired through classical conditioning, another stage 2 mechanism.‡[24]

The first signs recorded in young stage 2 infants are relatively simple behaviors that are characteristic of the chimpanzees' natural repertoire. The signs for "drink" (thumb extended from closed fist and brought to the mouth) and "more" (hands loosely open, with palms toward signer, fingertips brought together) are common primary circular reactions.

‡ The Gardners do not distinguish between classical and operant conditioning, nor between operant conditioning and the method of successive approximations, or shaping, They treat the "babbling" method as a separate teaching method, whereas I classify it under operant conditioning.

These observations are consistent with the Piagetian model, since only simple behaviors consisting of single motor patterns involving species-typical behavior are manifested until stage 4.

The Gardners report that Washoe learned some of her early signs through operant (instrumental) conditioning—a stage 3 process. For example, when Washoe was observed to spontaneously touch her nose with her index finger (a close approximation to the sign for funny), she was socially rewarded with laughter. Gradually she learned to make this sign in amusing contexts. Most signs required shaping (the method of successive approximations, which is a more elaborate form of operant conditioning) as well as contextual operant conditioning, since the spontaneous behaviors from which the signs were derived usually were only poor approximations to the desired signs. For example, poor approximations of the sign for "open"—flat hands placed side by side, palms down, then drawn apart and rotated palms up—were made spontaneously by Washoe as she banged her flat palms on a door to indicate she wanted to go out through the door. By being rewarded for successively closer approximations, she eventually learned the correct "open" sign.[24]

The Gardners found that molding, or guidance, and later imitation were more effective teaching methods than operant conditioning, since it was impractical to wait for the spontaneous approximations of signs to occur in appropriate contexts.

The first two-sign combinations are reported to occur during stage 4, the stage of coordinations, when the first naturally occurring behavioral combinations also appear.§ Stage 4 occurs from approximately 6 to 11 or 12 months in chimpanzees, and from 6 or 7 to 14 months in gorillas. The Gardners' infant chimpanzee, Tatu, was making two-sign combinations at 6 months, and Moja, Pili, and Dar were combining signs by 7 months.[28] Koko, whose signing instruction did not begin until she was 1 year of age, made her first two-sign combination at 14 months, late stage 4 in gorillas, and was making spontaneous two-sign combinations regularly by 15 months.[30]

Following the Piagetian model, one would predict that until the fourth stage, signs would function only as conditioned responses to specific objects or situations (stage 2), or as means to reinstate interesting spectacles (stage 3). However, as the infant enters stage 4 and becomes capable of applying familiar behaviors to new situations, and to performing goal-oriented behaviors, and as he reaches stage 5 and

§ Nim is reported to have made his first two-sign combination at 3½ months,[2] but Terrace has the impression that his first combination was really two independent signs that happened to occur very close to each other in time (personal communication).

becomes capable of figuring out how to use signs as tools to attain goals, one would predict that signs would no longer be context-bound, and one would expect to see the generalization of signs to new situations. Washoe began to show generalization at 14 to 20 months (early stage 5), as she began to use the sign "more" in contexts other than to reinstate tickling.[24] Koko began to manifest generalizations at 13 months (late stage 4 or early stage 5) as she signed "drink" for fruit drinks as well as for her formula.[29] Koko also made over-generalizations and these also began at 13 months, as she signed "drink" for a piece of fruit. Her over-generalizations were most common between 13 months and 3½ years. Terrace reports that Nim also made generalizations and over-generalizations. Citing Clark,[31] Patterson points out that over-generalizations are common in human children between 1 and 2½ years of age.

During stage 5, which occurs from about 11 or 12 months to 4 years in chimpanzees, as the apes begin to experiment with and to coordinate object–object, object–space, and object–force relationships to solve problems (e.g., tool use) and in play, apes instructed in signing also begin to use signs in similar ways. They begin to initiate and use signs "as tools" to effect person–object, person–action, and person–person changes. This use of signs by apes is extremely common, and, in fact, most of the ape signing in all of the studies constitutes requests for objects (e.g., "drink," for "you give me drink") or requests for actions (e.g., "hug," "come," or "open"). During this stage, they will also figure out how to use new iconic signs that are actually parts of the instrumental acts, but do not require mental representation. At 18 months Nim figured out that by rubbing his hands together, he could request hand cream.

Apes also will use signs to name objects or to comment on objects or their properties, although this is less common than the requests. At about 23 months Nim signed "hat" while looking at a billboard picture of a cowboy wearing a hat, and he signed "red" at about 21 months as he passed a red flower. Apes also use signs to express names and concepts in play. Washoe signed "in" and "out" in different contexts as she played by herself.[32]

Finally, during stage 6, which occurs between about 18 months to about 7 years in apes, nonsigning apes begin to solve problems through mental representation or insight, rather than through experimentation, or action. They become capable of mentally figuring out how objects can be used as tools, and they become capable of mentally figuring out subversive means (deception and "lying") to attain goals. Similarly, during stage 6 signing apes figure out through mental representation that they can use signs "as tools" to request objects that are not present. For example, Terrace reports that Nim made requests in the absence of desired objects.

Displacement in time even more saliently demonstrates mental representation. Signing apes are evidently capable of comprehending and expressing events displaced in time. For example, at 5 years 6 months, Koko had made a mess of her room and it had taken her teacher, Penny Patterson, almost an hour that morning to clean it up:

> Patterson: Do you remember what happened this morning?
> Koko: Penny clean. [Reference 29, page 281.]

Again, at 5 years 6 months, the day after Koko bit a companion:

> Patterson: What did you do yesterday?
> Koko: Wrong wrong.
> Patterson: What wrong?
> Koko: Bite. [Reference 29, page 282.]

During stage 6, through mental representation, signing apes also invent new signs to solve problems. Between 2 years 2 months and 2 years 8 months, Washoe invented the iconic sign for "bib"—drawing the outline of a bib on her chest—in order to get someone to put on her bib.[23,27,33] Koko invented novel iconic signs for "bite"—open mouth contacting side of hand held palm down (at 2 years 10 months), "stethoscope"—index finger placed in each ear (at 3 years 9 months), and "eyeglasses"—index fingers trace line from eye to back of ears (at 6 years 1 month).[29]

The use of single signs by the older more advanced ape signers likewise suggests the use of mental combinations. If one examines these single signs contextually, in dialogues, they show "vertical" grammar similar to that of two-sign combinations (see below). For example:

> Washoe: Please.
> Person: What you want?
> Washoe: Out.

> Washoe: Come.
> Person: What you want?
> Washoe: Open.

> Washoe: More.
> Person: More what?
> Washoe: Tickle.

> Washoe: Out.
> Person: Who out?
> Washoe: You.
> Person: Who more?
> Washoe: Me.

Washoe: Go.
Person: Where?
Washoe: In. [Reference 23, page 172.]

Similar sequences have been recorded for human infants:[34-36]

(Brenda at 18 months)
Brenda: Car.
Adult: What?
Brenda: Go. ¶[Reference 37, page 96.]

(Allison at 16 months has just eaten a cookie.)
Allison (looking around): Cookie.
Adult: Where is the cookie?
Allison: Gone. [Reference 36, page 155.]

These dialogues show that by about 18 months of age, neither the apes' signs nor the children's words can be attributed to simple stimulus-response learning; the signs and words are not just context-appropriate names. On the contrary, they stand for ideas. Brenda's "car" means "car go," and it may even mean "car goes fast," or "Daddy's car goes." This indicates that rich interpretations for the single utterances of 18 + -month-old apes and children are appropriate — although it is sometimes difficult to verify whether specific interpretations are correct.

Signing apes also appear to mentally figure out new two-sign combinations and use them innovatively as compound names for new objects. Some psychologists have proposed that these so-called novel combinations, such as Washoe's "water bird," meaning duck, merely represent the apes signing two context-appropriate but unrelated combinations.[1-3] However, many of the innovative combinations that have been recorded do not appear to be as well-explained in this way. For example, Koko's term "white tiger," signed at 5 years 1 month, for a zebra,[29,30,38] can hardly be interpreted as a "white" and a "tiger." Furthermore, the combination probably represents a grammatical phrase, for the gorilla must have intended that the sign "tiger" follow the sign "white," since "white" alone was not contextually relevant. [It should be recalled that goal-oriented behavioral combinations are already possible at stage 4 (TABLE 1).] Similarly, at 4 years 10 months her "elephant baby" for a Pinocchio doll,[29,30,38] and at about 3 years 9 months Washoe's "dirty good" for a toilet and "clothes food eat" for a bib[39] seem most reasonably interpreted as compound signs. Patterson points out that these novel compound names are similar to those that have been recorded for children, such as a deaf child's signing "fireplace wall shelf" for a mantelpiece[40]

¶ Repetitions of signs and words are omitted from both the ape and child transcriptions.

and a 2- to 3-year-old Russian-speaking child's "giraffe bird" for an ostrich.[41]

Besides creating new signs and innovative compound signs, apes evidently are capable of producing phrases of 2 or more signs.[23,29,42] If one looks over the lists of 2- and 3-sign combinations that have been produced by the apes, including Nim's combinations, the signs that are strung together are contextually reasonable as groups. For example, Koko has been observed in unrehearsed sessions to make logical context-appropriate responses to questions. Some of her logical answers to questions on I.Q. tests administered when she was 4 years 3 months were:

Q.: What do we get in bottles?
A.: Drink sweet.

Q.: Name two animals.
A.: Cow, gorilla.

Q.: What lives in water?
A.: Tadpole good. [There recently had been aquaria with fish and tadpoles next to Koko's cage. Reference 28, page 226.]

In some cases, Koko's appropriate answers to questions could not have been anticipated, much less cued. For example, when Patterson experimented with asking incorrect "wh-" questions, made up of random wh-signs and other words, she obtained the following logical replies:

(asked at 7+ years)
Q.: Where color?
A.: There red [indicating a nearby object].

Q.: What good?
A.: Hug good. [Reference 29, page 216.]

In addition, these apes often make unsolicited context-appropriate comments. For example, at 3 years 6 months Koko commented "listen quiet" when an alarm clock in the next room stopped ringing; and at 5 years "chin red" after Patterson bumped her chin.[29]

Contrary to Terrace's interpretation of his data—that Nim's multisign combinations do not provide new information—an examination of his data shows that a large proportion of the compound signs do expand upon the meanings expressed by single signs. This is true for at least 92% of Nim's 25 most frequent 2-sign combinations (which represent approximately 3,000 phrases) and of at least 36% of his 25 most frequent 3-sign combinations (Reference 2, page 212). For example, "play me" expresses far more information than just "play" or "me" alone. Likewise with "tickle me," "eat Nim," "more eat," "me eat," "grape eat Nim," "banana Nim eat," "me more eat," and so forth. Expansion of meaning is greatly reduced, and possibly insignificant, in Nim's four or more multisign combinations. For example, the combinations "drink me drink me drink me drink tea," "drink me eat me eat me eat me eat me eat Nim," and "banana eat banana eat Nim"[43] do not expand on the meanings that could have been expressed by two or three of the signs. An examination of the corpus of Nim's combinations[43] shows that most of his 4 + -sign combinations are of this nature. This evidently is not the case for many of Washoe's and Koko's longer combinations, as one can see from the following examples: "breakfast eat some cookie eat," signed by Koko at 5 years 6 months[29] and "please tickle more, come Roger tickle," "you me go peekaboo," and "you me go out hurry," signed by Washoe at about 3 years 9 months.[23,39] Besides providing new information, the structures of these phrases (like those of the novel compound names) imply that they are intentionally planned sequences. In the last phrase, for instance, "you," "me," "go," "out," and "hurry" are not separate relevant signs. "You" alone would have been meaningless. Furthermore, Washoe would not have started the sequence with "you me" unless she had planned to add "go" or "out."

Patterson,[29] Terrace,[23] and Fouts[42] all have found grammatical structure in the apes' 2- and 3-sign sequences. Patterson and Terrace have both reported that the combinations show statistical regularities. For example, "more" followed by "X" occurs with greater frequency than "X" followed by "more." Terrace was quite certain that these regularly occurring sequences were not due to imitation. Terrace also found that Nim regularly placed transitive verbs before nouns in his 2-sign combinations. For example, "give apple" and "give eat" were more common than "apple give" and "eat give." In a more complicated task, Fouts found that Ally correctly ordered signs "subject, preposition, location" 77%–85% of the time, as in "ball on table" or "ball under table," when asked where specific objects were.

Deception, "lying," and joking are all behaviors that logically are dependent upon mental combinations, or symbolization, and, like other stage 6 behaviors, they cannot be cued. As mentioned above, deception, lying, and joking all appear in stage 6 in nonsigning apes, and I have observed this kind of behavior both nonlinguistically and in conjunction with signing in the gorilla Koko during this stage. Consequently, I have no reason to doubt, as some authors have,[1,44] Patterson's reports that Koko tells lies and jokes.[29,30,38]

Besides lying and joking, the gorilla Koko also has been recorded to argue with and correct others. Arguing and correcting are dependent upon comparing two viewpoints of a situation—existing conditions with nonexisting ones—and therefore require mental representation.

Consequently, they would be classified as stage 6 behaviors. Clearly they cannot be cued. An argument between the gorilla and a teacher (Koko is 6 years 3 months):

Koko: Key key time.
Teacher: No, not yet time.

Koko: Yes time.
Teacher: No time. [Reference 29, page 294.]

Corrections have been recorded since Koko was 4 years 5 months. Koko at 6 years 4 months:

Teacher (after Koko has pointed to some squash on her plate): Potato.
Koko: Wrong squash.
Koko (pointing again): Do hurry that.
Koko: Squash eat hurry. [Reference 29, page 292.]

One day when a teacher was talking about Koko to a visitor:

Teacher (to the visitor in vocal speech): No, she's not an adolescent yet, she is still a juvenile.
Koko (who understands vocal speech): No, gorilla. [Reference 29, page 292.]

Thus it appears that while only sensorimotor stage 2 is involved in the Clever Hans phenomenon and stages 2 to 5 are involved in cuing, ape signing is based on all six stages of the Sensorimotor Intelligence Series (FIGURE 4).

Stage 6 sensorimotor abilities enable apes, through mental representation, to insightfully and symbolically make up new iconic signs, to invest new innovative 2-sign compound names for objects, to produce contextually relevant new 2+-sign combinations (phrases), and to deceive, lie, joke, argue, and correct others.

Imitative Development of Ape Signing

Fewer data are available on imitative development of ape signing than on sensorimotor development. The Gardners report that the first sign Washoe learned by direct imitation was "sweet" (extended index and second finger touches lower lips) acquired between 14 and 20 months (stage 5).[23] The first manifestation of the sign was inaccurate, Washoe grabbing her tongue with her thumb and index finger. It took more than a month before the sign was formed properly. This indicates that the sign was learned gradually, through matching, as new motor behaviors are characteristically learned during stage 5.

The Gardners report that Washoe learned her first sign through delayed imitation ("toothbrush") between 18 and 24 months (early stage

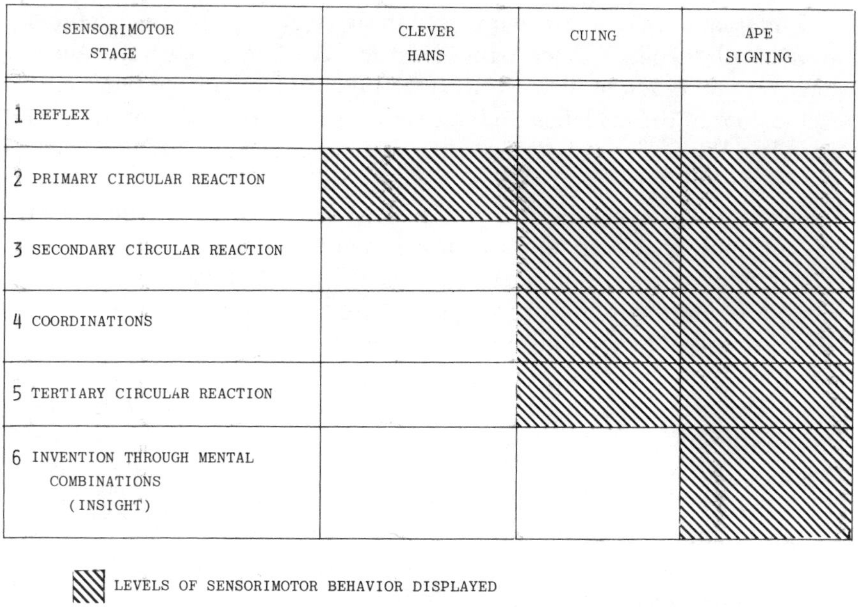

SENSORIMOTOR STAGE	CLEVER HANS	CUING	APE SIGNING
1 REFLEX			
2 PRIMARY CIRCULAR REACTION	▨	▨	▨
3 SECONDARY CIRCULAR REACTION		▨	▨
4 COORDINATIONS		▨	▨
5 TERTIARY CIRCULAR REACTION		▨	▨
6 INVENTION THROUGH MENTAL COMBINATIONS (INSIGHT)			▨

▨ LEVELS OF SENSORIMOTOR BEHAVIOR DISPLAYED

FIGURE 4. Sensorimotor comparison of the Clever Hans phenomenon, cuing, and ape signing.

6).[24] The sign for "smoke" was also learned by delayed imitation between 3 years and 2 months and 3 years 8 months.[23]

However, while the Gardners report that Washoe succeeded in learning and performing a few simple gestures through direct (stages 3 to 5) and indirect (stage 6) imitation before she was 2 years old,[24] Fouts reports that even at 3 or 4 years old she rarely succeeded in imitating complicated signs involving new motor patterns, and success generally occurred only after many attempts at matching (stage 5).[45] Not until she was 5 to 8 years old was she observed to regularly imitate new motor patterns on the first try (stage 6).[26]

The limited data indicate that stage 5 and 6 imitations are involved in ape sign learning. Stage 3 and 4 imitations are probably also involved. Imitation of behaviors already in the repertoire, as occurs in stage 3 and 4 imitation, would be more difficult to detect than stage 5 and 6 imitation of new motor patterns. But the Gardners have noted that Washoe imitated familiar motor patterns far more readily than new motor patterns.[23]

Imitation does seem to be involved in sign learning; to what extent is it involved in cuing learned signs in day-to-day signing? Several authors have proposed that imitation is the major determinant of ape signing.[1-3] Let us look at the evidence. In the case of one author,[2,3] this proposal is based on the analysis of 3½ hours of videotaped sessions of Nim's signing, recorded between 2 years 2 months and 3 years 8 months. This analysis revealed that 34% of Nim's signs were imitations. These data do not imply that even Nim's signing is merely imitation, since 66% of his signs were not imitations. Patterson also has examined the frequency of imitations in Koko's signing. Based on 11 hours of videotaped sessions collected monthly between 2 years 7 months and 3 years 6 months, 7% of Koko's signs were imitations—and 93% were not (Reference 29, page 191). In addition, Washoe and Koko have both been observed signing when they were each alone, when they could not have been imitating and could not have been cued. For example, at 5 years 3 months, Koko, by herself and nesting, picks up one of her blankets, smells it, and signs "that stink."

In this context, one should note that attributing ape sign learning and day-to-day signing partially to imitation does not reduce the cognitive significance of the signing phenomenon. Only a few mammals have the ability to imitate new motor patterns, and this ability is one of the manifestations of high cognitive capacity. In addition, it is one of the major determinants of human vocal and sign language learning, and is also a determinant of day-to-day speech in human children. Bloom *et al.* report that imitation cued 12%–23% of 4 children's utterances at about 21 months.[46]

In summary, imitation does not appear to be involved in the Clever Hans phenomenon. Imitative cuing can occur on levels 3 to 6 of Piaget's Imitation Series, but stage 6 delayed imitation is not involved in cuing. Ape signing evidently incorporates stages 3 to 6 of the Imitation Series, including delayed imitation (FIGURE 5). While all levels of imitation are incorporated in ape sign learning, imitative cuing is not a major determinant in the manifestation of learned signs.

The instruction of signing to apes is often called "training." In view of the differences in the definitions of training (drill) and teaching (to instruct by example, and to cause to think independently), only the early stages (sensorimotor stages 2 to 5) of instructing apes to sign are properly termed "training." Any instruction that involves imitation or stage 6 sensorimotor learning is properly called "teaching."

Ape Signing: An Orderly Development of Increasingly Complex Cognitive Functioning

The above discussion suggests that ape signing develops though an

IMITATIVE STAGE	CLEVER HANS	CUING	APE SIGNING
1 REFLEXIVE			
2 SPORADIC "SELF-IMITATION"	?	?	?
3 PURPOSEFUL "SELF-IMITATION"	?	▓	▓
4 ATTEMPTS TO IMITATE NEW MOTOR PATTERNS		▓	▓
5 IMITATES NEW MOTOR PATTERNS THROUGH MATCHING		▓	▓
6 FIRST TRY IMITATION OF NEW MOTOR PATTERNS		▓	▓
DELAYED IMITATION			▓

 LEVELS OF IMITATIVE BEHAVIOR DISPLAYED

FIGURE 5. Imitative comparison of the Clever Hans phenomenon, cuing, and ape signing.

orderly progression of increasingly complex cognitive functioning. This progression appears to be similar to the development of vocal language in human infants,[7] which also appears to develop hand-in-hand with sensorimotor and imitative cognitive development.[47]

Such an interpretation implies that signs will have different meanings to the infants at different developmental stages. The sign "more" is simply a conditioned response to a specific, familiar object or context to the stage 2 infant. To the stage 3 infant, "more" becomes a means to reinstate a pleasurable or interesting activity, and the infant may delight in his ability to control the reenactment of this particular activity. For example, after being tickled the infant may sign "more" in order to reinstate the tickling. To the stage 4 infant, "more" becomes a means for achieving a predetermined goal. For example, upon seeing his bottle, the infant may sign "more" in order to obtain it. To the stage 5 infant signs become "tools" that can be used to obtain goals, and a single sign often can signify complex ideas involving person–object or person–person relationships. The stage 5 infant, looking first at her teacher and then at her bottle and signing "more," may mean "Roger give me my bottle." To the stage 6 infant, signs become symbolic representations.

Following such a model, it becomes evident that simple interpretations* of signs, as some psychologists have proposed, are appropriate during the early stages of cognitive development. But later, as stages 5 and 6 are achieved, more complex, and eventually symbolic, interpretations are most likely correct.

Furthermore, this model provides a methodological framework for interpreting the discrepant results of Terrace's study of Nim's signing and the studies of the Gardners, Fouts, and Patterson on Washoe's and Koko's signing. Terrace collected his corpus of Nim's data, the 19,203 multisign combinations, when Nim was between 18 months and 3 years 2 months of age. My studies on cognitive development have shown that while apes begin to manifest stage 5 behavior at 1 year and begin to manifest stage 6 behavior at 18 months, they do not regularly and effectively manifest advanced stage 5 behavior until they are 4 years old, or stage 6 behavior until they are more than 4 years of age (probably 7 years).† Therefore, it is likely that Nim was functioning at early to advanced stage 5 levels during the 20 months that Terrace's corpus of data was collected. During this period, one would expect Nim to have been using signs instrumentally, as tools, to request person–object and person–action changes and to name objects—these things he did. He also made up a few iconic signs that did not require mental representation. At this stage one would also expect a few, though occasional, stage 6 behaviors. His requests for objects that were not present are examples of his early stage 6 signing behaviors.

Washoe at 15 years and Koko at 9 years are now adults. Their stage 6 innovative compound names for new objects, such as Washoe's proverbial "water bird" and Koko's "white tiger" began to appear in their signing after they were around 4 years of age. Although these data have not yet been analyzed in detail, their inventions of compound names began to occur at low frequencies when they were about 4 years of age. They

---

* *Interpreting Behavior:* Some ethologists and linguists say that one cannot and ought not attempt to interpret behavior of other species or nonverbal children at all. I believe that if we do not attempt to interpret behavior we will never learn all its causes or functions. Interpretations can be *proposed* (not necessarily proven) *by examining the contexts* in which specific behaviors repeatedly occur. For example, if a specific open-mouthed facial expression is followed by attacks 90% of the time, it can be interpreted as an anger or threat expression.

† Both in humans and in apes, each stage unfolds gradually. Some abilities characteristic of a particular stage appear early, and others appear later. So a stage should not be conceived of as a single level of functioning, but rather as a group of substages. At the beginning of stage 5, an ape will first throw balls, then bounce, and later spin balls as he experiments with object–force relationships, but he will not begin to catch balls until late in the stage. Similarly, he will begin to poke objects with sticks early in this stage as he experiments with object–object relationships, but he will not use a stick as a tool to extract honey from a small-mouthed jar until late in the stage.

rose in frequency until they were 6 or 7 years. When Koko was 6 years old, she was recorded to sign six to nine new innovative names in a month.[29] After 6 or 7 years, the frequencies of inventing new names dropped.[26,48] Similarly, most of Koko's impressively creative multisign combinations (e.g., her displacements in time, her logical answers to I.Q. test questions, her responses to incorrect "wh-" questions, her arguments and corrections) were recorded when she was 4 to 7 years of age.

Besides being older, and therefore cognitively more advanced than Nim, it appears that by 3½ to 4 years Koko was also a more competent signer than Nim. As discussed above, her 4 + -sign phrases show expansion of meaning and planned word ordering, while Nim's generally do not. Data on Nim's and Koko's vocabulary sizes and their mean lengths of utterances (MLU) also suggest that Koko is more advanced. At 3 years 10 months Koko's vocabulary included about 180 signs,[29] while Nim's included 125 signs.[2] While both Nim's and Koko's mean utterance lengths were 1.2 signs at 2 years 2 months, by 3 years 9 months Nim's MLU had increased to only 1.6 signs,[3] while Koko's had increased to about 2.0 signs.[29] Perhaps the conditioning methods used by Nim's teachers (which are ideal for teaching lower stage abilities) were inappropriate for advanced sign learning, and may have been partially responsible for his less proficient signing at an older age.

Chimp-to-Chimp Sign Learning: The Ultimate Evidence Against Clever Hans

On March 24, 1979, Washoe, whose natural infant had recently died, adopted the 10-month-old male infant, Loulis. The experiment, set up by Roger Fouts, was designed to discover whether this infant, who was signed to by no one except Washoe, would learn to sign.

Within one month, Loulis had learned 6 signs. By the end of the sixth month he had learned 7 signs, and by the end of the ninth month he had learned 11 signs.[49] His sign learning was comparable to Washoe's. Her signing instruction had begun when she was about the same age, and by the end of the tenth month of schooling she had learned 12 signs.[50] Furthermore, Loulis probably had learned more signs in this period than Koko had learned, since Koko, whose signing instruction had begun at 1 year, acquired 13 signs in 17 months of schooling.[29]

Like the other signing apes, Loulis was soon using combinations of signs. His first 2-sign combination, "drink that" occurred at 16 months. But his combinations occurred only rarely until he was 18 months, at which time they became frequent as he signed: "come hurry," "come tickle," "come eat," "me food," and "drink me."[49] Loulis' development of 2-sign combinations was also comparable to Washoe's, which she first began to produce between 18 and 24 months of age.[23]

Loulis learned his signs mainly by imitation—as human infants learn language. However, Washoe has been observed to intentionally try to teach signs to the infant. On one occasion, Washoe repeatedly (five times) modeled the sign for chair, looking at the infant and picking up a chair and signing "chair" each time. On another occasion, Washoe was signing to a human for food. She looked at the infant, who was not signing, and molded his hand into the "food" sign.[26]

Even though the infant is learning his signs primarily through imitation, a contextual analysis reveals that most of his learned signs occur spontaneously, and are not imitations. During a period of about 2 months, during stage 5, when he was between 16 and 18 months of age, 214 signs and their contexts were recorded. Of these signs, less than 2% were imitations of Washoe's signs, and more than 98% were spontaneous. (During the period from 18 months to 2 years of age, less than 1% of his signs were imitations, and more than 99% were not.) Most signs (71%) occurred in appropriate contexts. The rest (29%) occurred during play.[49]

This experiment conclusively proves that ape signing is not simply a group of conditioned responses that the animals are drilled to display as circus animals are trained to perform. It also demonstrates that apes' signing of *known* signs cannot be attributed merely to imitation.

Conclusions

Recent proposals, that signing in apes can be attributed to the Clever Hans phenomenon, or to cuing, appear to be unfounded. A cognitive analysis of the Clever Hans phenomenon, cuing, and ape signing shows that the ape signing involves more advanced cognitive processes than the other two phenomena.

The Clever Hans phenomenon is based only on lower stage, sensorimotor stage 2 cognitive processes, and could account for only the early rudiments of signing. Cuing refers to several distinct phenomena, based on several levels of cognitive complexity (sensorimotor stages 2 to 5, imitation stages 3 to 5, and some aspects of stage 6 imitation). Theoretically, cuing *could* account for a major portion of ape signing. However, the data show that while cuing is a major determinant of sign learning, it does not appear to be a major determinant in eliciting known signs. Furthermore, neither the Clever Hans effect nor cuing can account for the most advanced manifestations of ape signing, which are based on stage 6 sensorimotor and stage 6 imitative abilities. Only by incorporating cognitive abilities more advanced than those involved in cuing are apes able to make up new iconic signs, to invent new innovative 2-sign compound names for objects, to produce new 2 + -sign combinations to solve problems, to show displacement in time, to deceive, lie, joke, argue, or correct, or to manifest delayed imitation.

Several psychologists have proposed that one should not use complex interpretations of ape signing if simpler interpretations are possible. This analysis shows that simple interpretations do not fit the data. Apes manifest advanced cognitive process nonlinguistically, and since they appear to manifest them in their signing, it is illogical to attribute their signing to the simpler cuing and Clever Hans phenomenon.

Of the three phenomena examined, ape signing is the most complex, and, like verbal and sign language of 2- to 3-year-old children, incorporates the highest levels of sensorimotor ability.

References

1. UMIKER-SEBEOK, J. & T. A. SEBEOK. 1980. Questioning apes. *In* Speaking of Apes: A Critical Anthology of Two-Way Communication with Man. T. A. Sebeok & J. Umiker-Sebeok, Eds. pp. 1–59. Plenum Press, New York, N.Y.
2. TERRACE, H. S. 1979. Nim. A. A. Knopf, New York, N.Y.
3. TERRACE, H. S., L. A. PETITTO, R. J. SANDERS & T. G. BEVER. 1979. Can an ape create a sentence? Science **206**: 891–902.
4. PIAGET, J. 1951. Play, Dreams and Imitation in Childhood. Translated by C. Gattegno & F. M. Hodgson. W. W. Norton & Company, Inc., New York, N.Y.
5. PIAGET, J. 1952. The Origins of Intelligence in Children. Translated by Margaret Cook. International Universities Press, Inc., New York, N.Y.
6. PIAGET, J. 1954. The Construction of Reality in the Child. Translated by Margaret Cook. Ballatine Books, New York, N.Y.
7. CHEVALIER-SKOLNIKOFF, S. 1976. The ontogeny of primate intelligence and its implications for communicative potential: A preliminary report. Ann. N.Y. Acad. Sci. **280**: 173–211.
8. CHEVALIER-SKOLNIKOFF, S. 1977. A Piagetian model for describing and comparing socialization in monkey, ape, and human infants. *In* Primate Bio-Social Development: Biological, Social, and Ecological Determinants. S. Chevalier-Skolnikoff & F. E. Poirier, Eds. pp. 159–187. Garland Publishing, Inc., New York, N.Y.
9. PARKER, S. T. 1977. Piaget's sensorimotor series in an infant macaque: A model for comparing unstereotyped behavior and intelligence in human and nonhuman primates. *Ibid.* pp. 43–112.
10. CHEVALIER-SKOLNIKOFF, S. 1979. A primatological perspective: Commentary on Brainerd, C. J. (1978) The stage question in cognitive-developmental theory. Behav. Brain Sci. **2**: 139–140.
11. PFUNGST, O. 1904 (reprinted 1965). Clever Hans: The Horse of Mr. von Osten. R. Rosenthal, Ed. Holt, Rinehart & Winston, Inc., New York, N.Y.
12. PAVLOV, I. P. 1927 (1960). Conditioned Reflexes. Translated & edited by G. V. Anrep. Oxford University Press, New York, N.Y.

13. SKINNER, B. F. 1938. The Behavior of Organisms. Appleton-Century Co., New York, N.Y.
14. HEDIGER, H. 1955. Studies of the Psychology and Behaviour of Captive Animals in Zoos and Circuses. Translated by Geoffrey Sircom. Criterion Books, New York, N.Y.
15. HILGARD, E. R. 1957. Introduction to Psychology. Harcourt, Brace & Company, Inc., New York, N.Y.
16. CHEVALIER-SKOLNIKOFF, S. 1978. Intelligence in wild cebus monkeys, and its implications for the evolution of tool use. Paper delivered at the 77th Annual Meeting of the American Anthropological Association, Los Angeles, Calif., November.
17. TAYLER, C. K. & G. S. SAAYMAN. 1973. Imitative behaviour by Indian Ocean bottlenose dolphins (*Tursiops aduneus*) in captivity. Behavior 44: 286–298.
18. REDSHAW, M. 1978. Cognitive development in human and gorilla infants. J. Hum. Evol. 7: 133–141.
19. CHEVALIER-SKOLNIKOFF, S. 1979. Kids: Zoo research reveals remarkable similarities in the development of human and orang-utan babies . . . and one very special difference. Animal Kingdom 82: 11–18.
20. WEBB, N. G. 1977. Symbolic thinking in dolphins. Search 1: 38–44.
21. CHEVALIER-SKOLINIKOFF, S. 1975. The Ontogeny of Intelligence in Gorillas. Super 8 mm film, color, silent. 10 min.
22. HAYES, K. J. & C. HAYES. 1952. Imitation in a home-raised chimpanzee. J. Comp. Physiol. Psychol. 45: 450–459.
23. GARDNER, B. T. & R. A. GARDNER. 1971. Two-way communication with an infant chimpanzee. *In* Behavior of Nonhuman Primates. A. Schrier & F. Stollnitz, Eds. pp. 117–184. Academic Press, New York, N.Y.
24. GARDNER, R. A. & B. T. GARDNER. 1969. Teaching sign language to a chimpanzee. Science 165: 664–672.
25. FOUTS, R. S. 1973. Talking with chimpanzees. *In* Science Year: The World Book Science Annual, 1974. pp. 34–49. Field Enterprises Educational Corporation. Chicago, Ill.
26. FOUTS, R. S. 1980. Personal communication.
27. GARDNER, R. A. & B. T. GARDNER. 1974. Early signs of language in child and chimpanzee. Science 187: 752–753.
28. GARDNER, R. A. & B. T. GARDNER. 1978. Comparative psychology and language acquisition. Ann. N.Y. Acad. Sci. 309: 37–76.
29. PATTERSON, F. G. P. 1979. Linguistic Capabilities of a Lowland Gorilla. Ph.D. Dissertation, Stanford University, Stanford, California. University Microfilms International. Ann Arbor, Mich.
30. PATTERSON, F. 1978. Linguistic capabilities of a lowland gorilla. *In* Sign Language and Language Acquisition: New Dimensions in Comparative Pedolinguistics. F. C. C. Peng, Ed. pp. 161–201. Westview Press, Boulder, Colo.
31. CLARK, E. V. 1973. What's in a word? On the child's acquisition of semantics in his first language. *In* Cognitive Development and the Acquisition of Language. T. E. Moore, Ed. Academic Press, New York, N.Y.

32. GARDNER, R. A. & B. T. GARDNER. 1973. Teaching sign language to the chimpanzee Washoe (16 mm sound film). Psychological Cinema Register, State College, Pa.
33. GARDNER, B. T. & R. A. GARDNER. No date. Development of behavior in a young chimpanzee: 5th summary of Washoe's diary.
34. MOSKOWITZ, B. A. 1978. The acquisition of language. Sci. Am. **239**: 92–108.
35. SCOLLON, R. 1976. Conversations with a One Year Old: A Case Study of the Developmental Foundation of Syntax. University Press of Hawaii and The Research Corporation of the University of Hawaii, Honolulu, Hi.
36. BLOOM, L. 1973. One Word at a Time: The Use of Single Word Utterances Before Syntax. Mouton, The Hague.
37. SCOLLON, R., quoted in Moskowitz, Reference 34.
38. PATTERSON, P. 1978. The gestures of a gorilla: Language acquisition in another Pongid. Brain and Lang. **5**: 72–97.
39. GARDNER, B. T. & R. A. GARDNER. No date. Development of behavior in a young chimpanzee: 8th summary of Washoe's diary.
40. SCHLESINGER, H. S. 1976. The acquisition of sign language. *In* Current Trends in the Study of Sign Languages of the Deaf. I. M. Schlesinger & L. Namir, Eds. Mouton, The Hague.
41. CHUKOVSKY, K. 1963. From Two to Five. Translated & edited by M. Morton. University of California Press, Berkeley, Calif.
42. FOUTS, R. S., J. B. COUCH & C. R. O'NEIL. 1979. Strategies for primate language training. *In* Language Intervention from Ape to Child. R. L. Schiefelbusch & J. H. Hollis, Eds. pp. 295–323. University Park Press, Baltimore, Md.
43. TERRACE, H. S., L. A. PETITTO, R. J. SANDERS & T. G. BEVER. 1980. On the grammatical capacity of apes. *In* Children's Language. Vol. 2. K. Nelson, Ed. Halsted Press, New York, N.Y.
44. GARDNER, M. 1980. Monkey business. New York Review of Books (March 20): 3–4.
45. FOUTS, R. S. 1972. Use of guidance in teaching sign language to a chimpanzee (*Pan troglodytes*). J. Comp. Physiol. Psychol. **80**: 515–522.
46. BLOOM, L., L. ROCISSANO & L. HOOD. 1976. Adult–child discourse: Developmental interaction between information processing and linguistic knowledge. Cogn. Psychol. **8**: 521–552.
47. BATES, E., L. BENIGNI, I. BRETHERTON, L. CAMAIONI & V. VOLTERRA. 1977. *In* Interaction, Conversation, and the Development of Language. M. Lewis & L. A. Rosenblum, Eds. pp. 274–307. John Wiley & Sons, New York, N.Y.
48. PATTERSON, P. 1980. Personal communication.
49. FOUTS, R. S., G. H. KIMBALL, D. L. DAVIS & R. L. MELLGREN. 1979. Washoe and Her Children. Paper presented at the Psychonomics Meeting, Phoenix, Ariz., November.
50. GARDNER, B. T. & R. A. GARDNER. No date. Development of behavior in a young chimpanzee: 4th summary of Washoe's diary.

A Report to an Academy, 1980*

H. S. TERRACE

Department of Psychology
Columbia University
New York, New York 10027

THE FIRST ACCOUNT of an ape who learned to talk appears to be fictional. In 1917, Franz Kafka wrote a tale about a chimpanzee who acquired the gift of human language. Recent research appears to have confirmed Kafka's sense of what it would take to induce an ape to speak:

> . . . there was no attraction for me in imitating human beings. I imitated them because I needed a way out, and for no other reason . . . And so I learned things, gentlemen. Ah, one learns when one needs a way out; one learns at all costs. [F. Kafka, "A Report to an Academy."]

During the 63 years that have elapsed since the publication of Kafka's short story, much has been written about man's presumably unique capacity to use language and attempts to show that apes can master some of its features. Linguists, psychologists, psycholinguists, philosophers, and other students of human language have yet to capture its many complexities in a simple definition. They do agree, however, about one basic property of all human languages, that is, the ability to create new meanings, each appropriate to a particular context, through the application of grammatical rules. Noam Chomsky[1] and George Miller,[2] among others, have convincingly reminded us of the futility of trying to explain a child's ability to create and understand sentences without a knowledge of rules that can generate an indeterminately large number of sentences from a finite vocabulary of words.

The dramatic reports of the Gardners,[3] Premack,[4] and Rumbaugh[5] that a chimpanzee could learn substantial vocabularies of words of visual languages and that they were also capable of producing utterances containing two or more words, raise an obvious and fundamental question: Are a chimpanzee's multi-word utterances grammatical?

In the case of the Gardners, one wants to know whether Washoe's

* The research reported in this article was, in part, funded by grants from the W.T. Grant Foundation, the Harry Frank Guggenheim Foundation, and The National Institutes of Health (RO1MH29293). Portions of this article appeared previously in Terrace, H.S. 1979. How Nim Chimpsky Changed My Mind. Psychol. Today **13** (6): 65–76.

signing *more drink* in order to obtain another cup of juice or *water bird*, upon seeing a swan, were creative juxtapositions of signs. Likewise one wants to know whether "Sarah," Premack's main subject, was using a grammatical rule in arranging her plastic chips in the sequence, *Mary give Sarah apple*, and whether "Lana," the subject of a related study conducted by Rumbaugh, exhibited knowledge of a grammatical rule in producing the sequence, *please machine give apple*.

In answering these questions, it is important to remember that a mere sequence of words does not qualify as a sentence. A rotely learned string of words presupposes no knowledge of the meanings of each element and certainly no knowledge of the relationships that exist between the elements. Sarah, for example, showed little, if any, evidence of understanding the meanings of *Mary*, *give*, and *Sarah* in the sequence, *Mary give Sarah apple*. Likewise, it is doubtful that, in producing the sequence *please machine give apple*, Lana understood the meanings of *please machine* and *give*, let alone the relationships between these symbols that would apply in actual sentences.[6] There is evidence that Sarah or Lana could distinguish the symbol *apple* from symbols that named other reinforcers. This suggests that what Sarah and Lana learned was to produce rote sequences of the type ABCX, where A, B, and C are nonsense symbols and X is a meaningful element. That conclusion is supported by the results of two studies, one an analysis of a corpus of Lana's utterances, the other an experiment on serial learning by pigeons.

Thompson and Church[7] have recently shown that a major portion of a corpus of Lana's utterances can be accounted for by three decision rules that dictate when one of six stock sentences might be combined with one of a small corpus of object or activity names. The decision rules are (1) did Lana want an ingestible object, (2) was the object in view, and (3) was the object in the machine. For example, if the object was in the machine, an appropriate stock sequence was *please machine give*; if it was not in the machine an appropriate stock sequence would be *please move object name into machine*, and so on.

A recent experiment performed in my laboratory[8] showed that pigeons could learn to peck four colors presented simultaneously in a particular sequence. Such performance is of interest as evidence of the memorial capacity of pigeons. It does not, of course, justify interpreting the sequence of the colors ($A \rightarrow B \rightarrow C \rightarrow D$) as the production of a sentence meaning *please machine give grain*.

While the sequences, *please machine give grain* and *please machine give apple*, are logically similar, they are not identical. It has yet to be shown that pigeons can learn ABCX sequences of the type that Sarah and Lana learned (where X stands for different reinforcers) or that pigeons can learn to produce different sequences for different rein-forcers. But given the relative ease with which a pigeon can master an A→ B→ C→ D sequence, neither of these problems seem that difficult, *a priori*. And even if a pigeon could not perform such sequences or, as would probably be the case, a pigeon learns them more slowly than a chimpanzee, we should not lose sight of the fact that learning a rote se-quence does not require any ability to use a grammar.

Utterances of apes who were not explicitly trained to produce rote sequences pose different problems of interpretation. The Gardners report that Washoe was not required to sign sequences of signs nor was she differentially reinforced for particular combinations. She never-theless signed utterances such as *more drink* and *water bird*. Before these and other utterances can be accepted as creative combinations of signs, combinations that create particular meanings, it is necessary to rule out simpler interpretations.

The simplest nongrammatical interpretation of such utterances is that they contain signs that are related solely by context. Upon being asked what she sees when looking in the direction of a swan it is appro-priate for Washoe to sign *water* and *bird*. On this view, if Washoe knew the sign for sky, she might just as readily have signed such less in-teresting combinations as *sky water, bird sky, sky bird water*, and so on.

Even if one could rule out context as the only basis of Washoe's com-binations, it remains to be shown that utterances such as *water bird* and *more drink* are constructions in which an adjective and a noun are com-bined so as to create a new meaning. In order to support that interpreta-tion, it is necessary to show that she combined adjectives and nouns in a particular order. It is, of course, unimportant whether Washoe used the English order (adjective + noun) or the French order (noun + adjec-tive) in creating combinations in which the meaning of a noun is quali-fied by an adjective. But it is important to show that all, if not most, presumed adjectives and that all, if not most, presumed nouns are com-bined in a consistent manner so as to create particular meanings.

It is, of course, true that sign order is but one of many grammatical devices used in sign language and that it is less important in sign lan-guage than it is in spoken languages such as English. At the same time, sign order is one of the easiest, if not the easiest, grammatical device of sign language to record. It also provides a basis for demonstrating an awareness of such simple constructions as subject–verb, adjective–noun, verb–object, subject–verb–object, and so on.

With only two minor exceptions,[9,10] the Gardners have yet to publish any data on sign order. Accordingly, the interpretation of combinations such as *water bird* and *more drink* remains ambiguous. One has too little information to judge whether such utterances are manifestations of a simple grammatical rule or whether they are merely sequences of contextually related signs.

PROJECT NIM

In order to distinguish between these interpretations it is necessary to examine a large body of utterances for regularities of sign order. The initial goal of Project Nim, a project I started in 1973, was to provide a basis for deciding whether a chimpanzee could use one or more simple grammatical rules. Our immediate aim was to collect and to analyze a large corpus of a chimpanzee's sign combinations for regularities of sign order. Such regularities could be used as evidence of a chimpanzee's ability to use a finite-state grammar. As puny as such an accomplishment might seem from the perspective of a child's acquisition of language, it would amount to a quantal leap in the linguistic ability of nonhumans. As we shall see, showing that a chimpanzee can learn a mere finite-state grammar proved to be an elusive goal.

Socialization and Training

The subject of our study was an infant male chimpanzee, named "Nim Chimpsky." Nim was born at the Oklahoma Institute for Primate Studies in November 1973 and was flown to New York at the age of two weeks. Until the age of 18 months, he lived in the home of a former student and her family; subsequently he lived in a University-owned mansion in Riverdale where he was looked after by four students.

At the age of 9 months, Nim became the sole student in a small classroom complex I designed for him in the Psychology Department of Columbia University. The classroom allowed Nim's teachers to focus his attention more easily than they could at home and it also provided good opportunities to introduce Nim to many activities conducive to signing, such as looking at pictures, drawing, and sorting objects. Another important feature of the classroom was the opportunity it provided for observing, filming, and videotaping Nim without his being aware of the presence of visitors and observers who watched him through a one-way window or through cameras mounted in the wall of the classroom.

Nim's teachers kept careful records of what he signed both at home and in the classroom. During each session, the teacher dictated into a cassette recorder as much information as possible about Nim's signing and the context of that signing. Nim was also videotaped at home and in the classroom. A painstaking comparison of Nim's signing at home and in his classroom revealed no differences with respect to spontaneity, content, or any other feature of his signing that we examined. In view of comments attributed to other researchers of ape language,[11] that Nim was conditioned in his nursery school like a rat or a pigeon in a Skinner box, it should be emphasized that his teachers were just as playful and spontaneous in the Columbia classroom as they were at home. (A detailed description of Nim's socialization and instruction in sign language can be found in *Nim*.[12])

Nim's teachers communicated to him and amongst themselves in sign language. Although the signs Nim's teachers used were consistent with those of American Sign Language (ASL or Ameslan), their signing is best characterized as "pidgin" sign language. This state of affairs is to be expected when a native speaker of English learns sign language. Inevitably, the word order of English superimposes itself on the teacher's signing, at the expense of the many spatial grammatical devices ASL employs. An ideal project would, of course, attempt to use only native or highly fluent signers as teachers, teachers who would communicate exclusively in Ameslan. There is, however, little reason to be concerned that this ideal has not been realized on Project Nim, or for that matter, on *any* of the other projects that have attempted to teach sign language to apes. The achievements of an ape who truly learned pidgin sign language would be no less impressive than those of an ape who learned pure Ameslan. Both pidgin sign language and Ameslan are grammatically structured languages.

Nim was taught to sign by the methods developed by the Gardners[3] and Fouts:[13] molding and imitation. During the 44 months he was in New York he learned 125 signs, most of which were common and proper nouns; next frequent were verbs and adjectives; least frequent were pronouns and prepositions.[12]

Combinations of Two or More Signs

During a 2-year period, Nim's teachers recorded more than 20,000 of his utterances that consisted of 2 or more signs. Almost half of these utterances were 2-sign combinations, of which 1,378 were distinct. One characteristic of Nim's 2-sign combinations led me to believe that they were primitive sentences. In many cases Nim used particular signs in either the first or the second position, no matter what other sign that sign was combined with.[12,14] For example, *more* occurred in the first position in 85% of the 2-sign utterances in which *more* appeared (such as *more banana, more drink, more hug,* and *more tickle*). Of the 348 2-sign combinations containing *give*, 78% had *give* in the first position. Of the 946 instances in which a transitive verb (such as *hug, tickle,* and *give*) was combined with *me* and *Nim*, 83% of them had the transitive verb in the first position.

These and other regularities in Nim's two-sign utterances[12] are the first demonstrations I know of a reliable use of sign order by a chimpanzee. By themselves, however, they do not justify the conclusion that they were created according to grammatical rules. Nim could have simply imitated what his teachers were signing. That explanation seemed doubtful for a number of reasons. Nim's teachers had no reason to sign many of the combinations Nim had produced. Nim asked to be tickled long before he showed any interest in tickling; thus, there was no reason for the teacher to sign *tickle me* to Nim. Likewise, Nim requested various objects by signing *give* + X (X being whatever he wanted) long before he began to offer objects to his teachers. More generally, all of Nim's teachers and many experts on child language learning, some of whom knew sign language, had the clear impression that Nim's utterances typically contained signs that were not imitative of the teacher's signs.

Another explanation of the regularities of Nim's two-sign combinations that did not require the postulation of grammatical competence was statistical. However, an extensive analysis of the regularities observed in Nim's two-sign combinations showed that they did not result from Nim's preferences for using particular signs in the first or second positions of two-sign combinations.[15] Finally, the sheer variety and number of Nim's combinations make implausible the hypothesis that he somehow memorized them.

The analyses performed on Nim's combinations provided the most compelling evidence I know of that a chimpanzee could use grammatical rules, albeit finite-state rules, for generating two-sign sequences. It was not until after our funds ran out and it became necessary to return Nim to the Oklahoma Institute for Primate Studies that I became skeptical of that conclusion. Ironically, it was our newly found freedom from data-collecting, teaching, and looking after Nim that allowed me and other members of the project to examine Nim's use of sign language more thoroughly.

Differences Between Nim's and a Child's Combinations of Words

What emerged from our new analyses was a number of important differences between Nim's and a child's use of language. One of the first facts that troubled me was the absence of any increase in the length of Nim's utterances. During the last two years that Nim was in New York, the average length of Nim's utterances fluctuated between 1.1 and 1.6 signs. That performance is like what children do when they begin combining words. Furthermore, the maximum length of a child's utterances is related very reliably to their average length.[16] Nim's showed no such relationship.

As children get older, the average length of their utterances increases steadily. As shown in FIGURE 1, this is true both of children with normal hearing and of deaf children who sign.[14] After learning to make ut-

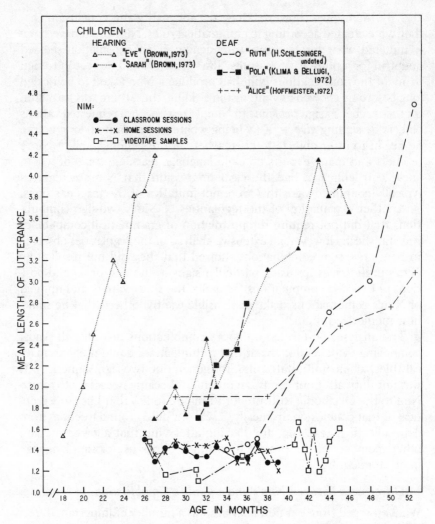

FIGURE 1. Mean length of signed utterances (MLU) of Nim and three deaf children and mean length of spoken utterances of two hearing children. The functions showing Nim's MLU between January 1976 and February 1977 (age, 29 to 39 months) are based on data obtained from teachers' reports; the function showing Nim's MLU between February 1976 and August 1977 (age, 27 to 45 months) is based upon video transcript data.

terances relating a verb and an object, as, for example, *eats breakfast*, and utterances relating a subject and a verb, as, for example, *Daddy eats*, the child learns to link them into longer utterances relating the subject, verb, and object, for example, *Daddy eats breakfast*. Later, the child

TABLE 1

MOST FREQUENT 2- AND 3-SIGN COMBINATIONS

2-Sign Combinations	Frequency	3-Sign Combinations	Frequency
play me	375	play me Nim	81
me Nim	328	eat me Nim	48
tickle me	316	eat Nim eat	46
eat Nim	302	tickle me Nim	44
more eat	287	grape eat Nim	37
me eat	237	banana Nim eat	33
Nim eat	209	banana eat Nim	26
finish hug	187	eat me eat	22
drink Nim	143	me Nim eat	21
more tickle	136	hug me Nim	20
sorry hug	123	yogurt Nim eat	20
tickle Nim	107	me more eat	19
hug Nim	106	more eat Nim	19
more drink	99	finish hug Nim	18
eat drink	98	banana me eat	17
banana me	97	Nim eat Nim	17
Nim me	89	tickle me tickle	17
sweet Nim	85	apple me eat	15
me play	81	eat Nim me	15
gum eat	79	give me eat	15
tea drink	77	nut Nim nut	15
grape eat	74	drink me Nim	14
hug me	74	hug Nim hug	14
banana Nim	73	play me play	14
in pants	70	sweet Nim sweet	14

learns to link them into longer utterances such as *Daddy didn't eat breakfast,* or *when will Daddy eat breakfast?*

Despite the steady increase in the size of Nim's vocabulary, the mean length of his utterances did not increase. Although some of his utterances were very long, they were not very informative. Consider, for example, his longest utterance, which contained 16 signs: *give orange me give eat orange me eat orange give me eat orange give me you.* The same kinds of run-on sequences can be seen in comparing Nim's 2-, 3-and 4-sign combinations. As shown in TABLE 1, the topic of Nim's 3-sign combinations overlapped considerably with the apparent topic of his 2-sign combinations (compare TABLE 2).

Of Nim's 25 most frequent 2-sign combinations, 18 can be seen in his 25 most frequent 3-sign combinations, in virtually the same order in which they appear in his 2-sign combinations. Furthermore, if one ignores sign order, all but 5 signs that appear in Nim's 25 most frequent two-sign combinations *gum, tea, sorry, in,* and *pants* appear in his 25

TABLE 2

MOST FREQUENT 4-SIGN COMBINATIONS

4-Sign Combinations	Frequency	4-Sign Combinations	Frequency
eat drink eat drink	15	drink eat me Nim	3
eat Nim eat Nim	7	eat grape eat Nim	3
banana Nim banana Nim	5	eat me Nim drink	3
drink Nim drink Nim	5	grape eat me Nim	3
banana eat me Nim	4	me eat drink more	3
banana me eat banana	4	me eat me eat	3
banana me Nim me	4	me gum me gum	3
grape eat Nim eat	4	me Nim eat me	3
Nim eat Nim eat	4	Nim me Nim me	3
play me Nim play	4	tickle me Nim play	3
drink eat drink eat	3		

most frequent three-sign combinations. We did not have enough contextual information to perform a semantic analysis of Nim's 2- and 3-sign combinations. However, Nim's teachers' reports indicate that the individual signs of his combinations were appropriate to their context and that equivalent 2- and 3-sign combinations occurred in the same context.

Though lexically similar to 2-sign combinations, the 3-sign combinations do not appear to be informative elaborations of 2-sign combinations. Consider, for example, Nim's most frequent 2- and 3-sign combinations: *play me* and *play me Nim*. Combining *Nim* with *play me* to produce the 3-sign combination, *play me Nim*, adds a redundant proper noun to a personal pronoun. Repetition is another characteristic of Nim's 3-sign combinations, for example, *eat Nim eat*, and *nut Nim nut*. In producing a 3-sign combination, it appears as if Nim is adding to what he might sign in a 2-sign combination, not so much to add new information, but instead to add emphasis. Nim's most frequent 4-sign combinations reveal a similar picture. In children's utterances, in contrast, the repetition of a word, or a sequence of words, is a rare event.

Now that we have seen what Nim signed about in 2-, 3-, and 4-sign combinations, it is instructive to see what he signed about with single signs. As shown in TABLE 3, the topics of Nim's most frequent 25 single-sign utterances overlap considerably with those of his most most frequent multisign utterances. (Most of the exceptions are signs required in certain routines, e.g., *finish*, when Nim was finished using the toilet; *sorry*, when Nim was scolded, and so on.) In contrast to a child, whose longer utterances are semantically and syntactically more complex than his shorter utterances, Nim's are not. When signing a combination, as opposed to signing a single sign, Nim appears to be running on with his hands. It appears that Nim learned that the more he signs the better his chances for obtaining what he wants. It also appeared as if Nim made no effort to add informative, as opposed to redundant, signs in satisfying his teacher's demand that he sign.

The most dramatic difference between Nim's and a child's use of language was revealed in a painstaking analysis of videotapes of Nim's and his teacher's signing. These tapes revealed much about the nature of Nim's signing that could not be seen with the naked eye. Indeed they were so rich in information that it took as much as one hour to transcribe a single minute of tape.

Terrace *et al.*[14] showed that Nim's signing with his teachers bore only a superficial resemblance to a child's conversations with his or her parents. What is more, only 12% of Nim's utterances were spontaneous. That is, only 12% were not preceded by a teacher's utterance. A significantly larger proportion of a child's utterances is spontaneous.

In addition to differences in spontaneity, there were differences in

TABLE 3

25 MOST FREQUENT SIGNS*

Sign	Tokens
hug	1,650
play	1,545
finish	1,103
eat	951
dirty	788
drink	712
out	615
Nim	613
open	554
tickle	414
bite	407
shoe	405
pants	372
red	380
sorry	366
angry	354
me	351
banana	348
nut	328
down	316
toothbrush	302
hand cream	301
more	301
grape	239
sweet	236

* July 5, 1976–February 7, 1977.

creativity. As a child gets older, the proportion of utterances that are full or partial imitations of the parent's prior utterance(s) decreases from less than 20% at 21 months to almost zero by the time the child is 3 years old.[17] When Nim was 26 months old, 38% of his utterances were full or partial imitations of his teacher's. By the time he was 44 months old, the proportion had risen to 54%. A doctoral dissertation by Richard Sanders[18] showed that the function of Nim's imitative utterances differed from those of a child's imitative utterances. Bloom and her associates[19] have observed that children imitated their parents's utterances mainly when they were learning new words or new syntactic structures. Sanders found no evidence for either of these imitative functions in Nim's imitative utterances.

As children imitate fewer of their parents' utterances, they begin to expand upon what they hear their parents say. At 21 months, 22% of a child's utterances add at least one word to the parent's prior utterance; at 36 months, 42% are expansions of the parent's prior utterance. Fewer than 10% of Nim's utterances recorded during 22 months of videotaping (the last 22 months of the project) were expansions. Like the mean length of his utterances, this value remained fairly constant.

The videotapes showed another distinctive feature of Nim's conversations that we could not see with the naked eye. He was as likely to interrupt his teacher's signing as not. In contrast, children interrupt their parents so rarely (so long as no other speakers are present) that interruptions are all but ignored in studies of their language development. A child learns readily what one takes for granted in a two-way conversation: each speaker adds information to the preceding utterance and each speaker takes a turn in holding the floor. Nim rarely added information and showed no evidence of turn-taking.

None of the features of Nim's discourse—his lack of spontaneity, his partial imitation of his teacher's signing, his tendency to interrupt—had been noticed by any of his teachers or by the many expert observers who had watched Nim sign. Once I was sure that Nim was not imitating *precisely* what his teacher had just signed, I felt that it was less important to record the teachers' signs than it was to capture as much as I could about Nim's signing: the context and specific physical movements, what hand he signed with, the order of his signs, and their appropriateness. But even if one wanted to record the teacher's signs, limitations of attention span would make it too difficult to remember all of the significant features of *both* the teacher's and Nim's signs.

Once he knew what to look for, the contribution of the teacher was easy to see, embarrasingly enough, in still photographs that we had looked at for years. Consider, for example, Nim's signing the sequence *me hug cat* as shown in FIGURE 2. At first, these photographs (and many others) seemed to provide clear examples of spontaneous and meaningful combinations. But just as analyses of our videotapes provided evidence of a relationship between Nim's signing and the teacher's prior signing, a re-examination of FIGURE 2 revealed a previously unnoticed contribution of the teacher's signing. She signed *you* while Nim was signing *me* and signed *who* while he was signing *cat*. While Nim was signing *hug*, his teacher held her hand in the "n"-hand configuration, a prompt for the sign *Nim*. Because these were the only photographs taken of this sequence, we cannot specify just when the teacher began her signs. It is not clear, for example, whether the teacher signed *you* simultaneously or immediately prior to Nim's *me*. It is, however, unlikely that the teacher signed *who?* after Nim signed *cat*. A few moments before these photographs were taken the teacher repeatedly quizzed Nim as to the contents of the cat box by signing *who?* At the very least, Nim's sequence, *me hug cat*, cannot be interpreted as a spontaneous combination of three signs.

FIGURE 2. Nim signing the linear combination, *me hug cat*, to his teacher (Susan Quinby). (Photographed in classroom by H.S. Terrace.)

There remain other differences between Nim's and a child's signing that need to be explored. One important question is how similar are the contents of a chimpanzee's and a child's utterances, whether or not they are structured grammatically? A cursory examination of corpora of children's utterances[20] indicates that chimpanzees and children differ markedly with respect to the variety of the utterances they make. While children produce certain routine combinations, they are relatively infrequent. Nim's utterances, on the other hand, showed a high frequency of routine combinations (e.g., *play me, me Nim*; see TABLE 1 for additional examples).

It also remains to be shown that the patterns of vocal discourse observed between a hearing parent and a hearing child are similar to the patterns of signing discourse that obtain between a deaf parent and a deaf child. Videotapes of such discourse are now available from a number of sources.[21-23]

CRITICISMS OF PROJECT NIM

Of more immediate concern is the generality of the conclusions drawn about Nim's signing. This issue can be approached in two ways. One is to consider the methodological weaknesses of Project Nim and to pursue their implications. The other is to recognize that, methodological weaknesses notwithstanding, it is reasonable to ask if all signing apes sign because they are coaxed to do so by the teachers and how much sign-for-sign overlap exists between the teacher's and the ape's signing?

Consider first some of the questions raised about Project Nim. It has been said that Nim was taught by too many teachers (60 in all), that his teachers were not fluent enough in ASL, that terminating Nim's training at the age of 44 months prevented his teachers from developing Nim's full linguistic competence, and that Nim may simply have been a dumb chimp (e.g., quotes of other researchers studying language in apes as reported by Bazar[11]).

Aside from the speculation that Nim may have been a dumb chimp, I believe that all of these criticisms are valid. And, if questioning Nim's intelligence is simply a nasty way of asking whether an N of 1 is an adequate basis for forming a general conclusion about an ape's grammatical competence, I would readily admit that it is not. A case can be made, however, that most of the methodological inadequacies of Project Nim have been exaggerated and, in any event, that they are hardly unique to Project Nim. Though Nim was taught by 60 teachers, he spent most of his time in the presence of a core group of 8 teachers: (Stephanie LaFarge, Laura Petitto, Amy Schachter, Walter Benesch, Bill Tynan, Joyce Butler, Dick Sanders, and myself). As described in *Nim* (see especially Appendix B),[12] many of Nim's 60 teachers served as occasional playmates rather than as regular teachers. Nevertheless, they were each listed as a teacher. The Gardners have yet to publish a full list of the teachers who worked with Washoe. Roger Fouts, however, estimates that their number was approximately 40. Of greater importance is Fouts's observation that, during Washoe's 4 years in Reno, Nevada, she was looked after mainly by a small group of 6 core teachers. Francine Patterson's doctoral dissertation presents data provided by 20 teachers during the 3 year period in which they worked with "Koko." This is hardly surprising. It was difficult for an ape's teachers to sustain the energy needed to carry out lesson plans, to engage its attention, to stimulate it to sign, and to record what it signed for more than 3 to 4 hours a day. A 3- to 4-hour session with Nim also entailed an additional 1 or 2 hours needed to transcribe the audio casette on which the teacher dictated information about Nim's signing and to write a report. At least 6 full-time people would be needed to carry out such a schedule on a 16-hour/day, 7-day/week basis. As far as I know, no project has been able to afford the salaries of such a staff. Accordingly, it is necessary to make do with a large contingent of part-time volunteers.

Both Patterson and Fouts speak English while signing with their apes. Films of the Gardners and Fouts signing with Washoe, of Fouts signing with "Ally" and "Booee" (two resident chimpanzees of the Oklahomas Center for Primate Research), and of Francine Patterson signing with Koko make clear that none of these researchers use Ameslan. As mentioned earlier, pidgin sign seems to be the prevalent form of communication on *all* projects attempting to teach apes to use sign language.

Washoe is now 15, Koko is 9, and Ally is 9 years old. I know of no evidence that their linguistic skills increased as they became older. An ape's intelligence undoubtedly increases after infancy. One must, however, also keep in mind that as an ape gets older, its ability to master its environment by physical means also increases. An ape's increasing strength and its recognition that it can get its way without signing should result in *less* motivation to sign and a reduction in its teacher's dominance. I am therefore skeptical of conjectures that an ape's increasing intelligence would manifest itself in a more sophisticated use of language.

THE GENERALITY OF THE CONCLUSIONS OF PROJECT NIM

Whatever the shortcomings of Project Nim, it should be recognized that they are irrelevant to the following hypothesis about an ape's use of signs. An ape signs mainly in response to his teachers' urgings, in order to obtain certain objects or activities. Combinations of signs are not used creatively to generate particular meanings. Instead, they are used for emphasis or in response to the teacher's unwitting demands that the ape produce as many contextually relevant signs as possible. The validity of this hypothesis rests simply on the conformity of data obtained from other signing apes.

Finding such data has proven difficult, particularly because discourse analyses of other signing apes have yet to be published. Also, as mentioned earlier, published accounts of an ape's combinations of signs have centered around anecdotes and not around exhaustive listings of all combinations. One can, however, obtain some insight into the nature of signing by other apes by looking at films and videotape transcripts of their signing. Two films made by the Gardners of Washoe's signing, a doctoral dissertation by Lyn Miles[24] (which contains four videotape transcripts of the two Oklahoma chimps, Ally and Booee), and a recently released film, "Koko, the Talking Gorilla," all support the hypothesis that the teacher's coaxing and cuing have played much greater roles in so-called "conversations" with chimpanzees than was previously recognized.

In "The First Signs of Washoe," a film produced for the "Nova" television series, Beatrice Gardner can be seen signing *What time now?*, an utterance that Washoe interrupts to sign, *time eat, time eat*. A longer version of the same exchange shown in the second film, "Teaching Sign Language to the Chimpanzee Washoe," began with Gardner signing *eat me, more me*, after which Washoe gave her something to eat. Then she signed *thank you* — and asked *what time now?* Washoe's response *time eat, time eat* can hardly be considered spontaneous, since Gardner had just used the same signs and Washoe was offering a direct answer to her question.

The potential for misinterpreting an ape's signing because of inadequate reporting is made plain by another example in both films. Washoe is conversing with her teacher, Susan Nichols, who shows the chimp a tiny doll in a cup. Nichols points to the cup and signs *that*; Washoe signs *baby*. Nichols brings the cup and doll closer to Washoe, allowing her to touch them, slowly pulls them away, and then signs *that* while pointing to the cup. Washoe signs *in* and looks away. Nichols brings the cup with the doll closer to Washoe again, who looks at the two objects once more, and signs *baby*. Then, as she brings the cup still closer, Washoe signs *in*. *That*, signs Nichols, and points to the cup; *my drink*, signs Washoe.

Given these facts, there is no basis to refer to Washoe's utterance — *baby in baby in my drink* — as either a spontaneous or a creative use of "in" as a preposition joining two objects. It is actually a "run on" sequence with very little relationship between its parts. Only the last two signs were uttered without prompting from the teacher. Moreover, the sequence of the prompts (pointing to the doll, and then pointing to the cup) follows the order called for in contructing an English prepositional phrase. In short, discourse analysis makes Washoe's linguistic achievement less remarkable than it might seem at first.

In commenting about the fact that Koko's mean length of utterances (MLU) was low in comparison with that of both hearing and deaf children, Francine Patterson speculates, in her doctoral dissertation, that, "This probably reflects a species difference in syntactic and/or sequential processing abilities" (Reference 25, page 153). Patterson goes on to observe that, "The majority of Koko's utterances were not spontaneous, but elicited by questions from her teachers and companions. My interactions with Koko were often characterized by frequent questions such as 'What's this?'" (Reference 25, page 153).

Four transcripts appended to Lyn Miles' dissertation provided me with a basis for performing a discourse analysis of the signing of two other chimpanzees, Ally and Boee. Each transcript presents an exhaustive account of one of these chimps signing with one of two trainers: Rogert Fouts and Joe Couch. The MLU and summaries of the discourse analysis of each tape is shown in TABLE 4. TABLES 5 and 6 show exhaustive summaries of two conversations, one from a session with the highest MLU and one from a session with the highest percentage of adjacent utterances that were novel. The novel utterances are very similar to Nim's run-on sequences. They also overlap considerably with adjacent utterances that were expansions and with noncontingent utterances.

In his discussion of communicating with an animal, the philosopher Ludwig Wittgenstein cautions that apparent instances of an animal using human language may prove to be a "game" that is played by simpler rules. Nim's, Washoe's, Ally's, Booee's, and Koko's use of signs suggests a

TABLE 4

SUMMARY OF ALLY AND BOOEE TRANSCRIPTS

	Video Tape Number				
	3	4	5	6	Mean
Number of utterances	38	79	102	72	72.75
MLU	1.63	1.52	2.25	1.93	1.85
% Adjacent	76.3	93.7	77.4	86.1	83.4
% Imitations:	13.6	22.8	7.84	8.33	13.03
% Expansions	7.89	7.59	13.7	4.16	8.34
% Novel	55.3	63.3	55.9	73.6	62.3
% Noncontingent	23.7	6.3	2.6	3.9	16.6

TABLE 5

CONVERSATION NUMBER 4: ROGER AND ALLY

Adjacent Utterances (N = 74: C = 67, ? = 7)*

Novel (N = 50: C = 43, ? = 7)			
Roger	14	that	2
Roger tickle Ally	9	that that box	1
Roger tickle	3	that shoe	1
tickle Roger	1	Roger string George	1
tickle	1	comb	1
Roger Tickle Ally hurry	1	good	1
George	2	food-eat	2
Joe	7	Ally	2
string	1		
Expansions (N = 6: C = 6)			
George smell Roger	1	Roger tickle Ally	1
tickle hurry	1	Roger tickle Roger comb	1
Roger tickle	1	Roger tickle Roger	1
Imitations (N = 18: C = 18)			
Roger	5	baby	1
tickle	2	pillow	1
Roger tickle Ally	1	comb	2
shoe tickle	2	pull	2
shoe tickle	1		
Noncontingent utterances (N = 5)			
Joe	2	shoe	1
you tickle	1	pillow	1

* C denotes the number of utterances that followed teachers' commands; "?" denotes the number that followed teachers' questions.

type of interaction between an ape and its trainer that has little to do with human language. In each instance the sole function of the ape's signing appears to be to request various rewards that can be obtained only by signing. Little, if any, evidence is available that an ape signs in order to exchange information with its trainer, as opposed to simply demanding some object or activity.

In a typical exchange the teacher first tries to interest the ape in some object or activity such as looking at a picture book, drawing, or playing catch. Typically, the ape tries to engage in such activities without signing. The teacher then tries to initiate signing by asking questions such as *what that?*, *what you want?*, *who's book?*, and *ball red or blue?*

The more rapidly the ape signs, the more rapidly it can obtain what it wants. It is therefore not surprising that the ape frequently interrrupts the teacher. From the ape's point of view, the teacher's signs provide an excellent model of the signs it is expected to make. By simply imitating

TABLE 6

CONVERSATION NUMBER 6: JOE AND BOOEE

*Adjacent utterances (N = 79: C = 31, ? = 47)**

  Novel (N = 57: C = 21, ? = 35)

food-eat Booee that	3	more that Booee	1
food-eat Booee	3	that Booee	1
food-eat	3	more Booee	1
you-food-eat Booee that	1	Booee	2
Booee food-eat more Booee	1	you Booee	1
food-eat Booee hungry	1	more	1
food-eat me fruit Booee	1	you	3
food-eat fruit Booee	1	Booee hurry	1
that food-eat Booee	1	there you gimme	1
fruit food-eat you	1	you Booee you Booee hurry	1
food-eat fruit hurry	1	more Booee hurry	1
gimme food-eat	1	tickle Booee	3
gimme	1	tickle you me	1
fruit more Booee	1	you there	1
gimme fruit hurry	1	hurry tickle Booee hurry	1
there more fruit	1	you there that there	1
hurry	4	that Booee there Booee	1
fruit	1	baby that	1
that Booee over there	1	more baby	1
that	4		

  Expansions (N = 14: C = 4, ? = 10)

hurry tickle Booee hurry Booee hurry	1	that Booee	1
Booee you Booee hurry gimme	1	that more baby	1
tickle Booee gimme	1	there that there	1
me tickle	1	that tickle	1
me tickle hurry	1	there you	1
food-eat Booee that	1		

  Imitation (N = 8: C = 6, ? = 2)

that	4	tickle	1
you	2	baby	1

Noncontingent Utterances (N = 23)

tickle Booee	6	more fruit there	1
tickle	2	that	1
more Booee	1	that that Booee gimme	1
hurry that tickle	1	gimme	2
tickle gimme	1	Booee	1
food-eat	1	hurry gimme hurry	1
food-eat Booee	1	that there	1

* C denotes utterances that followed teachers' commands; "?" denotes utterances that followed teachers' questions.

a few of them, often in the same order used by the teacher, and by adding a few "wild cards"—general purpose signs such as *give, me, Nim,* or *more*—the ape can produce utterances that appear to follow grammatical rules. What seems like conversation from a human point of view is actually an attempt to communicate a demand (in a nonconversational manner) as quickly as possible.

FUTURE RESEARCH

It might be argued that signing apes have the potential to create sentences but did not do so because of motivational rather than intellectual limitations. Perhaps Nim and Washoe would have been more motivated to communicate in sign language if they had been raised by smaller and more consistent groups of teachers, thus sparing them emotional upheavals. It is, of course, possible that a new project, administered by a permanent group of teachers who are fluent in sign language and have the skills necessary for such experiments, would prove successful in getting apes to create sentences.

It is equally important for any new project to pay greater attention to the function of the signs than to mastery of syntax. In the rush to demonstrate grammatical competence in the ape, many projects (Project Nim included) overlooked functions of individual signs other than their demand function. Of greater significance, from a human point of view, are the abilities to use a word simply to communicate information and to refer to things which are not present. One would like to see, for example, to what extent an ape is content to sign *flower* simply to draw the teacher's attention to a flower with no expectation that the teacher would give it a flower. In addition one would want to see whether an ape could exchange information about objects that are not in view in order to exchange information about those objects. For example, could an ape respond, in a nonrote manner, to a question such as, *What color is the banana?* by signing *yellow* or to a question such as *Who did you chase before?* by signing *cat*. Until it is possible to teach an ape that signs can convey information other than mere demands it is not clear why an ape would learn a grammatical rule. To put the question more simply, why should an ape be interested in learning rules about relationships between signs when it can express all it cares to express through individual signs?

The personnel of a new project would have to be on guard against the subtle and complex imitation that was demonstrated in Project Nim. In view of the discoveries about the nature of Nim's signing that were made through videotape analyses, it is essential for any new project to maintain a permanent and unedited visual record of the ape's discourse with its teachers. Indeed, the absence of such documentation would make it impossible to substantiate any claims concerning the spontaneity and novelty of an ape's signing.

Requiring proof that an ape is not just mirroring the signs of its teachers is not unreasonable; indeed, it is essential for any researcher who seeks to determine, once and for all, whether apes can use language in a human manner. Nor is it unreasonable to expect that in any such experiment, ape "language" must be measured against a child's sophisticated ability. That ability still stands as an important definition of the human species.

While writing "A Report to an Academy," Kafka obviously had no way of anticipating the numerous attempts to teach real apes to talk that took place in this country and in the U.S.S.R.[3-13,25-28] Just the same, his view that an ape will imitate for "a way out" seems remarkably telling. If one substitutes for the phrase, "a way out," rewarding activities such as being tickled, chased, hugged, and access to a pet cat, books, drawing materials, and items of food and drink, the basis of Nim's, Washoe's, Koko's, and other apes' signing seems adequately explained. Much as I would have preferred otherwise, a chimpanzee's "Report to an Academy" remains a work of fiction.

REFERENCES

1. CHOMSKY, N. 1957. Syntactic Structures. Mouton, The Hague.
2. MILLER, G. A. 1964. The psycholinguists. Encounter **23** (1): 29–37.
3. GARDNER, B.T. & R.A. GARDNER. 1969. Teaching sign language to a chimpanzee. Science **162**: 664–672.
4. PREMACK, D. 1970. A functional analysis of language. J. Exp. Anal. Behav. 4: 107–125.
5. RUMBAUGH, D.M., T.V. GILL & E.C. VON GLASERSFELD. 1973. Reading and sentence completion by a chimpanzee. Science **182**: 731–733.
6. TERRACE, H.S. 1979. Is problem-solving language? J. Exp. Anal. Behav. **31**: 161–175.
7. THOMPSON, C.R. & R.M. CHURCH. 1980. An explanation of the language of a chimpanzee. Science **208**: 313–314.
8. STRAUB, R.O., M.S. SEIDENBERG, T.G. BEVER & H.S. TERRACE. 1979. Serial learning in the pigeon. J. Exp. Anal. Behav. **32**: 137–148.
9. GARDNER, B.T. & R.A. GARDNER. 1974. Comparing the Early Utterances of Child and Chimpanzee. pp. 3–23. University of Minnesota Press, Minneapolis, Minn.
10. GARDNER, B.T. & R.A. GARDNER. 1974. Teaching sign language to a chimpanzee, VII: Use of order in sign combinations. Bull. Psychonomic Soc. **4**: 264–267.
11. BAZAR, J. 1974. Catching up with the ape language debate. Am. Psychol. Assoc. Monitor **11**: 4–5, 47.
12. TERRACE, H.S. 1979. *Nim*. A. Knopf, New York, N.Y.
13. FOUTS, R.F. 1972. Use of guidance in teaching sign language to a chimpanzee. J. Comp. Physiol. Psychol. **80**: 515–522.
14. TERRACE, H.S., L.A. PETITTO, R.J. SANDERS & T.G. BEVER. 1979. Can an ape create a sentence? Science **206**: 891–902.
15. TERRACE, H.S., L.A. PETITTO, R.J. SANDERS & T.G. BEVER. 1980. On the Grammatical Capacity of Apes. Gardner Press, New York, N.Y.

16. BROWN, R. 1973. A First Language: The Early Stage. Harvard University Press, Cambridge, Mass.
17. BLOOM, L.M., L. ROCISSANO & L. HOOD. 1976. Adult-child discourse: Developmental interaction between information processing and linguistic knowledge. Cogn. Psychol. 8: 521–552.
18. SANDERS, R.J. 1980. The Influence of Verbal and Nonverbal Context on the Sign Language Conversations of a Chimpanzee. Ph.D. dissertation, Columbia University.
19. BLOOM, L., L. HOOD & P. LIGHTBOWN, 1974. Imitation in language development: If, when, and why. Cogn. Psychol. 6: 380–420.
20. BLOOM, L. 1970. *Language development: Form and function in emerging grammars.* M.I.T. Press, Cambridge, Mass.
21. BELLUGI, U. & E.S. KLIMA. 1976. The Signs of Language. Harvard University Press, Cambridge, Mass.
22. HOFFMEISTER, R.J. 1978. The Development of Demonstrative Pronouns, Locatives and Personal Pronouns in the Acquisition of American Sign Language by Deaf Children of Deaf Parents. Ph.D. dissertation, University of Minnesota.
23. MCINTIRE, M.L. 1978. Learning to take your turn in ASL. Working paper. Department of Linguistics, UCLA.
24. MILES, H.L. 1978. Conversations with Apes: The Use of Sign Language by Two Chimpanzees. Ph.D. dissertation, University of Connecticut.
25. PATTERSON, F.G. 1979. Linguistic Capabilities of a Lowland Gorilla. Ph.D. dissertation, Stanford University.
26. KELLOGG, W.N. 1968. Communication and language in the home-raised chimpanzee. Science 182: 423–427.
27. HAYES, C. 1951. The Ape in Our House. Harper & Row, New York, N.Y.
28. TEMERLIN, M.K. 1975. Lucy: Growing up Human: A Chimpanzee Daughter in a Psychotherapist's Family. Science and Behavior, Palo Alto, Calif.

Ape Signing: Problems of Method and Interpretation

MARK S. SEIDENBERG*
Center for the Study of Reading
University of Illinois at Urbana-Champaign
Champaign, Illinois 61820

Bolt, Beranek & Newman, Inc.
Cambridge, Massachusetts 02138

LAURA A. PETITTO
Department of Human Relations
Harvard University Graduate School of Education
Cambridge, Massachusetts 02138

REPORTS IN THE LAST DECADE that apes can learn to communicate with humans using a rudimentary form of sign language[1-5] have been received with great interest. For many anthropologists, linguists, philosophers, and psychologists, research with apes such as Washoe, Koko, and others appeared to provide decisive evidence bearing on several fundamental issues. For example, the belief that language is species-specific could not be sustained if apes acquired some linguistic facility, as these projects seemed to indicate. The apes' performance also appeared to rescue learning-theoretic views of language acquisition from critical analyses such as Chomsky's.[6] The methods used in the ape sign language projects were derived from such theories, and some researchers were self-consciously behavioristic in orientation.[7] Whatever the logical merit of such criticism of empiricist theories of language acquisition, it might be waived aside if the apes acquired linguistic skills through shaping and differential reinforcement of responses. This would at the same time undermine strongly nativist positions.

This research has gained renown outside the academic community through accounts in the print and broadcast media, documentary and educational films, and popularized texts. As might be expected, the mass media accepted the ape researchers' claims uncritically, focusing instead on speculation. If apes can talk, what will they say? Will they be

* Current address: Department of Psychology, McGill University, Montreal, PQ, Canada H3A 1B1.

0077-8923/81/0364-0115 $01.75/0 © 1981, NYAS

able to provide a new perspective on the human race? Will they demand their civil rights? If language is the only capacity that distinguishes humans from other species, and apes possess that capacity, should they not be treated as human? The ape researchers themselves appeared all too willing to exploit media interest in such nonscientific issues. At the point when they were raised, the question, "What *evidence* have these projects provided concerning the linguistic capacities of apes?" was obscured. Even among scholars with a more dispassionate interest in this research, discussion was largely concerned with the *implications* of linguistic skills in lower primates, rather than evidence that they *possessed* them.

Recently, the claims on behalf of the linguistic abilities of signing apes have come under close scrutiny as a result of research on two fronts. First, a team at Columbia University (Terrace, Petitto, Sanders & Bever) attempted to replicate the experiments of the Gardners, Fouts, and Patterson.[8,9] The Columbia team interpreted the behaviors of their subject, Nim Chimpsky, not as evidence for linguistic skill, but rather as the result of nonlinguistic processes such as imitation. We shall consider this work below. Second, the ape researchers' claims were questioned by Seidenberg and Petitto, who in a series of papers[10-13] performed a close analysis of the methods and procedures of the original studies. We concluded that the available data could not sustain the radical claims on the apes' behalf, since they are compatible with and suggestive of much weaker nonlinguistic interpretations. Let us summarize those findings briefly.†

(1) The primary data in a study of ape language must include a large corpus of utterances, a substantial number of which are analyzed in terms of the contexts in which they occurred. No corpus exists of the utterances of any ape for whom linguistic abilities are claimed. The absence of this fundamental source of data means that it is impossible to evaluate the character of the apes' behavior (e.g., is it context-appropriate? imitative? spontaneous? reactive? rule-governed? repetitive?). Most importantly, its absence affects the status of the examples from the apes' behaviors (e.g., that Washoe signed *water bird* for a duck, and Koko signed *cookie rock* for a stale sweet roll), which are the primary evidence offered to this time. In conjunction with a corpus, these might constitute representative examples of the apes' behaviors. Lacking a corpus, their typicality cannot be established, nor can alternate, nonlinguistic interpretations be eliminated, for example, that they were the output of a random sign-generation process which produced a few inter-

---

† These issues are discussed in greater detail in the cited references (10–13).

pretable sequences amid a large amount of noise. As such, they are simply anecdotes. A small number of anecdotes are repeatedly cited in discussions of the apes' linguistic skills. However, they support numerous interpretations, only the very strongest of which is the one the ape researchers prefer, i.e., that the ape was signing "creatively." These anecdotes are so vague that they cannot carry the weight of evidence which they have been assigned. Nonetheless, two important claims—that the apes could combine signs creatively into novel sequences, and that their utterances showed evidence of syntactic structure—are based exclusively upon anecdote.

(2) Belief that apes can use signs to name entities is founded primarily on the Gardners' "double blind" vocabulary tests.[2,14,15] These are the tests in which Washoe named objects or pictures of objects in a box. Washoe's performance on these tests is probably the most widely known and cited aspect of her behavior. The tests have been portrayed as providing hard, objective evidence of Washoe's naming skills, and her performance has been witnessed by many millions of persons through television and film. It is startling to realize how little information has been provided either about the methods and procedures used in these tests, or about Washoe's performance. Nowhere have the experimenters provided a complete description of the stimuli used, the manner in which they were presented, the scoring procedures, Washoe's training for the test, or her performance on multiple tests. It is not even clear which of her vocabulary signs were tested. The examples cited are simple objects such as *shoe* or *apple*. But she is credited with knowing many other signs—such as *good, mine, quiet,* and *sorry*—that are not exemplified by simple objects. It is unclear how such signs could be tested under this procedure. Rather than providing an account of her performance on several testing sessions, the Gardners provide examples of her performance on individual sessions. As with the anecdotes discussed above, these examples entail a sampling problem. It cannot be determined whether the cited behavior is typical of her performance, whether, for example, it was sustained across multiple testing sessions.

Much has been made of these tests because, it is asserted, the "double blind" procedures ensured that the subject's responses could not be "cued." However, this is surely false. Blind procedures merely ensure that the experimenters could not cue responses themselves. Other aspects of the test—the order in which stimuli were presented, the number of times a particular stimulus was presented, the manner in which trials were blocked, the nature of feedback, and so forth—could provide information that would increase the probability of a correct response. This is not to say that the tests necessarily provided such information, only that an adequate report would provide the relevant procedural details, and that in their absence, the cited examples are subject to a variety of interpretations.‡

Perhaps the published reports are schematic because it was believed that films of Washoe performing this task provided sufficiently compelling evidence of her skill.[16,17] It is certainly true that the filmed examples are provocative. However, a documentary film is not a substitute for a scientific report. The method of a documentary film is to select the best examples of a given phenomenon. In presenting the vocabulary test, the phenomenon of interest was, "Washoe successfully naming objects in the box," which the filmed examples represent well. The scientific, rather than cinematic, question is different: on what proportion of all trials, with which stimuli, presented and scored in what manner, could Washoe successfully produce the correct sign? In a scientific report, all behaviors, both desired and undesired, photogenic and otherwise, must be documented and analyzed. The Washoe films would serve a useful purpose if there were a complete and unedited characterization of Washoe's performance on the tests in a published source. However, no such record exists. In the published discussions of both these tests and other aspects of ape signing, the ape researchers follow the method of the documentary filmmaker, presenting isolated examples which, however compelling, cannot be evaluated outside the context of the ape's behavior in general.

(3) Some of the strongest claims on behalf of the linguistic abilities of apes derive from Washoe's performance on a test of her ability to answer "wh-" questions.[3] It is said that she could answer questions correctly and that her performance exceeded that of a child at Stage III (as defined by Brown[18]). Consider for a moment the procedure used in this test. Washoe's response to a question was scored as correct not if its meaning was correct in a particular context but rather if it contained a sign from a predesignated target category. For example, the question *what's that* took responses from the category *noun*; thus, if the Gardners held up an orange and asked this question, a response containing *any* noun was correct, not merely those containing *orange*. No data are included concerning the proportion of trials on which the semantically correct sign was produced. Furthermore, in contrast to the vocabulary tests, the experimenters did not consider only the first sign in a response. Thus, Washoe could answer questions "correctly" by emitting signs from her vocabulary in random order until she produced one from the target category.

---

‡ Given the results of the Rumbaughs (Reference 36), it seems likely that apes can be taught to perform this task, at least with simple objects. Clearly establishing this would motivate further experiments exploring what the ape must know in order to perform in this manner.

Other aspects of this test are also disturbing. Although the test consisted of 500 trials, only a few examples of Washoe's actual replies to each question are included. This is problematical because her replies were transformed one or more times before they were scored. Eliminated from Washoe's answers were all signs that repeated those in a question, signs that appeared more than once in a reply, and signs from a class termed "markers." The latter is a heterogeneous category including *gimme, please, want, hurry, more, can't,* and others. Signs were eliminated from 46% of her replies. The effect of these transformations is to turn long, redundant utterances into strings that more closely resemble human utterances in their superficial form. In this way, her utterances are altered radically, before they have been fully documented.

Such methodological problems are not limited to this test. Fouts *et al.*[5] report a test of the ability of another signing chimpanzee, "Ali," to use the sign *in, on* and *under.* They constructed a test in which Ali had to describe whether one object was *in, on,* or *under* another object. Performance on one test is given as 49.1% correct, significantly above chance for a three-choice test. The crucial aspect of their procedure is that a given sign was never correct on two successive trials. If the ape learned about this contingency during the long course of training and testing, chance level responding was closer to 50% correct, rather than 33%. With the guessing probability correctly calculated, Ali is merely performing at chance levels.

(4) The existing reports are largely concerned with the apes' production of signs. There have been no tests of their abilities to comprehend. Nonetheless, all of the ape researchers assert that the apes possessed good comprehension skills. For example, the Gardners state that "There were 132 signs of Ameslan in Washoe's expressive vocabulary and a much larger number of signs in her receptive vocabulary."[3] Fouts *et al.* remark that "Ali has a sign language vocabulary of approximately 130 signs and a comprehension of English vocabulary at least equal to that" (Reference 5, page 167). Terrace *et al.*[9] provide a list of 199 signs that Nim could comprehend; these range from *banana* and *hat* to *attention* and *false.* In each of these cases, the basis for concluding that the apes could understand signs or words is at best informal tests of some lexical items, and at worst the subjective impressions of the researchers. It is clear from the child language literature that extreme caution must be exercised before assuming that a child "understands" a word or sign. Utterances occur in rich contexts that provide the perceiver with multiple sources of information. This nonlinguistic information may permit enactment of an "appropriate" response without comprehension of the utterance's meaning. For this reason, young children's comprehension skills appear to exceed their ability to produce utterances, and their

receptive skills are frequently overestimated by untrained observers.[19-21] Substantive evidence of comprehension could only be provided by tests that include controls over contextual cuing far more subtle than those in informal experiments or observations.

In concluding that the apes could comprehend words and signs, as in judging their other capabilities, the ape researchers ascribe control over complex linguistic skills to their subjects without a systematic characterization of the behavior in question and the circumstances under which it was elicited.

(5) In the absence of systematic data concerning the apes' production and comprehension of signs, it must be questioned whether there was an empirical basis on which to assign them specific meanings and grammatical functions. These attributions appear to have been made on the basis of the experimenters' *intentions* in teaching signs, not on the basis of the apes' actual behaviors. The question here is: What is the behavioral basis for deciding that a sign meant *banana, good, silly, false, large, sorry*? Inclusion of very abstract signs in lists of the apes' vocabularies underscores a serious question of over-attribution.

(6) Finally, despite a widespread assumption to the contrary, there is no evidence that the apes had acquired facility in American Sign Language (ASL or Ameslan). An ape could be said to have learned ASL only to the extent that its signing showed evidence of the grammatical structures and expressive devices characteristic of that language. There is no indication that any of the apes' signing behavior had these properties. A weaker claim would be that the apes learned lexical items of ASL—i.e., the citation forms of signs—but none of its grammar. The extent to which this is true is also unclear, since many of the apes' signs were either modified from the ASL equivalents or created especially for them.[2,4]

The extent to which an ape could learn ASL is limited by the degree to which ASL was provided as input. Discussions of the apes include clear descriptions of ASL as the language medium.[2-4,14] For example, the Gardners stated,

> We chose American Sign Language (ASL) because it is so widely used by humans in the United States. An alternative would have been a special language based on those gestures most easily performed by a chimpanzee. This alternative becomes much less attractive when one considers the formidable technical problems involved in the invention of a synthetic language. Moreover, it would be necessary to prove that our synthetic system was a language. [Reference 2, page 121.]

Many other statements imply that ASL is the language the apes were learning (e.g., "Washoe's exposure to Ameslan did not begin until she was nearly one year old."[31] "Washoe's exposure to verbal communication other than ASL was kept to a minimum."[2] "American Sign Language was chosen as the primary mode of communication [with Koko]. . . ."[4]). The degree to which the apes were in fact exposed to ASL is uncertain, however. As the Gardners acknowledge,

> Perhaps the greatest discrepancy between Project Washoe as we originally conceived it and as we were able to execute it lies in the weakness of the adult models that we could provide. . . . For the most part, Washoe's human companions and adult models for signing could be described as hearing people who had acquired Ameslan very recently and whose fluency, diction, and grammar were limited because they had little practice except with each other and with Washoe. [Reference 31, page 42.]

Questions as to which sign code the apes were exposed to could only be answered by systematic analysis of the teachers' actual input. These questions assume importance because instead of characterizing the structure of the apes' utterances, the researchers merely invoke ASL, a language whose properties are independently known.§[22-25] Given the absence of any evidence that their utterances possessed the structure of ASL, and unanswered questions as to whether they were exposed to ASL, it is inaccurate to term their behavior "signing in ASL." Of course, the apes might have acquired facility in a sign language that was not ASL, e.g., a pidgin sign. It would then be necessary, as the Gardners observed, to establish that their signing, while not ASL, nonetheless had interesting linguistic properties. At this time, however, we have no detailed information concerning the nature of their sign code.¶

THE NIM PROJECT

The data of Terrace et al.[8,9] on Nim are more robust than those offered by other ape researchers. Although their data are limited in several

---

§ They are also important because they affect the selection and interpretation of behavioral data. For example, discussions of linear orderings in the apes' utterances as evidence for syntax[2-4,8,9] presume that the apes learned a code more similar to English than to ASL.

¶ The published reports exhibit considerable confusion over this issue. In one paper, the Gardners state, "We do not assert that our chimpanzee subjects use the signs of Ameslan. . . .,"[31] but elsewhere they freely describe their behaviors as such (e.g., "There were 132 signs of Ameslan in Washoe's expressive vocabulary. . . ."[3]). Patterson states that, "In this study, simultaneous communication, or the use of sign language accompanied by spoken English, was employed. American Sign Language was chosen as the primary mode of communication. . . ."[4] However, the structure of English is so different from that of ASL that it is almost impossible to use both languages simultaneously. "Simultaneous communication" typically consists of both spoken and signed English, not ASL.[26] Hence, it is not clear that the ape was exposed to ASL, nor can this be determined elsewhere in the report.

respects, they are the only systematic data on any signing ape. They recorded a large number of Nim's utterances and tabulated them according to frequency of occurrence (but not, however, according to their contexts of occurrence). The single most striking fact about this corpus is that Nim relied extremely heavily upon a small set of signs. In fact, seven signs — *Nim, me, you, eat, drink, more*, and *give* — account for a very high proportion of the signs in their published corpus. For example, they provide a listing of Nim's 25 most frequently used single signs during five periods of recording.[9] In each of these five periods, the seven signs account for between 43% and 47% of the data. Nim continued to rely upon these signs even though the number of signs in his putative vocabulary increased from 33 to 157. The proportion of Nim's entire output that was composed of these seven signs may in fact be much higher if his repetitions were predominately from this class. These repetitions are deleted from the corpus, however. Similarly, these seven signs account for 84% of the signs in the 25 most frequent 2-sign utterances, and 83% of the signs in the 25 most frequent 3-sign utterances. Note that the pool of favorite signs has an interesting property: they are semantically "appropriate" in nearly every context. That is, signs such as *Nim* or *eat* could always be interpreted as correct. These signs present a difficult problem for the teacher attempting to differentially reinforce "appropriate" and "inappropriate" signs. In producing these signs, Nim appears to have developed a very simple strategy for signing "correctly" regardless of the context.

The data of Terrace *et al.* cannot provide evidence that Nim could consistently name objects or actions because the corpus is not analyzed with respect to contexts of occurrence. Thus, although Nim's behaviors are glossed as lexical items, there is no evidence that they possessed specific meanings or grammatical functions, or that they were consistently elicited by particular stimuli. Much of the remainder of their discussion is devoted to a search for evidence of syntactic organization in Nim's multisign sequences. Their strategy is to determine whether there are any statistical regularities in the combinations. Indeed, there are; for example, Nim produced the sequence *give* + X more often than X + *give* (where X is any sign); similarly, *more* + X exceeded X + *more* and *me* + X was more frequent than X + *me*. The critical fact is that although in each case there is a preference for one order, a high proportion of the tokens exhibiting this order is accounted for by a very small number of types. Thus, *give* + X occurs more than twice as often as X + *give*, but 44% of the *give* + X tokens are accounted for by only three combinations, *give eat, give me*, and *give Nim*. Similarly, 54% of the *more* + X combinations consist of *eat, drink*, and *tickle*, as do 54% of the X + *more* combinations. Note also that determining whether Nim had learned syntactic rules for combining a sign such as *give* with other signs depends upon contextual information that is not available. Either *give* + X or X + *give* could be appropriate depending upon the context.

In comparing Nim's multisign utterances and mean length of utterances (MLU) to those of children, it is important to realize that all contiguous repetitions were deleted. In this respect, Terrace *et al.* follow the practice established by the other ape researchers. The repetitions in ape signing constitute one of the primary differences between their behavior and the language of deaf and hearing children, yet they have always been eliminated from analyses. Given their ubiquity and importance, it is essential that they be evaluated eventually. The other major conclusions of Terrace *et al.* are that Nim interrupted and imitated his teachers more often than children do in conversation. The first point is subject to further confirmation, since close comparisons between the interruptions of child and ape have not been performed. The second point is supported by a comparison which shows that Nim imitated more often than the children in a study by Bloom, Rocissano, and Hood.[27] If the imitation hypothesis is true, then it must be concluded that Nim's teachers provided input largely consisting of the small number of signs that occur with very high frequency in the corpus. An alternate interpretation is that Nim did *not* imitate heavily, but rather had a general strategy of producing signs from his pool of favorites. In responding to Nim, and attempting to engage him, his teachers may also have produced these signs in abundance. In effect, Nim's signing may have cued his teachers' responses, rather than vice versa. These alternatives could be distinguished empirically in several ways, e.g., by determining whether Nim imitated signs that were not from the favored pool. However, the mere fact that the same signs tended to appear in both Nim's signing and that of his teachers does not establish whether imitation was occurring or who was imitating whom.

OTHER VIEWPOINTS

The ape researchers' responses to objections to their methods and conclusions have taken several forms. Primarily, they have pointed out deficiencies in Terrace *et al.*'s project.[28-30] The fact, it is argued, that Nim failed to acquire linguistic skills does not necessarily mean that apes such as Koko and Washoe lacked them as well. Nim's behavior (or lack thereof) could be a consequence of factors specific to that project. Fouts believes that Terrace *et al.*'s teaching methods were inadequate.[28] Patterson believes that their project was too short.[30] The Gardners and Fouts assert that Nim had too many teachers.[28,30] Premack states that "it is a repetition of a non-optimal experiment."[28] Perhaps Nim was simply less clever than the other apes. Rather than defend the Nim project, it will suffice to note two points. The first is that the experiment was not a "failure to replicate" earlier findings. So far as can be determined from available information, Nim's behavior closely resembled that of Washoe and other signing apes. He learned to form signs, combine them into sequences, and produce them in various contexts. Terrace *et al.* merely came to a different interpretation of these behaviors, by considering aspects of the subject's performance and that of his teachers which the other researchers ignored. In light of the similarities among the behaviors of all the signing apes—and deficiencies common to all the projects—questions as to who used the best methods are largely moot.

The second point is that criticism of the Nim project does nothing to validate claims on behalf of Washoe and Koko. We have argued that these claims are suspect because they are not supported by data. This criticism can only be countered by providing such data, not by attacking the merits of another experiment.

Another line of defense is to suggest that critics have applied more stringent criteria to the evaluation of the apes' behavior than are applied to children. The Gardners in particular believe that their critics have applied a "rubber ruler."[29,30] They fault Terrace *et al.* for using the method of "rich interpretation"[18] with respect to children but not apes. This argument is orthogonal to the primary problem, however, which lies in the absence of adequate data on which to base any interpretation. Both Patterson and the Gardners have consistently misinterpreted the "method of rich interpretation" to mean, "apply the richest interpretation to the data" without evaluating weaker alternatives.*

The Gardners' comment raises an important point. It is clear that many studies of child language suffer from the same problems of under-substantiation and over-attribution as do the ape studies. One value of the ape research has been to emphasize the methodological issues in the study of language acquisition. However, it is also clear that evaluation of the apes' behaviors would be advanced immeasurably by having available the kinds of evidence seen in many studies of child language. Beyond this there is a limited sense in which it may be justified to require documentation of the apes' behaviors that is both more complete and more rigorous than that in most studies of children. There are several reasons why this might be so. The first is that while the general

* This reaches the heart of the matter. Researchers in this field appear content to establish that aspects of the apes' behaviors are *compatible* with a "rich" interpretation. With the exception of Terrace *et al.*, they do not actively attempt to evaluate simpler interpretations.

character of child language is well known, the same cannot be said of ape signing. Access to children for the purpose of obtaining data on their linguistic performance is easy; access to apes is not. The ape researchers' primary responsibility was to provide a full characterization of their behavior regardless of the interpretation assigned to it. This was especially important because the behavior had not been observed previously, and it could be observed independently only by a few persons at enormous expense. This is not a responsibility that typically falls on students of child language, whose data and interpretations can be verified by others.

Second, it might be argued that questions about ape language should be resolved using the most robust data that can be obtained under existing procedures. It is not sufficient simply to apply any method that has been used in a study of child language. Surely it would be wrong to decide these issues on the basis of observational and experimental data that are themselves questionable. Stringent criteria of scientific investigation are rightly invoked in the face of an unusual, implausible, or controversial claim (as, for example, in recent debates over ESP and heritability of intelligence). The nature of the language acquisition process makes it difficult to develop methods that offer the rigor available in other areas of psychology (and other sciences). In light of this, it may be that the best strategy is to seek converging evidence from a variety of sources, i.e., multiple measures of a wide range of behaviors.

It is also important to note differences in the goals which the ape researchers have set and those of persons studying child language since these affect the types of data that are cited. Psycholinguists typically do not ask of children's behavior, as some have asked of the apes' behavior, "Is it language?"[31] Rather, they attempt to track the developmental sequence by which humans move from an initial state (lack of language) to a steady state (adult competence). Most questions in the study of child language are diachronic: How does language behavior change over time, and how do such changes reflect the growth of knowledge? The ape researchers are concerned with a synchronic question: Do the behaviors that the ape has come to produce possess certain linguistic properties? It is not that a "rubber ruler" is being applied, but rather that different kinds of evidence are necessary to decide these questions. Evaluating the apes' behavior turns heavily on the issue of nonlinguistic response strategies, which is an important issue in the study of child language, but one that is irrelevant to many other topics.

There is another important difference between research on apes and children. In interpreting child language data, researchers are able to draw upon the fact that in the absence of pathology, children learn languages and retain this skill through maturity. Although it may not be possible to determine exactly when a child has acquired a particular type of linguistic knowledge—indeed, this may not be an empirical question—one knows that the child's fragmentary utterances are evidence for an emerging linguistic skill. We cannot have the same confidence about the apes' utterances, both because they have a very different character from those children, and because the apes do not achieve this higher competence.†

A NONLINGUISTIC INTERPRETATION OF APE SIGNING

The available information concerning ape signing—derived from published reports, films, and our own experiences with a signing ape—leads us to an interpretation that differs radically from that offered by the ape researchers. What the apes appear to have learned is not the meanings or grammatical functions of signs, or how to combine them using productive rules, or how to decode what was signed to them, but rather that signing behavior was very important. The activity of producing signs was highly valued by their teachers. Under this circumstance, they learned that the mere behavior of producing signs could be used to effect certain outcomes, e.g., getting food, social approval, release from work time, and the like. This behavior is similar to that of a very young child who has learned to say things it does not understand. Consider the child who learns to say something very adult, perhaps mildly obscene. It learns that this utterance will produce certain behaviors in the adult audience: they will laugh or titter, pat it on the head. The child understands the consequence of saying the phrase, even though it has no knowledge whatsoever of the meanings or grammatical functions of the elements, or any control over rules for combining these elements. All it has is the very general knowledge that saying the phrase is highly valued—at least until the audience tires of it.

In the service of effecting a variety of positive outcomes, the signing ape learns response strategies of various sorts. If the Nim data are representative, then it is clear that one strategy is "sign from the pool of favorites." In the Gardners' test of Washoe's ability to answer questions, the strategy, "sign until the experimenter terminates the trial," would have worked effectively, as would, "randomly select signs from the

† Some observers have detected another kind of bias against the ape research. Failure to accept the ape researchers' conclusions has been attributed to anthropocentric bias, an irrational attachment to belief in the "uniqueness" of man and the inferiority of other species.[32,33] Those who criticize the ape research are portrayed as objecting not to the methods, procedures, or conclusions of the ape researchers, but rather to the mere idea that apes can talk. This move trivializes the issues, and does nothing to advance the debate.

vocabulary of responses until you hit the correct one." Being able to respond in these ways is not an uninteresting skill to have learned. In a very limited sense, it concerns the pragmatics of language use, that is, the relation between an utterance, the context in which it occurs, and the effects it has on the perceiver. This is a form of communication, but not one that relies upon shared linguistic knowledge. This behavior places the ape at a more advanced level than children with certain language disorders. For example, in some extreme forms of echolalia, children have control over the production of linguistic forms without possessing knowledge of their meanings and grammatical functions, or knowledge of the ways in which the act of producing utterances can be used to affect another. This is performance, so to speak, without either grammatical or communicative competence. Comparing the apes' behaviors to those of autistic children who have acquired primitive signing skills might also be revealing.[34,35]

CONCLUSIONS

We believe that the above considerations establish that there is no basis on which to conclude that signing apes acquired linguistic skills. Simply stated, the existing claims are not supported by evidence. There are numerous methodological problems with this research; perhaps the most damaging is uncritical over-reliance on production data. Production is, of course, an important skill to evaluate, and it is a simple matter to record examples of such behaviors. It is far more difficult, however, to establish a systematic relationship between the utterances and their contexts of occurrence. In the absence of this information the data are of limited value.

We should close by emphasizing that evaluating the apes' behaviors is not beyond the limits of the scientific method. In this paper and elsewhere[10-13] we have described the kinds of observational and experimental data that would bear decisively on the important issues raised by this research. It would be unfortunate if the inadequacies of the existing research—and widespread publicity concerning this controversy—were to inhibit further work with these complex animals, whose behavior is so little understood.

ACKNOWLEDGMENT

We are grateful to William Brewer and Edward Lichtenstein for comments on an earlier draft.

REFERENCES

1. GARDNER, R. A. & B. T. GARDNER. 1969. Teaching sign language to a chimpanzee. Science. **165**: 664–672.
2. GARDNER, B. T. & R. A. GARDNER. 1971. Two-way communication with an infant chimpanzee. *In* Behavior of Non-Human Primates. A Schrier & F. Stollnitz, Eds. Vol. 4: 117–184. Academic Press, New York, N.Y.
3. GARDNER, R. A. & B. T. GARDNER. 1975. Evidence for sentence constituents in the early utterances of child and chimpanzee. J. Exp. Psychol. Gen. **104**: 244–267.
4. PATTERSON, F. 1978. The gestures of a gorilla: Language acquisiton in another pongid. Brain Lang. **5**: 56–71.
5. FOUTS, R., G. SHAPIRO & O'NEILL. 1978. Studies of linguistic behavior in apes and children. *In* Understanding Language Through Sign Language Research. P. Siple, Ed. pp. 163–186. Academic Press, New York, N.Y.
6. CHOMSKY, N. 1959. Review of Skinner's *Verbal Behavior*. Language **35**: 26–58.
7. GARDNER, R. A. & B. T. GARDNER. 1974. Review of Brown's A *First Language*. Am. J. Psychol. **87**: 729–736.
8. TERRACE, H. S., L. A. PETITTO, R. J. SANDERS & T. G. BEVER. 1979. Can an ape create a sentence? Science **206**:891–902.
9. TERRACE, H. S., L. A. PETITTO, R. J. SANDERS & T. G. BEVER. In press. On the grammatical capacities of apes. *In* Children's Language. K. Nelson, Ed. Vol. 2. Gardner Press, New York, N.Y.
10. SEIDENBERG, M. S. 1976. There is no evidence for linguistic abilities in signing apes. Columbia University Master's Thesis.
11. SEIDENBERG, M. S. & L. A. PETITTO. 1978. What do signing apes have to say to linguists? *In* Papers from the 14th Regional Meeting of the Chicago Linguistic Society. D. Farkas, W. M. Jacobsen & K. W. Todrys, Eds. pp. 430–444. Chicago Linguistic Society, Chicago, Ill.
12. SEIDENBERG, M. S. & L. A. PETITTO. 1979. Signing behavior in apes: A critical review. Cognition **7**:177–215.
13. PETITTO, L. A. & M. S. SEIDENBERG. 1979. On the evidence for linguistic abilities in signing apes. Brain Lang. **8**: 72–88.
14. GARDNER, B. T. & R. A. GARDNER. 1974. Comparing the early utterances of child and chimpanzee. *In* Minnesota Symposium on Child Psychology. A. Pick, Ed. vol. 8: 3–23. University of Minnesota Press, Minneapolis, Minn.
15. GARDNER, R. A. & B. T. GARDNER. 1972. Communication with a young chimpanzee: Washoe's vocabulary. *In* Modeles Animaux du Comportement Humain. R. Chauvin, Ed. pp. 241–264. Centre National de la Recherche Scientifique, Paris, France.
16. THE FIRST SIGNS OF WASHOE. 1976. WGBH-Nova. Time-Life Films, New York, N.Y.
17. TEACHING SIGN LANGUAGE TO THE CHIMPANZEE WASHOE. 1973. No. 16802. Psychological Cinema Register, University Park, Pa.
18. BROWN, R. 1973. A First Language. Harvard University Press, Cambridge, Mass.

19. CHAPMAN, R. 1977. Comprehension strategies in children. *In* Language and Speech in the Laboratory, School and Clinic. J. Kavanaugh & P. Strange, Eds. MIT Press, Cambridge, Mass.
20. BLOOM, L. & M. LAHEY. 1978. Language Development and Language Disorders. Wiley, New York, N.Y.
21. CLARK, E. 1973. Non-linguistic strategies and the acquisition of word meanings. Cognition **2**: 161–182.
22. KLIMA, E. & U. BELLUGI. 1979. The Signs of Language. Harvard University Press, Cambridge, Mass.
23. STOKOE, W. C., D. C. CASTERLINE & C. C. CRONEBERG, Eds. 1965. A Dictionary of American Sign Language on Linguistic Principles. Gallaudet College Press, Washington, D.C.
24. SIPLE, P., Ed. 1978. Understanding Language Through Sign Language Research. Academic Press, New York, N.Y.
25. LANE, H., P. BOYES-BRAEM & U. BELLUGI. 1976. Preliminaries to a distinctive feature analysis of American Sign Language. Cog. Psychol. **8**: 263–289.
26. MARMOR, G.S. & PETITTO, L. Simultaneous communication in the classroom: How well is English grammar represented? Sign Lang. Stud. **23**: 99–136.
27. BLOOM, L., L. ROCISSANO & L. HOOD. 1976. Adult-child discourse: Developmental interaction between information processing and linguistic knowledge. Cogn. Psychol. **8**: 521–552.
28. MARX, J. L. 1980. Ape-language controversy flares up. Science **207**: 1330–1333.
29. GARDNER, R. A. & B. T. GARDNER. 1980. Letter to the Editor. N.Y.U. Education Q. **11**: 33.
30. BAZAR, J. 1980. Catching up with the ape language debate. Am. Psychol. Assoc. Monitor **11**: 4–5.
31. GARDNER, R. A. & B. T. GARDNER. 1978. Comparative Psychology and Language Acquisition. Ann. N.Y. Acad. Sci. **309**: 37–76.
32. HAHN, E. 1978. Look Who's Talking! Crowell, New York, N.Y.
33. SAGAN, C. 1977. The Dragons of Eden. Random House, New York, N.Y.
34. KONSTANTAREAS, M., J. OXMAN & C. D. WEBSTER. 1977. Simultaneous communication with autistic and other severely dysfunctional nonverbal children. J. Commun. Disorders **10**: 267–282.
35. CREEDON, M. P. 1973. Language development in nonverbal autistic children using simultaneous communication system. Biennial Meeting of the Society for Research in Child Development.
36. RUMBAUGH, E. S. & D. M. RUMBAUGH. 1978. Symbolization, language and chimpanzees: A theoretical reevaluation based on initial language acquisition processes in four young Pan troglodytes. Brain Lang. **6**: 265–300.

Clever Hans: Training the Trainers, or the Potential for Misinterpreting the Results of Dolphin Research

JOHN H. PRESCOTT

New England Aquarium
Central Wharf
Boston, Massachusets 02110

RESEARCH WITH DOLPHINS, and particularly research concerned with the communication capabilities of dolphins, suffered in its early days. Early research results of efforts conducted between 1960 and 1965 appear to be fraught with pitfalls related to the misinterpretation of the results. Initially, dolphins were reported to have a language, and communication with this nonhuman species was elevated to a level of mystique bordering on religious fanaticism that we are only beginning to overcome. J. C. Lilly reported on the intelligence of dolphins and speculated upon their use of language.[1-3] Dr. Lilly reported that dolphins were capable of learning and repeating words, and speculated that the combination of words was contactual expression, defending this supposition on the basis that dolphins had a larger brain size than any other mammal, including man. Brain size was equated with intelligence, and intelligence extrapolated to language. Review and analysis of the results indicate that Dr. Lilly's initial results were no more than mimicry. Other investigators were unable to reproduce the results of his experiments, while the popular press shaped public impression regarding brain size, intelligence, and communication.

Nearly simultaneously, utilizing the same training techniques—a modified operant conditioning methodology using bridging, stimuli, and food as reinforcement—an obscure dolphin trainer stumbled upon the ability of dolphins to mimic human sounds. This trainer, at a marine exhibition center (Ocean World, Fort Lauderdale, Fla.), also developed humanlike sounds by shaping the vocalizations of the dolphins. Whereas Dr. Lilly's efforts were concentrated on the identification of people and objects, the Ocean World trainer pursued a response to satisfy audience needs, and the Ocean World dolphins mimicked phrases. Whenever a dolphin show began, the trainer would enter the stage, talk to the audience about the capabilities of dolphins, and simul-

0077–8923/81/0364–0130 $01.75/0 © 1981, NYAS

taneously inquire how many people in the audience were celebrating their birthday that day. Knowing that in a crowd there is always somebody present whose birthday is that day, the trainer then turned to the dolphin and asked, "What do you think about that?" In apparent response to the raising of hands in the audience, the dolphin responded, "Happy birthday." Later in the performance the dolphin was apparently noncooperative with the trainer, having been asked to swim out in the pool and jump before receiving his reward. Despite repeated signals, the dolphin refused to work and swam back to the trainer. Upon inquiry by the trainer about "what did he want?" the dolphin responded, "more fish." Upon receiving that reinforcement, the dolphin swam off to perform the requested activity. Unbeknownst to the audience, cues to swim away and return were distinct from cues to swim away, jump, and perform the required task. To the audience, "Happy birthday" and "more fish" were interpreted as a sign of intelligence and language.

Unlike Lilly, this trainer realized that he had shaped a dolphin's behavior to mimic human sounds and incorporated the result into a basic animal performance, leaving only the audience to misinterpret the results.

Awareness of the Clever Hans phenomenon, although not by that name, occurred about 1962 from experience gained in training dolphins for echo-ranging experiments and observation of animal training and learning associated with the popular performance of dolphins in captivity.[4,5] During that period, three questions were raised regarding dolphin training and interpretation of results. It was frequently unclear what we had asked dolphins to do, how we had asked them to do it, and, finally, whether we understood the response.

Three anecdotal illustrations can be used to illustrate and generate an awareness of past weaknesses in dolphin research in order to avoid misinterpretation. These anecdotal illustrations include observation of the inventiveness of dolphins, spontaneous learning, and the transfer or simultaneous use of multiple visual and auditory cues by the bottlenose dolphin (*Tursiops truncatus*).

The inventiveness of captive dolphins led to speculation about their intelligence. Prior to 1960, dolphins were often maintained in community exhibits. These exhibits frequently contained fishes, turtles, and birds. These community exhibits provided opportunities that no longer exist, since there has been a subsequent isolation of dolphins in monospecific exhibits. Community pools allowed researchers to observe the multispecific interactions of the inhabitants and watch the games that dolphins play. Dolphins are quite inventive. In these pools it was often observed that dolphins played with feathers, sea weed, or any other object (animate and inanimate) available. It was also obvious that their boredom with confinement led to many "barnyard" behaviors such as mating with other species and even other classes of animals. If we review the early literature on this subject, barnyard behaviors were not recognized and it was often implied that other sexual characteristics such as homosexuality were involved in dolphin behavior.[6-8] Since this was a nondiscussed subject, the terminology of homosexuality was inferred by the lay public, press, and even scientists as another expression aligning dolphins with man.

To illustrate the inventiveness of dolphins, we must refer to some of the early reports. Brown[6] and Brown and Norris[7] discussed play activities in dolphins utilizing feathers, sea weed, and turtles. It was not unusual, in a community tank, to observe a dolphin playing with a feather or a piece of sea weed, first balancing it on its nose and then, while still swimming at high speed, let loose of the feather, catch it on its dorsal fin, swim on, roll, dislodge the feather from its dorsal fin only to entrap it on its tail flukes and continue to swim at high speed. This game was repeated over and over with the feather being finally dislodged from the tail fluke; the dolphin would then roll up and pick it up again on its nose or trade it off to a second dolphin playing the game. Dolphins were also noted to pursue fish, turtles, and many other aquatic inhabitants of the tank. On one occasion, a moray eel living in the tank had apparently learned to avoid dolphin play and had found a recess from which a single dolphin could not extricate the eel. During one observation period, the eel was observed to be dislodged from its hiding place by the overt actions of two dolphins. While one kept the moray eel occupied and staying in its hole, the second swam about the tank, captured a very spiny fish and returned. Holding the fish in its mouth, it stuck the fish's spine into the eel, whereupon the eel swam out of its hole to be captured by the second dolphin, thence to be included in a game of balance, breaching, and "homosexual" activity.

Similar examples of inventiveness in game playing have been transferred to the experimenter, with the frequent result being that the results are anthropomorphized and translated into an almost mystical and even superhuman intellect.[2] However, dolphins left alone or in specific isolation develop sterotypic behavior patterns not unlike the pacing of captive bears and lions and the playing with feces and urine by apes in isolation. Without an understanding for sterotypic behaviors and the transference of behavior by captive animals towards their captors, misinterpretation of results may result.

A second anecdotal illustration can be used relative to spontaneous or observational learning. Dolphins are capable, as are most animals, of learning the technique of learning, and they become aware that a response is required by the trainer whenever the trainer is present. In the early 1960's, operant conditioning was not understood by all trainers, and some believed they communicated with dolphins in mystic ways. However, most trainers were aware that dolphins learn faster or catch on with experience. In this respect, how we ask a question may be unclear and may lead to transference in our interpretation of the results.

Two examples by a single individual can be used to illustrate the phenomenon of spontaneous learning, a phenomenon that prompts the question: Are we sure we know what we are asking an animal to do? During the mid-1960's a false killer whale (*Pseudorca crassidens*) named "Swifty" was maintained at the exibition center Marineland of the Pacific. After capture and appropriate quarantine conditioning to the captive environment, the whale was introduced into the main show pool with two pilot whales and five dolphins. During subsequent months Swifty was trained to perform, and, being a particularly tractable animal, her activities were generally focused upon the more elaborate and spectacular behaviors, such as breaching, bowing, and hurdle jumping. After more than three years in captivity, Swifty demonstrated that she knew the game better than anticipated. On the first occasion of this demonstration of observational learning, a single Pacific whitesided dolphin (*Lagenorhynchus obliquidens*) was being taught to backflip. The trainer had spent several days attempting to initiate the behavior and coordinate it with a signal. In this particular instance, the dolphin frequently backflipped spontaneously in the early morning and the trainer was present to capture the activity by using a bridging stimulus to reinforce the activity. During one of these sessions, Swifty departed from her staging position, swam off, and backflipped, not only surprising the trainer but others present. Seeing a 12-foot whale backflip was more than anticipated. The whale subsequently did backflips to a cue from the trainer in three sessions. A remarkable learning experience.

Several months later the same whale participated in probably one of the most unusual occurrences of spontaneous learning. During the regular performance of shows at Marineland, "Bubbles," a pilot whale (*Globicephala scammoni*) was the star performer. During one performance, however, Bubbles performed her opening act and then refused to participate further in the show. Swifty swam over, took Bubbles' position, and upon the trainer's command, Swifty accepted all the cues for Bubbles' performance and completed the entire show. More than half of the behaviors had never before been performed by nor asked of Swifty.

This kind of response can be explained as mystical or intellectual, but we must consider that the appearance of a trainer was a cue for the whale. In this case the performance was not what the trainer expected but the whale, stimulated by the trainer's presence and receiving reinforcement, responded. It was apparent that we had not anticipated this learning process and that the dolphins and whales learn faster than previously expected.

In addition to inventiveness and spontaneous learning, it must be acknowledged that how we ask a dolphin to perform a specific task may be unclear to dolphin and man. Our perceptions are biased. Humans are visual species and results of our actions may often appear to be "Clever Hansing" by the cetacean. There is a particular need for human awareness of the sensoral capabilities, not only of dolphins but of all experimental animals. It is particularly necessary to understand the dolphin sensoral capabilities, and the following anecdote illustrates a potential for misunderstanding.

In 1959 an experiment was developed to demonstrate the echo-ranging capability of dolphins.[5] In this experiment, a dolphin was blindfolded with suction latex eyecups that totally eliminated its visual capability. This behavior was subsequently incorporated in nearly every dolphin exhibit/performance in the United States. Upon one occasion in 1966, at Marineland of the Pacific, one of the trainers called to report that he had a dolphin that was echo-ranging through the air/water interface and answering signals using echo-ranging when his head was out of water and blindfolded. Because of the physical characteristics of sound and its transmittal through the air/water interface, a careful observation process was initiated to determine if dolphins can echo-range through the air/water interface and echo-range in air alone.

On this occasion the trainer reported that a dolphin, when blindfolded, took a visual command, swam around the pool and jumped over a hurdle suspended six feet above the surface of the pool. In effect, an applicable demonstration that dolphins can echo-range in air as well as under water. However, because of the physical characteristics of sound, this was a highly unlikely phenomenon. The regular routine of the dolphin for the previous seven years had been to perform a high hurdle jump during the show. At that time the hurdle was regularly raised to the elevation of six feet above the surface of the pool and the dolphin commanded to leave the stage, swim about the pool, and leap. Upon observation, undoubtedly the dolphin did take the command, swim about the pool, and leap over the hurdle while blindfolded. Subsequent observation revealed, however, that if the hurdle was elevated or lowered, the dolphin jumped at a pre-determined height. It is assumed that the dolphin was capable of echo-ranging under water and of determining his position, take-off point, and trajectory prior to the leap. If the elevation of the hurdle was altered, the dolphin always jumped at the predetermined height.

There still remained, however, the question of accepting the command. In the normal training process, visual cues are used; signals consist of waving motions, and elevated, cupped, clasped, and directional thrusts of the hands are frequently used. In this instance the dolphin, with its head above water, was perceived to take a waved arm motion command prior to take-off and jumping over the hurdle. However, it was determined that the initial interpretation of the results was incorrect, for observation of the trainer indicated that not only did he give the dolphin a visual cue, which he perceived to be the cue for jumping the hurdle, he also gave an inadvertent acoustic cue. Dolphins have very acute hearing, in response to the utilization of underwater sounds for communication as well as echo-ranging. In this particular instance, it was discovered that not only did the trainer wave his arm to stimulate the dolphin's response, he also stepped forward and it is assumed that the dolphin, upon hearing the step coincidental with the arm waving motion, took both as a cue to respond and jump the hurdle. The blindfolded dolphin was subsequently asked to jump the hurdle using the visual arm motion cue while the trainer was stationary. In every trial the dolphin simply sat in front of the trainer, maintaining its position without the appropriate response once the auditory portion was eliminated. In this case the auditorial link with the visual cue was not understood by the trainer and led to misinterpretation of results and capability of the dolphins and perhaps, without close control, would have led to a new inference of intelligence, contributing to the mystique of these animals.

These observations are offered not to serve as conclusions relative to cetacean research in learning, communication, or language, but rather to raise cautions and awareness. We must recognize the adaptability of the species with which we work, that dolphins are auditory animals, and that our interpretation of results may be biased because of our own preconceived desire to communicate. Current work recognizes some of these pitfalls, but care must be exercised and researchers must recognize that "noise" may bias the research if we do not understand how and what we ask the dolphins. Because of the difference in environment and physiological capabilities, and because dolphins are capable of learning the game, interpretation of results may not be possible.

REFERENCES

1. LILLY, J. C. 1961. Man and Dolphin. Doubleday & Co., Garden City, N.Y.
2. LILLY, J. C. 1967. The Mind of the Dolphin, A Non-Human Intelligence. Doubleday & Co., Garden City, N.Y.
3. LILLY, J. C. 1978. Communication Between Man and Dolphin, The

Possibilities of Talking with Other Species. Crown Publishers, New York, N.Y.

4. NORRIS, K. S. & J. H. PRESCOTT. 1961. Observations on Pacific Cetaceans of Californian and Mexican Waters. Univ. Calif. Press Publ. Zool. **63**: 291–402

5. NORRIS, K. S., J. H. PRESCOTT, P. V. ASA-DORIAN & P. PERKINS. 1961. An experimental demonstration of echo-location behavior in the porpoise, *Tursiops truncatus* (Montague). Biol. Bull. **120**: 163–176

6. BROWN, D. H. 1960. Behavior of a captive Pacific pilot whale. J. Mamm. **41**(3):342–349

7. BROWN, D. H. & K. S. NORRIS. 1956. Observations of captive and wild cetacea. J. Mamm. **37**(3): 120–145

8. BROWN, D. H., D. K. CALDWELL & M. C. CALDWELL. 1966. Observations on the behavior of wild and captive false killer whales, with notes on associated behavior of other genera of captive Delphinids. Los Angeles County Museum Contributions in Science (95): 1–32

Why Porpoise Trainers Are Not Dolphin Lovers: Real and False Communication in the Operant Setting

KAREN PRYOR*

Department of Zoology
Rutgers University
Newark, New Jersey 07102

DOLPHIN LOVERS ABOUND. Everyone who has ever seen "Flipper" on television, visited an oceanarium, or read John Lilly's books thinks that dolphins are cute, playful, friendly, harmless, and affectionate to each other and to man, that they save drowning people, and that they are possessed of extraordinary intelligence and a rich communication system comparable perhaps to our own.

Porpoise trainers know otherwise (many prefer the word "porpoise" to "dolphin" because it differentiates the mammals from a pelagic fish, *Coryphaenus hippurus*, which is also called "dolphin.")

The novice trainer quickly learns that porpoises can be very aggressive. They are highly social animals, to which rank order is a matter of considerable importance. Aggressive interactions between porpoises, usually during dominance disputes, include striking, raking with the teeth, and ramming with the beak or rostrum, sometimes with serious consequences, such as broken ribs or vertebrae, or punctured lungs, in the rammed animal. A dolphin that has become accustomed to humans shows no hesitation in challenging the human for dominance, by means of threat displays and blows; a person who is in the water with an aggressive porpoise is at a dangerous disadvantage. The sentimental view that these animals are harmless stems at least in part from the fact that they are usually in the water and we are usually on boats or dry land; they can't get at us.

Interaction with porpoises in a training situation, usually with the trainer at tank-side, also brings their vaunted intelligence under pragmatic scrutiny. A poll of experienced trainers[1] reveals that some trainers, after working with many individual animals of several species for five years or more, were not willing to place porpoise intelligence levels above that of dogs; the majority of the respondents, however, agreed

---

* Mailing address: 28 East Tenth St., New York, N.Y. 10003.

0077-8923/81/0364-0137 $01.75/0 © 1981, NYAS

with the dictum that porpoise intelligence is "between the dog and the chimpanzee, and nearer to the chimpanzee" (by no means, however, does it appear to trainers to be equal to or superior to that of the great apes.)

A confounding effect of this question is that most of the porpoises kept in captivity are Atlantic bottlenose porpoises, *Tursiops truncatus*, a coastal species which is highly adaptable, plastic in its behavior, and an opportunistic feeder, showing a marked tendency to play with and to manipulate objects, and some tolerance for solitude. Other genera of Delphinidae, such as pilot whales, belugas, and the small, pelagic, white-sided, spinner and spotter porpoises, exhibit quite different behavioral profiles. Spinner porpoises (*Stenella longirostris*), for example, show very little tendency to play and high avoidance of foreign objects. They have difficulty negotiating barriers or obstacles, seldom learn to tolerate (much less solicit) human touch, and become inappetent and in fact rapidly moribund if kept in isolation from species mates. It is perhaps unfortunate that the popular view of porpoises is based on the genus *Tursiops*, the bottlenose, which is in fact a rather anomalous member of the Delphinidae, behaviorally.

Nevertheless, whether the subject is an inactive pilot whale, a very active but timid spinner, or a bold, aggressive bottlenose, the trainer can train it: not by loving it, or even liking it — one may find oneself cordially despising a particular individual — but by operant conditioning.

People often ask a working trainer about "communication," John Lilly having established an apparently ineradicable mystique by holding that there is something unusual about dolphin communication.[2] The flippant answer is that anyone can communicate just fine with a whistle and a bucket of fish.

In fact what is "different" about porpoises, compared to other frequently trained animals, is the manner in which they are trained. Aversive methods are virtually unavailable. The porpoise trainer cannot use the choke chain, the spur, the elephant hook, the cattle prod, or even a fist, on an animal that can swim away if alarmed. As they cannot "get at" us, so we cannot "get at" them.

The laboratory psychologist's mind may at once turn to devising some arcane method of punishment. In fact, punishment is unnecessary. Other than the mild negative reinforcement of a brief interruption of the training session, (a "time-out"), porpoise trainers achieve highly disciplined and complex responses entirely with positive reinforcement.

The proliferation of trained porpoise shows, as well as some rather limited use of trained porpoises as domestic animals working in the open sea, has paralleled in time the promulgation and increasing public awareness of the laws of operant conditioning described by B.F. Skinner and others. Unlike traditional animal trainers, porpoise trainers are not only at the mercy of these laws, they are aware of them, and use them consciously. The jargon of the porpoise trainer is the jargon of the laboratory: successive approximation, conditioned stimuli, variable schedules of reinforcement, and so on. Unlike the shaper working under laboratory conditions, however, the porpoise trainer, in addition to being largely limited to positive reinforcement, is interacting with the animal; he or she can see the animal, the animal can see him or her, and both can introduce changes in the training process, at will. It is a situation both rigorous and admitting of spontaneity: a game.

The game has "rules" on both sides: "I will reinforce you only for jumps in which you do not touch the hoop as you pass through it;" "I have come to expect a fish for at least every three or four jumps, and will stop jumping if you let eight or ten responses go unrewarded." It is a game in which challenge is always present, as the trainer, mindful of the various techniques for maintaining response levels, raises criteria or introduces new criteria; but at least in the hands of a skilled trainer, it is a game that the animal always, eventually, wins.

A dog or a horse learns responses because it must do so to avoid aversive stimuli; a pigeon in a Skinner box must work, because it is hungry; and when we train people, we generally use a mix of positive and negative, of praise and coercion, though sometimes covert. There is little coercion, however, in the porpoise–trainer interaction (even food deprivation is hazardous, and seldom used) and this has an effect upon both trainer and animal. The animal is, as it were, training the trainer to give fish, and thus is shaped towards finding new ways to elicit fish; it is shaped in fact towards innovative response. The trainer in turn may become a very skilled and imaginative user of Skinner's laws. Porpoise trainer shoptalk (as opposed, let us say, to racehorse trainer shoptalk) is generally concerned with ingenious shaping programs, or novel use of operant conditioning techniques and not with the personalities or achievement of individual animals. It is the fascination of the game that keeps porpoise trainers in their strenuous, low-paying jobs, year after year, and not the fascination of the animals themselves. Many trainers in fact come to prefer the more reliably conditionable pinnipeds, and some greatly enjoy working with birds, another group that cannot be trained aversively.

This is not to say that every porpoise trainer is a walking compendium of Skinnerian laws. The less educated trainer, or the self-taught trainer working in isolation, may be full of superstitious behaviors ("You have to wear white; dolphins like white") or may be unable to say how he is cuing his animals, and thus fall victim to Clever Hans phenomenon, maintaining, for example, that his animals respond telepathically.

The animals of course quickly develop superstitious behavior too; for example, only responding to trainers in white clothes.

The nontrainer, interacting with dolphins, is also apt to misinterpret, especially in the matter of social signaling. He may, for example, interpret the gaped jaw, a threat display, as a "smile," or touches and jostling as affectionate play, when they are often dominance challenges. Dr. Lilly made much of anecdotes concerning a male porpoise making sexual advances to a human female,[3] but male bottlenosed porpoises in captivity may exhibit sexual behavior towards almost anything; and it is a behavior which, after all, we do not find intelligent or endearing in male dogs.

The interactive, positive-reinforcement training setting is an excellent way to become acquainted with the nature and function of social signals in an unfamiliar species. You do not need what George Schaller describes as 5000-hour eyeballs (months or years of observation) to discover which gestures, postures, and acts are aggressive, which affiliative, and so on. For example, in spinner porpoises, an extremely loud echo-location click-train is a threat display. This may not be obvious the first time you swim with spinners; if it is immediately followed by a sharp blow of a dorsal fin to your upper arm, you will recognize it the second time you hear it, and prepare to take evasive action.

In the operant setting, most large mammals quickly direct their intraspecific social signals at the trainer. They are not begging; begging does not work; they are exhibiting frustration, making submissive or aggressive displays, and so on, both giving and garnering information. One of the commonest trainer-directed social signals is sudden eye contact, which can be described metaphorically as the "Am I on the right track?" eye contact. Verifying that the trainer is indeed watching, the animal then escalates the vigor or duration of the response, and thus earns reinforcement. This is not a behavior seen only in porpoises, although they make eye contact more often than many other mammals; I have experienced this specific social interchange of information in an operant conditioning setting with an elephant, a wolf, a hyena, several polar bears, and primates.

The wise trainer makes use of whatever social signaling he feels he can accurately interpret. The animal can make use of this communication link too. Porpoises, for example, probably do not care what we think of them, and, according to Gish,[4] do not necessarily, in their own acoustic social signaling, increase volume to add emphasis; nevertheless porpoises can learn that human increased volume—yelling—means "I mean it!" and respond appropriately, not from fear or a desire to please, as a dog might, but from having gleaned the appropriate information in training interactions.

The richness and detail of information available in the operant setting enables communication to occur on a level considerably exceeding
that of the usual interactions between man and beast. The porpoise
trainer, for example, can change his tankside location to indicate when
he wants previously conditioned responses, and when the animal is at
liberty to earn reinforcement through new responses. A porpoise can indicate through a series of totally wrong responses that the quality of the
fish reward is not satisfactory; this device is not uncommon in research
animals being fed from feeding machines, in which fish may dry out or
spoil. The porpoise, through actions, and with eye contact, may deliberately test the trainer's criteria: take, for example, this episode:[5]

Two false killer whales (*Pseudorca crassidens*) have been trained to
jump a hurdle simultaneously, in opposite directions. The behavior used
in daily public performances, has deteriorated, due to trainer careless-
ness. One whale now always jumps too late, spoiling the effect of the
mid-air crossing of trajectories. A corrective training session was held, as
follows:

(1) Trainer presents cue (an underwater sound). Both animals approach
hurdle. Animal A jumps from the left; the conditioned reinforcer (a whistle) is
sounded, and the cue is turned off; animal B then jumps from the right.
Animal A receives a handful of fish. Animal B returns to trainer but is not
reinforced.

(2) Trainer presents cue. Both animals approach the jump, jump simultaneously in opposite directions, hear the whistle, the cue is turned off, and
both are rewarded with 2 lbs. of fish, many small fish ($\sim$ 10) dumped directly
into the animal's large mouths. These are very large animals, and that constitutes the usual reinforcement.

(3) Trainer presents cue, and the first episode is repeated, with animal B
jumping late, after the cue is off, getting no whistle, and no fish.

(4) Trainer presents cue and animal B does something quite unprece-
dented; it switches sides and jumps in synchrony with animal A, but from the
left or same side, hearing the cue, and the whistle, but getting no fish.

(5) Trainer presents cue. Both animals jump, and from opposite directions, and animal B is just slightly late. Animal A receives 2 lbs. of fish, and
animal B gets one little 2-ounce smelt. Animal B physically startles, and
makes eye contact with the trainer.

(6) Trainer presents cue. Animal B increases swimming speed and makes a
perfect jump, opposite to and in synchrony with animal A. Both animals are
given 4-pound rewards. Both animals henceforth perform the response correctly, eight times a day.

Does this demonstrate these large delphinids' "intelligence"? Perhaps: One would not expect such methodical testing of the criteria by a
spinner porpoise, which has a behavioral repertoire generally more
limited and rigid than that of *Pseudorca*. However, the anecdote may

demonstrate the kind of communication that can arise purely through operant conditioning and through using positive reinforcement flexibly. Nothing in the "rules" suggests that half a reinforcement should convey the information, "You're about half right." But whether or not that interpretation represents what truly happened, there was information in the unusually tiny reinforcement, information possibly accessible to an animal that had experienced the earning of many consistently larger reinforcements, information that the animal, to all pragmatic intents and purposes, recognized and made note of.

Gregory Bateson has stated that operant conditioning is a method of communicating with an alien species.[6] Others have suggested that the various language acquisition experiments with apes are nothing more than glorified operant conditioning.

Whether what the apes do is related to language, as we use language, is beside the point to a porpoise trainer. Like the porpoises, the apes have experienced very elaborate operant conditioning programs, in a setting conducive to interchange of social signals and a setting which, while rigorous, is admitting of spontaneity on both sides. It is a training circumstance that is rather rare in the world at large. What seems evident, and is taken now almost for granted by many researchers, is that at least chimpanzees are capable of assimilating enormous numbers of signs—conditioned stimuli, if you will—and of attaching correct meanings to these signs. A signing chimp, or even an ape that merely recognizes some signs (such as some orangutans and gorillas now do at the National Zoo) is capable both of giving and receiving information that is far more subtle than that normally conveyed between a person and a pet animal or a caged specimen. Something is developed; it may or may not be language; it is certainly heightened communication.

Innovative responses, and increased communication, thus may be not so much an indication of unusual or near-human capabilities in a species, but rather an artifact of the effect of advanced techniques ("glorified," if you like) of operant conditioning in opening pathways for communication, including two-way and unpremeditated communication, between other species—perhaps many other species—and man.

REFERENCES

1. DEFRAN, R. H. & K. PRYOR. 1980. The behavior and training of cetaceans in captivity. *In* Cetacean Behavior. L. Herman, Ed. Wiley-Interscience, New York, N.Y.
2. LILLY, J. 1961. Man and Dolphin. Doubleday and Co., New York, N.Y.
3. LILLY, J. 1978. Communication Between Man and Dolphin. Crown, New York, N.Y.

4. Gish, S. L. 1979. Quantitative Analysis of Two-Way Acoustic Communication between Captive Atlantic Bottlenose Dolphins (*Tursiops truncatus* Montague). Ph.D. dissertation. Univ. of California at Santa Cruz.
5. Pryor, K. 1975. Lads Before the Wind. Harper & Row, New York, N.Y.
6. Bateson, G. Personal communication.

Conversational Strategies*

STARKEY DUNCAN, JR.
Committee on Cognition and Communication
University of Chicago
Chicago, Illinois 60637

SO FAR TODAY we have been treated to discussions of communication, or the lack thereof, between humans and horses, chimpanzees, and porpoises. And we look forward to hearing about communication with and through ghosts, mice, and inert chemical substances. At the risk of deviance from the dominant themes of this conference, let us briefly consider something thoroughly prosaic, one of the most common forms of communication that humans indulge in: ordinary, everyday conversation. Consistent with a theme of this conference, we may view interaction from the special perspective of strategies used within conversations and the effects these strategies may have on the course of the conversation. Let us keep in mind that ordinary interaction may, among other things, be useful as a kind of benchmark against which more unusual interactions may be evaluated.

A hazard in discussing ordinary interaction is that it is a phenomenon on which each of us is an expert, a qualified native informant. As expert interactants, perhaps we could readily agree on the following: that individuals may have identifiable personal styles or strategies of interacting with others in specified contexts; that, at the same time, individuals may be observed to change their interactional styles in some respects as they move from one interaction to another; that various tactics may be taken—and changed—in the course of an interaction; that, one way or another, interactants may influence the behavior of their partners, either intentionally or unintentionally; and that intentional attempts at such influence may be more or less successful. Having congratulated ourselves as smugly as possible on these expert insights, permit me to venture some remarks on the nature of interaction strategy and the interactional process through which influence may be mediated. I recognize that I cannot hope to touch on anything that is not already well understood in some manner by each interactional expert present.

In speaking of influence, I shall not be concerned with power imbalances between participants in an interaction (an obvious source of in-

* This research was supported in part by National Institute of Mental Health Grant MH-30654 to Starkey Duncan, Jr. and Donald W. Fiske.

fluence), nor with the use of deception, another obvious source. I shall not indulge in speculation as to whether or not efforts to influence are intended, a rather difficult empirical issue. And I shall not venture to guess whether our human participants achieved their interactional competence legitimately or artifactually. I shall limit my discussion strictly to resources within the process of interaction itself that participants may, or in some cases must, use with possible influential effect. What does it mean to have a strategy? What is the connection between strategy and influence?

In these comments I shall draw on the insights of investigators of interaction process such as Erving Goffman,[1,2] Adam Kendon,[3] Norman McQuown,[4] and Emmanuel Schegloff[5] and on my own experience with a program of research on the process of two-person conversations between adults, on which Donald Fiske and I have collaborated for some years. For the past two years, we have been concerned specifically with studies of interaction strategy in such conversations.

First, what does it mean to have a strategy? I take it that human, face-to-face interaction, like language, is an activity based on convention. Participants are operating within a "communication system"[1] that can be described in terms of signals, rules, and other elements. Thus, the process of interaction is seen as rule-governed; my perspective on understanding interaction is essentially linguistic. Of particular relevance to interaction strategy is what Goffman terms "constraint to play" (Reference 1, page 114) imposed by the rules. Goffman discusses this constraint in terms of Harry, his protagonist-victim:

> Harry is not faced with a vast choice of moves, each a little different from the others, or with the possibility of creating variations and modifications; he is faced with a finite and often quite limited set of possibilities, each of which is clearly different from the others. Harry's situation, in other words, is *structured*. [Reference 1, page 114, original emphasis.]

To the extent that interaction is seen as structured by rules and the like, serious work on the process of interaction—including work on interaction strategy—must begin with a careful description of the rules applying to the interactions under study. One simply cannot understand strategy without knowing the rules, whether one is observing a game or a social interaction.

While I regard the concern with interaction structure as being essentially linguistic in nature, there are some necessary differences between linguistics as it is traditionally defined and research on interaction. As in linguistic description, rules formulated for interaction are concerned with specifying appropriate sequences of events. However, in interaction research the linguistic emphasis on individual messages is expanded to include both participants (in a two-person interaction). A sequence of events describable by a rule would include, at a minimum, an action by a participant and a subsequent action by the partner. In addition, studies of interaction will consider actions in paralanguage, body motion, spatial proximity, and the like, as well as those speech actions typically defined as "linguistic." It may well be that in everyday interaction the traditional distinction between language and "nonverbal communication" is less fastidiously preserved than the almost non-overlapping relationship that research literatures suggest.

This position is entirely unexceptional, I believe, among investigators such as those mentioned above, all of whom have been concerned, in one way or another, with the structure of interactions. But it is worth stating because the prevailing practice is for investigators to proceed directly to questions of strategy (e.g., sex differences in gaze during conversations) without considering the rule structure of the interaction they are studying. I believe that in taking this attempted shortcut, the investigator becomes exposed to several hazards.

First, when hypotheses concerning interaction structure are not available, the definition of strategy variables may be much less precise. It is important, not just that an action such as a smile or an attempt to take the turn occurs in an interaction, but also where in the stream of interaction that action occurs. Most studies are largely confined to variables based on gross counts and timings of actions, rather than more subtle variables based on sequential relationships between the target action and other, rule-related actions.

Second, when interaction structure is ignored, there will be an enduring confusion between behavioral regularities due to structure, and those regularities due to strategy. In fact, results are likely to represent an undifferentiated mixture of these two types of effects. I believe that straightforward analytic procedures exist for distinguishing structure effects from strategy effects, where structure and strategy are defined in terms of action sequences involving both participants. Discussion of these procedures lies outside the scope of this paper.

Finally, I would hold that descriptive power is certain to be weakened when the conceptual apparatus provided by the structural approach is ignored.

Regardless of the validity of these remarks—and I expect many current investigators would take exception to them—it remains a fact that the notion of interaction structure plays no part at all in the methodology of most studies in the "nonverbal communication" area, the area in which most studies of interaction are published.

Let us examine more closely the relation between interaction structure and strategy. It will be immediately apparent that one cannot engage in a rule-governed interaction like a conversation or a baseball game without simultaneously engaging in a strategy. The notion of structure necessarily entails strategy. This is because strategy is concerned with the exercise of options. Two types of rule-defined options are available to participants in an interaction. Ever-present is the option to violate the rules. A straightforward, rule-abiding interaction is, among other things, the result of both participants' continual choices not to break the rules. The second type of option occurs when the rules specify legitimate alternative actions a participant may take. When such a rule is operating in interaction, the participant subject to the rule cannot avoid making a strategic choice between available alternatives. In baseball a batter confronting a pitched ball cannot avoid exercising the option of either swinging or not. In the rules hypothesized for some conversations, an auditor is presented with the alternative of taking the speaking turn or not when a certain signal is displayed by the speaker. Because the signal display provides the rule-defined alternatives, the auditor's remaining silent is as definite a choice of action as taking the turn.

As rules for a type of interaction are discovered and documented, investigators become able to carry out a moment-to-moment description of the interaction strategies of participants, just as a chess game can be described move-by-move.

Given that participants cannot avoid strategic choices of actions in rule-governed interaction, how may a participant's action affect the probability that the partner will subsequently exercise one option, as opposed to another? That is, how is influence mediated through strategy?

I assume that in the realm of social action, we cannot speak of causality in any of the classic senses of that term. Apart from direct physical coercion, a participant cannot cause the partner to do something. Thus, while one might cause the other to fall down by a sufficient shove, one cannot cause the other to smile, shake hands, nod his head, take the turn, or the like. One can only strive to give the other sufficient reason to choose to do so. For example, if one wishes the other to smile, he might do something that elicits optional smiling in most cases, or he might attempt to place the partner in a position in which failing to smile is violating a rule. However, neither tactic may elicit the desired smile. I take this principle to hold in much more drastic situations, just as it does in this mild-mannered example. Ultimately, an individual through his actions can always just say "no" in word or deed, regardless of the consequences. In this sense, one cannot be assured of control over the other. One can only seek to affect the probability of a desired response; but this effort may fail—a result with which most of us are well acquainted.

While lack of control over the partner's actions seems obvious in most cases, it is more fascinating to me to note that in interaction a participant also lacks absolute control over something else: the meaning of his or her own actions. Even though I would claim that a participant may exercise ultimate choice over what action to take at a given moment, the same absolute control does not extend to the meaning of the action chosen. Obviously, an action acquires full meaning in a context, and in interaction that context includes to some degree the preceding rule-relevant action by the partner. In baseball a batter's choice not to swing at a pitched ball is much more fateful if there are two strikes against him, than if it is the first pitch to him. In the hypothesized rules for certain conversations, while the auditor can choose to take the turn at any moment, the meaning of that action depends in part on whether or not the proper speaker turn signal has been displayed. In this manner, a given action by the partner can have sharply different interactional implications, depending on the participant's preceding action. This ability to control in part the meaning of the partner's subsequent action is an important source of potential influence for the participant.

An obvious way that a participant can affect the probability of some action by the partner is to behave in such a way that the partner's taking that action would constitute a rule violation. For example, in the conversations we have studied, the auditor's attempting to take the speaking turn would be a violation of the hypothesized rules if the speaker is currently gesticulating. In such situations the auditor's turn attempts are either eliminated or sharply reduced. Some speakers succeed in turning conversations into near monologs by extensive use of gesticulations, not necessarily continuously, but at those specific points at which the auditor might otherwise legitimately act to take the turn. (I would add that more subtle tactics are available.)

Similarly, in games such as chess or baseball a participant may attempt to induce the partner to take disadvantageous moves in order to avoid an even worse result. With the bases loaded, a pitcher may deliberately walk in a runner to avoid pitching to a dangerous hitter. In our conversations, after an outright interruption by the auditor the speaker may nevertheless yield the turn in order to avoid a prolonged conflict over possession of the turn.

As interaction experts we should be protesting at this point that games such as chess and baseball are imperfect analogies for interaction because in many interactions participants may have no particular inclination to "win" or dominate the interaction. Rather, participants in an interaction may evidence concerted efforts at balance, coordination, or reciprocity. But what is to be balanced or coordinated? Research has centered on such variables as gross talking time, or certain speech–silence factors in the pacing of speech. A corollary view is that the process of interaction involves continual exchange of information, both "verbally" and "nonverbally," on a variety of internal states of each participant. On the mention of internal states and "nonverbal communication," one might think first of emotions, an area intensively studied by Paul Ekman and his colleagues.[6] One can also imagine many other states, such as relative tension. With respect to conversation, an internal state of "transition readiness" has been hypothesized[7] with respect to the exchange of speaking turns. Results suggest that a speaker has a variety of cues at his or her disposal to indicate the degree of current readiness to yield the turn. The auditor can read these cues, integrate this information with his or her own degree of transition readiness, and arrive at an appropriate decision on whether or not to take the turn at a particular moment. For example, it appears that an auditor, only weakly inclined to take the turn, may nevertheless do so if the speaker signals sufficient transition readiness. In this connection we have found a strong relationship between the number of turn cues activated by the speaker and the probability of an ensuing smooth exchange of the turn. In this way, one may influence the actions of the partner simply by letting the partner know one's conversational state at the moment.

Notice that an overreliance on the contest analogy for interactions may lead one to misinterpret strategies in some cases. In one of our conversations a woman talked most of the time, gesticulating more than any other participant in our studies. Her male partner gestured little and spoke little. An immediate interpretation might be that the woman was adopting a highly aggressive strategy, virtually requiring her partner to interrupt in order to get a word in edgewise. More careful examination of that particular interaction revealed something quite different was happening. Early on, the man had indicated a disinclination to speak extensively, and the woman was apparently accommodating her actions to his indication. In a subsequent conversation with a different partner, the woman behaved quite differently, in this respect, gesturing moderately and speaking less than the partner.

The process of coordinating action in interaction provides one perspective for interpreting the process of experimenter bias. Research conducted in collaboration with Robert Rosenthal and with Milton Rosenberg and Jon Finkelstein[8,9] suggests that at least part of the bias effect in Rosenthal's picture-rating task stems from certain cues involving contrasting stress in intonation and paralanguage — cues that are an integral part of speech itself — when the experimenter is reading certain parts of the instructions to the subject. The bias effect appears to be mediated in part through a subtle process of skewing the location of these contrastive stresses in the stream of speech. This skewing presumably provides information on the experimenter's relative investment in the two response tendencies a subject may show on the task. The subject, in turn, may or may not be inclined to be biased by this information. That is, the subject may or may not choose to coordinate his or her experimental responses with the perceived internal state of the experimenter. It seems worth noting at this conference that in our studies experimenter bias was most effective when the skewing of stresses was moderate and inconspicuous, not blatant. The skewing itself occurs most dependably when the experimenter is unaware of it; conscious efforts to skew the stresses in a prespecified manner were consistently unsuccessful, often producing skewing contrary to the experimenter's intensions. It is the actual skewing in the message, not the experimenter's intentions, that produces the bias effect.

I have made two assertions with regard to strategy. First, within the framework of rule-governed interaction a participant cannot avoid choosing among alternative rule-specified actions, including the alternative of breaking the rules. That is, one is inevitably implicated in a strategy if one accepts a set of rules applying to one's interaction. Second, a participant's action at any given point provides part of the context, and thus part of the meaning, of the partner's subsequent action. For example, the speaker's immediately preceding actions determine whether an auditor's attempt to take the turn is interpretable as an interruption or as an appropriate response to the state of the speaker. In contributing to the meaning of the partner's subsequent action, a participant cannot avoid exerting potential influence on the partner. It remains only to describe the characteristic patterns of the participant's strategy, together with the characteristic patterns of the partner's responses, given the rule-defined alternatives — a task much easier said than done. The patterns of action and response define the nature and extent of influence. And of course the process is truly interactive because the partner's response contributes in turn to the meaning of the participant's subsequent action.

Our current work on strategy in conversations is very much in the data-analysis stage. Although based on extensive prior analysis of the conversation structure, the research is entirely exploratory with respect to strategy phenomena. We continue to struggle with basic issues: What kinds of variables best reflect participants' strategies? How does one identify and represent a participant's strategy? What are useful analytic techniques for tracking the consistencies and changes in a participant's strategy over the course of an interaction? As we become better able to characterize the conduct of individual participants, are there ways to begin characterizing groups of participants in terms of communalities in strategy?

Given provisional solutions to these problems, we are concerned with discovering principles underlying the strategic conduct of an individual or a group. For example, with respect to smiling, does a participant simply show a consistent gross amount of smiling from one conversation to the next, maintain a constant ratio of smiling with respect to the partners smiling, key smiling to some other action or actions by the partner, or vary smiling according to the stage of the conversation? We have preliminary evidence that many rule-related actions vary according to the type of interaction, such as asking questions, answering questions, and exchanging comments. In this case, an apparent change in strategy may reflect in part a change in the type of interaction current in the conversation.

Finally, we seem far from the goal of analyzing and representing the process of mutual adjustment that takes place, at least to some extent, in all interactions.

REFERENCES

1. GOFFMAN, E. 1969. Strategic Interaction. Univ. Pennsylvania Press, Philadelphia, Pa.
2. GOFFMAN, E. 1976. Replies and responses. Lang. Soc. 5: 257–315.
3. KENDON, A. 1977. Studies in the Behavior of Social Interaction. Indiana Univ. Press, Bloomington, Ind.
4. McQUOWN, N. A., Ed. 1971. The Natural History of an Interview. Microfilm collection of manuscripts on cultural anthropology. 15th series. Univ. Chicago Joseph Regenstein Library Department of Photoduplication, Chicago, Ill.
5. SCHEGLOFF, E. A., G. JEFFERSON & H. SACKS. 1977. The preference for self-correction in the organization of repair in conversation. Language 53: 361–382.
6. EKMAN, P., W. B. FRIESEN & P. ELLSWORTH. 1972. Emotion in the Human Face: Guidelines for Research and an Integration of Findings. Pergamon Press, New York, N.Y.
7. DUNCAN, S. D., JR. & D. W. FISKE. 1977. Face-to-face Interaction: Research, Methods, and Theory. Erlbaum Press, Hillsdale, N.J.
8. DUNCAN, S. D., JR. & R. ROSENTHAL. 1968. Vocal emphasis in experimenters' instruction reading as unintended determinant of subjects' responses. Lang. Speech 11: 20–26.
9. DUNCAN, S. D., JR., M. J. ROSENBERG & J. FINKELSTEIN. 1969. The paralanguage of experimenter bias. Sociometry 32: 207–219.

The Significance of Unwitting Cues for Experimental Outcomes: Toward a Pragmatic Approach*

MARTIN T. ORNE

Unit for Experimental Psychiatry
The Institute of Pennsylvania Hospital
Philadelphia, Pennsylvania 19139

Department of Psychiatry
University of Pennsylvania
Philadelphia, Pennsylvania 19139

THE CLEVER HANS PHENOMENON excited the scientific community of the day because it appeared to document a unique cognitive skill on the part of a horse. When Pfungst,[1] after extensive study, was able to document this fascinating case of apparently unwitting communication, the scientific community after an initial shock rapidly lost interest in what seemed like an almost trivial alternative explanation. It is well to remember that several phenomena, ranging from ESP to the inheritance of acquired characteristics, and from the especially powerful effects of new drugs—which must be used while the special power persists—to some effects ascribed to biofeedback, have been shown to relate directly to the same process that explains the famous "talking horse." In these and many other instances, the effects of unwitting, subtle communications have been responsible for what was initially thought to be an entirely different phenomenon being subjected to scientific scrutiny. From the point of view of the scientific community, the Clever Hans effect has often been seen as a troublesome artifact which might readily mislead some *other* investigator.

It was not until the social psychology of research itself was recognized as requiring systematic inquiry that there has been some serious concern about the process of unwitting communication and its potential effects on experimental findings.[2,3] Unfortunately, to appreciate the

* The substantive research discussed here was supported in part by Grant MH 19156 from the National Institute of Mental Health and by a grant from the Institute for Experimental Psychiatry.

Clever Hans phenomenon it is essential to recognize that an experiment, just like any other social interaction, does not occur in a vacuum. Accordingly, to evaluate the effect of subtle communication, it is necessary to understand the roles of the various participants and how they perceive them.†

In contrast to the experimental model of the physical sciences, which studies the effect of stimuli on inanimate objects, the psychological experiment involves an interaction of an experimenter with an active, sentient participant who has agreed to take part in an experiment and recognizes that his role is to appear as a passive responder. The participant, however, takes part for his own purposes, seeks to present himself in a desirable way, generally wants to assist the research, and tries to do the "right" thing. In some ways the psychological experiment is viewed by the subject as a problem-solving situation, where the rules of the game preclude his being given sufficient information about what is expected of him but where he is nonetheless supposed to respond appropriately by attending carefully to the various sources of information available to him.

I have tried to suggest that one way the psychological experiment may be viewed advantageously is as two conceptually distinct experiments: the first, the study that the investigator has designed and operationalized to test his hypothesis; the second, the study that the subject perceives he is participating in.[5] What I have called the demand characteristics of the experiment are the sum total of cues available to the subject before the experiment, the instructions during the experiment, the covert communications during the experiment, and the nature of the procedure itself that communicate the experimental purposes and the desired behavior.[6] Clearly, unwitting communications are a major determinant of these demand characteristics. The extent to which the experiment designed by the investigator and the experiment participated in by the subject are the same will determine much of the ecological validity of the procedure and, therefore, the generalizability of the findings beyond the immediate experimental context.

It should be emphasized that it is not possible to design an experiment without demand characteristics, and it is unlikely that an experiment can be carried out without unwitting, subtle communications. It is, however, both possible and necessary to determine the kinds of effects that the demand characteristics of the situation will have for the subject in order to assess the adequacy of the experimental procedure —

† Goffman[4] has conceptualized this as frame analysis.

that is, the extent to which the subject perceives the study as the investigator intends him to perceive it.‡

Once it is recognized that the effects of demand characteristics and the potential consequences of unwitting communication are problems that are likely to be with any working scientist for as long as he carries out research with man and that these cannot be controlled by statistical design, there has been a tendency to take one of two opposing points of view. The first is to view the whole matter as a tempest in a teapot, to argue that these are problems only when proper controls are not carefully carried out, that with well-designed studies and trained investigators there are no difficulties, that the cooperative, good subject is a myth, and any variance accounted for by any such individuals is corrected by others who, in the jargon of the day, show a "screw-you effect."[7] The other point of view tends to be no less extreme in that it views with alarm the problems of experimental social psychology, seriously questioning whether any experiment is ever worth doing, using concepts like the Clever Hans phenomenon, demand characteristics, evaluation apprehension,[7] and the like as spoiler variables rather than as conceptual tools to design more appropriate studies. Proponents of this view enjoy citing some of the more dramatic studies, as for example by Rosenthal,[2] Adair,[9] Silverman,[10] Page,[11] and myself,[12] put iconoclastic interpretations on the findings, and tend to argue against the experiment as a means of learning about the nature of man.

It is of course obvious that neither of these extreme views is acceptable. In evaluating the outcome of research, the problems inherent in working with subjects who like ourselves have an interest and, to some degree, a stake in the experiment will have varying degrees of effects on different experimental situations. I would like to emphasize that it is essentially impossible to predict the extent of these effects from an armchair. They must be evaluated empirically, using procedures along the lines of quasi-controls, suggested elsewhere.[6]

---

‡ I cannot resist addressing the issue of deception at this point. The use of deception is neither inherently good nor inherently evil. The task of the investigator is to make certain that no one leaves his laboratory with more troubles than he brings, and I seriously doubt that many individuals in a nontherapeutic context will be helped by pointing out their infantile anger and other inadequacies even though the observations may be accurate in the sense that they are based on the best available data. By the same token, some investigators proudly proclaim that they need not worry about the subject's perception of the experiment since they always are truthful. However, it turns out that it matters less whether the investigator is truthful and more whether what he says is plausible. Many an experiment is done for what—from the subject's point of view—are arcane and entirely unlikely reasons, and the subject will accordingly disbelieve them, leading to potential discrepancies between the investigator's intent and the subject's perception, even though the former is always being truthful. Thus, in any experiment involving human participants, one needs to assess what the subjects perceive as the experimental purposes and their own appropriate role in the procedure.

Though I have tended to talk about the effects of demand characteristics and subtle communications on experimental contexts, it should be clear that these are by no means the only contexts where effects of this kind can be documented. Since one of the would-be solutions proposed for the difficulties of the experiment is the naturalistic field study, it seems worthwhile to examine how these effects can play a powerful role in nonexperimental contexts. Field studies are important and desirable, but they are not a substitute for carefully controlled laboratory research. For example, it is far more difficult and vastly more expensive to simultaneously control several relevant parameters in the field than in the laboratory. From my perspective, an interplay between field and laboratory research in psychology and psychiatry is as important as the interplay between *in vivo* and *in vitro* studies in biology. As with laboratory settings, it must also be emphasized that no meaningful field investigation should be carried out without a thorough analysis and detailed understanding of the social roles of the participants in the particular social context under study.

For example, a "real life" context that shares a number of important aspects of the experimental situation is the one-to-one psychotherapeutic interaction. The patient, even more than the subject, is motivated to participate in the interaction for reasons of his own. In superficial contrast to the experiment, where the subject may hope to learn something about himself and psychology in general, the patient in psychotherapy hopes to obtain direct personal benefit. In the therapeutic situation, however, like in the experiment, the participant does not know precisely what to expect and has only general knowledge of the rules of the game. Though he is generally told some things about what to do, the patient typically believes that there is more that he has to discover and that it is very important for him to do the "right thing." The therapist, analogous to the experimenter in the experiment, defines the correct response of the patient by his own behavior — being pleased, displeased, bored, and so on.

By way of illustration, it is common that among a new group of psychiatric residents one or another male resident will come to supervision troubled because an attractive female patient has for some time been bringing in extremely lurid sexual dreams, which take up most of the hour and bring treatment to a virtual standstill. To remedy this difficulty is as simple as it is effective. One first discusses the resident's interest in his patient, exploring briefly the countertransference, and then simply instructs him that henceforth he should stop writing down any sexual dream material. Instead he should write assiduously whenever the patient begins to talk about her current life difficulties but put down the pencil as soon as dream material is brought up. By the end of the next session usually, and certainly by the end of the next session thereafter, the sexual dreams simply cease to be an issue in treatment. This illustration documents that writing serves a function for the patient analogous to that of headnodding for Clever Hans!

Again, in the therapeutic setting I had observed that many patients —particularly those who did not know anyone else in psychotherapy— fail to do well in treatment because they do not understand the rules of the game. Accordingly, we developed a technique referred to as anticipatory socialization for treatment, designed to teach the individual how to be an effective patient.[13] This procedure was tested in two independent studies[14,15] where it was shown that patients who had a single session of anticipatory socialization did significantly better in psychotherapy as assessed four months later.

What was communicated to the patient about his role and expected behavior was in these instances deliberate and purposive. He was chosen as the target of the anticipatory socialization interview because we assumed he would have a very high level of motivation to learn about the enterprise of psychotherapy. The modification of the patient's behavior made it more congruent with the therapist's expectations and probably served to modify the therapist's expectations and behavior, which in turn led to greater therapeutic improvement. The therapist then was responding to the patient's unwitting communication. Thus, it would appear that in the therapeutic context either the therapist or the patient is capable of altering the course of treatment by purposive behavior, which affects the unwitting behavior of the partner in the dyadic relationship.

Another context where unwitting communications are potent is the forensic context. Recently I have become interested in the use and (at times) abuse of hypnosis in attempts to enhance the recall of witnesses to or victims of crimes. There have been instances where hypnosis facilitates the recall of important information not previously available. The recall of several numbers of the license plate of the kidnapper's car in the case of the Chowchilla kidnapping of a bus load of children is a good example. Here, neither the bus driver, the authorities, nor the media had any notion of the correct license number. In other instances, however, where the authorities or the media had identified suspects, the probable impact of unwitting communication can be seen. Under such circumstances, witnesses who had previously been unable to identify assailants became able to do so when their memory was "refreshed by hypnosis." The key question here, however, is whether in such circumstances hypnosis serves to enhance actual recall or to create a pseudo memory where there is no recall. That is, by increasing the response to subtle cues, hypnosis helps to persuade the witness who has no recall that he had in fact seen the plausible suspect commit the crime.[16]

One example of such a situation is a court martial case that occurred several years ago in the Philadelphia Naval Base. Two sailors were sitting drinking coffee when a man walked by, pulled a gun, and fired at one of them. The victim fortunately jumped away and was only grazed. Subsequently the witness identified an individual as the assailant. The victim, however, was unable to identify this individual as his attacker. At a preliminary hearing, the witness again identified the defendant as the assailant whereas the victim said that he looked somewhat like the assailant but that he was not the right person.

The victim was then hypnotized in order to help him recall. After the first session he still had no memory of the original event. During the second session he relived the events and was given a posthypnotic suggestion to remember everything that happened. On awakening he was certain that he had actually seen the defendant attempt to kill him. During the court martial trial I testified about the problems of confabulation when hypnosis is used in this fashion, pointing out that the pressure to remember caused the victim to accept a plausible suspect as the assailant though in the wake state he had rejected this possibility. However, the pseudo memory created in hynosis was then confused with his earlier recollection. The judge advocate excluded the testimony based on hypnotically enhanced recall and the defendant was acquitted. Within several weeks two reliable witnesses returned from overseas and corroborated the defendant's alibi, thus proving that he could not have been the assailant. Interestingly, the effects of the hypnosis persisted and the victim still accepted his pseudo memory as real one year later.

This case is one of a considerable number where the hypnotic process was used to alter the recollection of witnesses or victims. I bring it up in this context because it illustrates a rather frightening consequence of the Clever Hans phenomenon where, in conjunction with hypnosis, unwitting communication may cause an honest person to compellingly lie.

I have tried to show that powerful effects of subtle communication are ubiquitous and by no means limited to the laboratory experiment. Field settings such as the therapeutic context or the context of hypnotic interrogation may also bring about precisely the same kinds of consequences that have been described for psychological experiments. If anything, the effects are stronger and more durable in nonlaboratory settings. Clearly, the notion of leaving the laboratory in order to escape from the methodological problems inherent in Clever Hans or demand characteristic effects is naive and ill founded. By all means let us conduct field research, but we must be clear that we are exchanging one social psychological context for another, and the failure to understand or to specify that context is not likely to advance science.

With varying degrees of comfort, psychologists have learned to live with the need to use statistical tests in order to appropriately assess the outcomes of research. In an analogous way, we will need to become comfortable with the realization that any research involving human beings occurs in a social context and is never carried out in a vacuum. The nature of that social context and the subject's awareness of his of his particular social role will help determine his responsivity to the Clever Hans phenomenon, as well as other aspects of the demand characteristics inherent in the situation.

The fact that such effects exist does not mean that they determine the experimental outcome. The question of the impact of unwitting cues needs to be raised and considered in relation to each piece of research. It always represents an alternative hypothesis to be considered. While this involves a certain amount of effort, it need not paralyze us. Once we recognize that research findings are obtained with varying degrees of control and can therefore be relied upon with varying degrees of certainty, we will begin to develop a pragmatic point of view toward our observations. Certainly, if a research finding is to be the basis of decisions that seriously affect the health and welfare of man, or involve great expenditures of valuable resources, the level of certainty demanded should be far greater than if one were dealing with a purely academic exercise. Many would define science as the approximation of truth, and if we take such a point of view seriously, the realization that different levels of certainty are required, depending upon the purpose of the research, should not make us too uncomfortable. It should, in fact, ultimately lead to a pragmatic view of empirical psychological research as one approach to an understanding of the real world.

Acknowledgments

I would like to thank my colleagues David F. Dinges and Kevin M. McConkey for their detailed substantive suggestions and Emily Carota Orne, William H. Putnam, and William M. Waid for their comments in the preparation of this manuscript.

References

1. Pfungst, O. 1965. Clever Hans (The Horse of Mr. Von Osten). [Edited by R. Rosenthal.] Holt, Rinehart & Winston. New York, N.Y.

2. ROSENTHAL, R. 1966. Experimenter Effects in Behavioral Research. Appleton-Century-Crofts, New York, N.Y.
3. ORNE, M. T. 1962. On the social psychology of the psychological experiment: With particular reference to demand characteristics and their implications. Am. Psychol. 17: 776–783.
4. GOFFMAN, E. 1974. Frame Analysis: An Essay on the Organization of Experience. Harper Colophon Books, New York, N.Y.
5. ORNE, M. T. 1973. Communication by the total experimental situation: Why it is important, how it is evaluated, and its significance for the ecological validity of findings. In Communication and Affect. P. Pliner, L. Krames & T. Alloway, Eds. pp. 157–191. Academic Press, New York, N.Y.
6. ORNE, M. T. 1969. Demand characteristics and the concept of quasi-controls. In Artifact in Behavioral Research. R. Rosenthal & R. L. Rosnow, Eds. pp. 143–179. Academic Press, New York, N.Y.
7. MASLING, J. 1966. Role-related behavior of the subject and psychologist and its effects upon psychological data. In Nebraska Symposium on Motivation. D. Levine, Ed. pp. 67–103. Univ. of Nebraska Press, Lincoln, Neb.
8. ROSENBERG, M. J. 1969. The conditions and consequences of evaluation apprehension. In Artifact in Behavioral Research. R. Rosenthal & R. L. Rosnow, Eds. pp. 279–349. Academic Press, New York, N.Y.
9. ADAIR, J. G. 1973. The Human Subject: The Social Psychology of the Psychological Experiment. Little, Brown & Company, Boston, Mass.
10. SILVERMAN, I. 1968. Role-related behavior of subjects in laboratory studies of attitude change. J. Pers. Soc. Psychol. 8: 343–348.
11. PAGE, M. M. & A. R. LUMIA. 1968. Cooperation with demand characteristics and the bimodal distribution of verbal conditioning data. Psychonom. Sci. 12: 243–244.
12. ORNE, M. T. 1970. Hypnosis, motivation and the ecological validity of the psychological experiment. In Nebraska Symposium on Motivation. W. J. Arnold & M. M. Page, Eds. pp. 187–265. University of Nebraska Press, Lincoln, Neb.
13. ORNE, M. T. & P. H. WENDER. 1968. Anticipatory socialization for psychotherapy: Method and rationale. Am. J. Psychiat. 124: 1202–1212.
14. HOEHN-SARIC, R., J. C. FRANK, S. D. IMBER, E. H. NASH, A. R. STONE, & C. C. BATTLE. 1964. Systematic preparation of patients for psychotherapy. 1. Effects on therapy behavior and outcome. J. Psychiat. Res. 2: 267–281.
15. SLOANE, R., A. CRISTOL, M. PEPERNIK, & F. STAPLES. 1970. Role preparation and expectation of improvement in psychotherapy. J. Nerv. Ment. Dis. 150: 18–26.
16. ORNE, M. T. 1979. The use and misuse of hypnosis in court. Int. J. Clin. Exp. Hypnosis 27: 311–341.

Progressions of Conversations and Conversational Actions

KARL E. SCHEIBE

Department of Psychology
Wesleyan University
Middletown, Connecticut 06457

I WANT HERE to suggest several directions in the evolution or development of conversations. I call these conversational progressions. I do not propose to examine internal progressions of ordinary conversations, for much excellent work has already been done on the details of dialogue, the rhythms of conversations, their transitions and characteristic trajectories, and the multiform cues that seem to regulate the pace and content of conversational exchanges.[1-4] Rather, I propose here a molar examination of conversational progressions. Suppose an ordinary conversation between two interacting persons to be a normal condition, using just the sort of criteria articulated by Donaldson[5] in her definitional essay, "How do we know when we're conversing?"*

By departing from the normal case, it is possible to consider what conversations were before they became conversations, what they lead to, and what they become when one or another criterion for their existence is either exaggerated or diminished. This should lead us to recognize some lines of continuity between ordinary conversations and other states or forms of human interaction that are not normally thought to be conversations. For example, by decreasing the normal number of participants in a conversation from two to one, dialogue becomes monologue. By increasing the number from two to three, to ten, to twenty, to fifty, to a thousand, to a million, and beyond, we move from conversation to discussion to a meeting, to mass communication, and finally to something like a universal understanding.

Obviously, there are many such conversational progressions that can

---

* These criteria are the following: (1) A minimum of two participants. (2) Necessity of taking turns. (3) Remarks of participants deal with roughly the same subject. (4) Remarks must contain some information, not be empty or random. (5) Interaction must not be primarily for an explicitly stated (business) purpose. (6) For the duration of the exchange, participants behave as equals, with neither acting as an authority. (7) Remarks must have some measure of spontaneity and nonpredictability. (8) Participants must display some degree of reciprocity. (9) Few imperatives appear. (10) Regular deletion processes apply.

0077-8923/81/0364-0160 $01.75/0 © 1981, NYAS

be identified — just as many progressions can be identified in music or in mathematics — and there is no claim here about the unique importance of the particular progressions I have chosen to discuss. But I can say why I have chosen to discuss some progressions rather than others; it is because some conversational progressions carry strong implications for the motivational or evaluative significance of human interaction, and it seems to me that the semiotic literature on conversation is conspicuously inadequate in the affective domain.

First Progression: From Solitude to Communion

Solitude is an elemental fact of the human condition. Solitude is a backward progression from conversation, and conversation is the essential means for achieving a sense of human community and as a sequel, according to Mead,[6] a sense of self. Ortega y Gasset[19] asserts that the human being, when freed from the yoke of toil, always and everywhere chooses to engage in one of four kinds of activities — to converse, to dance, to compete in sports, and to hunt. Note that the latter three activities can be seen as variants of the first: The dance is synchronous and cooperative conversation; sporting, gaming and racing are competitive conversations; and the hunt is a competitive conversation with the natural prey of the hunter in the role of alter. Note also that each of these three forms of conversational activity are of necessity episodic and that they occur against a backdrop of solitude. So the progression from solitude to communion is cyclical. One result of the successive engagement in conversation and conversational variants is the differentiation of the social self. Once the self is differentiated, the act of meditation, which is natural to solitude, can be *about things* rather than hermetically sealed in solitude.

The affective consequence of this progression is plain. Montaigne expressed the matter clearly: "The most fruitful and natural exercise for our minds is, in my opinion, conversation. I find the practice of it pleasanter than anything else in life; and that is the reason why, if I were at this moment forced to choose, I should, I believe, rather consent to lose my sight than my hearing or speech."[7] Conversation for Montaigne and for the rest of us is the chief means of formation for the social self, and as conversation moves from solitude to communion and back again, so are our social selves expanded and contracted.

Second Progression: From Explication to Implication

Something happens to conversations between people as they become well known to each other. Conversations between individuals who are

foreign to each other are both simple in content and elaborate in detail
— the sort of conversation that comes out of phrase books. At the other
extreme, individuals who are of kindred mind converse in a rapid and
telegraphic fashion, with a minimum of explicit content and a maximum of meaning. William James describes the latter case as follows:
"Their conversation is chiefly remarkable for the summariness of its
allusions and the rapidity of its transitions. Before one of them is half
through a sentence the other knows his meaning and replies. Such
genial play with such massive materials, such an easy flashing of light
over far perspectives, such careless indifference to the dust and apparatus that ordinarily surround the subject and seem to pertain to its
essence, make these conversations seem true feats for gods to a listener
who is educated enough to follow them at all. His mental lungs breathe
more deeply, in an atmosphere more broad and vast than is their wont.
On the other hand, the excessive explicitness and short-windedness of
an ordinary man are as wonderful as they are tedious to the man of
genius."[8]

This progression is clarified by referring to H. P. Grice's distinction
between implication and explication in conversation.[9] A conversation is
much more than an exchange of words. A transcribed dialogue is merely
a representation of a conversation, not a reproduction of it. Each party
in a conversation tries to understand not only what is said but also what
is meant. Grice identifies a level of implicative meaning in conversations, wherein what is said is intended to communicate something that
is not said. The normal conversation is regulated by the Cooperative
Principle, consisting of a series of tacit agreements between the participants to stay on the subject, not to say too much or too little, and so
forth. When these normal rules are violated, as they commonly are in
conversation, then the hearer's attention is directed to the implication
of what is being said rather than to a literal interpretation.

In order for implication to work as a means of communication, there
must be some sharing of the tacit rules of implicature; sharing the same
explicit language is not a sufficient condition for the staging of a full
conversation. On the other hand, if there is a very great mutual understanding of the rules of implicature, conversations can be carried on entirely by means of glance, gesture, occasional verbal expression, and
posture. Tolstoy's representation in *Anna Karenina* of the truncated
conversation between Levin and Kitty as they are discovering their love
is often cited as an example of communication by shared implication.
This romantic sort of progression can be described roughly as follows:

> After the unspoken glance comes the verbal greeting—a mere certification
> of what has already happened. There ensues a bilateral flow of words,
> gradually to be truncated as more is understood. With the mutual willingness

that is essential to love, a conversation may now become cryptic, abbreviated, and proceed to the form of interaction known as dance. Dancing pairs are engaged in an extension of conversation; nothing is spoken, much is understood. The embrace of sexual union is an expression of a mutual desire to risk the furtherance of knowing.[10]

Conversation and intercourse are not synonymous, but etymologically and psychologically they are easily extended to each other.

The progression from explication to implication is accompanied by a blurring of the ordinary alternating transfers or exchanges of speaker and listener roles that are characteristic of ordinary conversation. Conversational participation becomes simultaneous. It might also be said that the decoding rules become less conscious or sink into unconsciousness, if by this is meant the inacessibility of these rules to verbal explication. So then we can see the happy show of intellectual prowess by Herr von Osten and Clever Hans to be a synchronous conversation by implication, where neither the horse nor the man were aware of how their results were achieved. The sagacious thought-reader or clinical diagnostician is not necessarily able to explicate the cues antecedent to their conclusions. Subjects responding to demand characteristics or to the nonverbal cues in Rosenthal's PONS test are functioning largely at the level of implication.[2,11]

THIRD PROGRESSION: RAW UTTERANCE TO FORMAL TEXTUAL DIALOGUE

Something there is in science that does not love a mystery, that wants it down. Katz's commentary on the Clever Hans phenomenon is introduced with these words: "At the beginning of the century, reports of the singular intellectual accomplishments of a horse caused a great sensation all over the world."[12] But Pfungst's discovery of the means of communication between interlocutor and horse effectively reveals what was previously a mystery to be something more ordinary. Attempts by Rosenthal and his colleagues to identify the channels and cues by which nonverbal communication functions is similarly a movement toward clarifying and making explicit what is going on. Scientific attempts to identify cues, specify encoding and decoding rules, or to formulate general descriptive models of communication are parallel to the progression noted by David Olson in his essay, "From Utterance to Text," for the cues transmitted in utterances are multiform and mysterious, while the line of text presents information clearly and in a single channel.

Olson argues that the usual primitive linguistic modes — conversations, storytelling, and singing — in time are transformed into a literate tradition, in which the text comes to stand for what was previously an utterance. He argues ". . . that the transition from utterance to text both culturally and developmentally . . . can be described as one of increasing explicitness, with language increasingly able to stand as an unambiguous or autonomous representation of meaning."[13] Olson goes on to argue that Chomsky's theory of sentence meaning is based on the premise that meaning inheres in formal text, and hence is not a theory of language in the sense of utterance. Perhaps because so much of the writing and theory about language has to do with text rather than utterance, or because literary criticism has to do almost entirely with text at the neglect of utterance, Olson claims that the direction of evolution is ". . . both culturally and developmentally from utterance to text."[13] Doubtless there is descriptive truth here, for this conference and these papers are testimony to a continuing cultural/scientific effort to make our meanings clear, to make our conversations unambiguous and explicit. But note that this progression, so ardently pursued by analytic philosophers, positivistic psychologists, and communication scientists, is quite opposite in direction to the progression I have posited just before — from explication to implication. It is also, I assert, quite distinct from that progression in its affective accompaniment. It satisfies our curiosity to know precisely how the dance achieves its marvelous coordination, but this is nothing like the satisfaction of dancing. There *is* something basically antiromantic about scientific analysis. The pursuit of literacy and textual explicitness is full of fascination, but the achievement of the end is no fun, or as Gerald Graff[14] has suggested of modern literary critics, the end of hermeneutic analysis is fear and angst.

Can the scholar be made to dance? I expect that at the end of his analysis he will need to, even if he cannot bring himself to want to. The reasons for the antinomy of Dionysus and Apollo become more clear when we consider the effect of the transformation from utterance to text on the first progression — from solitude to communion. For the end of explication and the final objective of full textual analysis is not something modest like the potentiation of dyadic conversations — rather it is the achievement of universal knowledge, or universal understanding. To simplify the matter greatly but to come right to the point, in a world where everything is known, where there are no residual mysteries, and where universal education has tutored all of the globe's inhabitants to perfect understanding, there would no longer be anything to talk about. It is surely a relief to be rid of fairies and goblins, and all our primitive animism, where the world is replete with spirits, spirits who speak never in explicit text, but might always have been detected by implication of the slightest natural sign or movement of wind, earth, fire, water, or entrail. The end of analysis is the achievement of full communion, entailing the end of private reserve and public mystery, and carrying as a consequence the withering of conversation.

But it is one thing to describe theoretically such a progression as this and quite another to affirm that this is precisely what is happening, for I do not think that this progression is a good description of what is happening in the world. One can progress out of solitude and its inchoate meanings and self-contained meditations into the full bright world of logically complete text; but after much time in the library there is no reason we cannot return once again to our solitude, and in that state to regain the wherewithal for pleasurable conversation again, and I think this is what happens in the world.

FOURTH PROGRESSION: FROM QUOTIDIAN TO TRANSCENDENT

Nothing is more boring, more painful, or a more perfect strain on patience than obligatory polite conversation about nothing that matters. And yet nothing is more sublime than the achievement of common understanding in living conversation. This latter sort of understanding, I contend, differs from that accomplished through the reading of text. One can read Freud, or Joyce or Proust or Thomas Mann and experience a shock of recognition. You find yourself saying, "My God, he understands what I have been thinking and have never been able to say." Such understanding is unilateral. What I am speaking of differs also from your being able to elicit a positive response from some general audience, and thereby conclude that you are understood. Rather, in living conversation between human parties a form of mutual understanding is attainable that is *sui generis* — it is transcendent, its implicit structure cannot be made explicit without loss, and it is not a replicable event. (Recognize now that I am close to talking nonsense, if one maintains the twin premises that the only form of knowledge is scientific knowledge and that scientific knowledge is not attainable for a nonreplicable event.)

The sort of progression I have in mind now is not one of gradual development, but one of metamorphosis of state — from egg to larva to pupa to moth. This elevation of the state of conversation is admirably described by Emerson:

> Persons themselves acquaint us with the impersonal. In all conversation between two persons tacit reference is made, as to a third party, to a common nature. . . . In groups where debate is earnest, and especially on high questions, the company become aware that the thought arises to an equal level in all bosoms. . . . They all become wiser than they were. It arches over them like

a temple, this unity of thought in which every heart beats with nobler sense of power and duty, and thinks and acts with unusual solemnity. All are conscious of attaining to a higher self-possession. . . . The action of the soul is oftener in that which is felt and left unsaid than in that which is said in any conversation. . . . We know better than we do. We do not yet possess ourselves, and we know at the same time that we are much more. I feel the same truth now often in my trivial conversations with my neighbors, that somewhat higher in each of us overlooks this by-play, and Jove nods to Jove from behind each of us."[15]

Clearly I am pointing here to a human possibility to which the poets can attest, but about which science can offer little testimony. According to my First Progression, we can emerge out of our solitude to have commerce with our fellows, but that commerce can be either boring chatter or a truly elevating exercise. If Emerson is correct, the elevating conversation entails some sort of contact with an ineffable *tertium quid*, knowable by implication only. Be that as it may, I offer here only the speculation that living conversation can be the vehicle to the transitory grasping of a truth that cannot be preserved long enough to survive the transformation to explicit knowledge. The only sort of evidence for this sort of possibility is internal: ask yourself.

CONCLUDING COMMENT: THE PROGRESSION TO PSEUDOCONVERSATION

It is curious that in an age of precarious and difficult human communication we should direct major amounts of attention to the question of whether it is possible to carry on true conversations with computers or nonhuman animals. Let there be no doubt: It is possible to talk to a computer, to an ape, to a plant, or to a pet rock. And it is also possible to exercise one's talents at deciphering implicit meaning in such a way as to find substance in the replies of these entities, and thus evidence of "consciousness." I do not think, incidentally, that all the boasting about the practical applications of the technology used to enable chimpanzees to speak is anything but the flimsiest of public relations cover stories. The interest is more intrinsic. As David Katz asserted at the head of another chapter in his book, "The question of the emergence of consciousness among living things has intrigued men throughout the ages."[12] And now, of course, this intriguing question must be extended to include computers and other sophisticated pieces of hardware.[16] I do not belittle this concern, for it is fundamental to both epistemology and to metaphysics. But I do not hesitate to take a position myself on this question, in accord with Terrace[17] and with Sebeok,[18] that evidence or argument does not compel us to grant true conversational capacity to any other than human alters. To be sure, Dr. Doolittle will continue to converse with his menagerie, and solitary matrons will chatter to their poodles—there is no harm in it.

I am suggesting that all the attention devoted to the progression from conversation to pseudoconversation—mimicked verbal interactions with nonhuman alters—is less fruitful and interesting than would be an exploration of some of the other progressions I have described: Conversation moves us from solitude to communion; it carries us from glance through wordy explication to the silent synchrony of the dance; it moves us back again from the mysteries of Dionysus and confusion into the clear light of certain and universal understanding; and when the clear and the commonplace begin to pall, conversation is at least possibly a means for our attaining transcendent mutual understanding, or the illusion of it, which is a question I cannot here judge.

There is plenty to talk about. And unless we are forced forever out of our solitude, there always will be.

References

1. JAFFE, J. & S. FELDSTEIN. 1970. Rhythms of Dialogue. Academic Press, New York, N.Y.
2. ROSENTHAL, R., Ed. 1979. Skill in Nonverbal Communication. Oelgeschlager, Gunn & Hain, Cambridge, Mass.
3. MISHLER, E. G. 1975. Studies in dialogue and discourse: An exponential law of successive questioning. Language in Society 4: 31-51.
4. DUNCAN, S. & D. FISKE. 1977. Face-to-Face Interaction: Research, Methods, and Theory. Erlbaum, Hillsdale, N.J.
5. DONALDSON, S. K. 1979. One kind of speech act: How do we know when we're conversing? Semiotica 28: 259-299.
6. MEAD, G. H. 1934. Mind, Self and Society. University of Chicago Press, Chicago, Ill.
7. MONTAIGNE, M. 1958. Essays. p. 286. Penguin, New York, N.Y.
8. JAMES, W. 1890. Principles of Psychology. Vol. 2: 390. Holt, New York, N.Y.
9. GRICE, H. P. 1975. Logic and conversation. In Syntax and Semantics. P. Coe and J. L. Morgan, Eds., vol. 3. Academic Press, New York, N.Y.
10. SCHEIBE, K. E. 1979. Mirrors, Masks, Lies and Secrets. p. 128. Praeger, New York, N.Y.
11. ORNE, M.T. 1962. On the social psychology of the psychological experiment: With particular reference to demand characteristics and their implications. American Psychologist 17: 776-783.
12. KATZ, D. 1937. Animals and Men. pp. 13, 28. Penguin, London.
13. OLSON, D. 1977. From utterance to text: The bias of language in speech and writing. Harvard Educational Review 47: 257-281.
14. GRAFF, G. 1977. Fear and trembling at Yale. American Scholar 46: 467-478.

15. EMERSON, R. W. 1883. Emerson's Works. pp. 260–261. Riverside Press, Cambridge, Mass.
16. SCHEIBE, K. E. & M. ERWIN. 1979. The computer as alter. Journal of Social Psychology 108: 103–109.
17. TERRACE, H. S. 1979. Nim: A Chimpanzee who learned sign language. Knopf, New York, N.Y.
18. SEBEOK, T. A. & D. J. UMIKER-SEBEOK. 1979. Speaking of apes: A critical anthology of two-way communication with man. Plenum, New York, N.Y.
19. ORTEGA Y GASSETT, J. 1972. Meditations on Hunting. [Originally published in 1943.] Scribners, New York, N.Y.

The Psychic Reading

RAY HYMAN
Department of Psychology
University of Oregon
Eugene, Oregon 97403

MIKE, A FREE-LANCE WRITER, visited a number of psychic readers to gather material for an article. He obtained a reading from each one. He also interviewed me to see how a psychologist would react to his experiences. Only one of the psychics, a palm-reader called Barbara, impressed Mike. In his article he wrote that, "This woman was studying lines in my hands and telling me, with devastating accuracy, about my strengths and weaknesses, my obsessions and my yearnings, my talents and my needs. As I drove away . . . I felt ready to chuck skepticism forever."

Mike discussed with me his experience with Barbara shortly after his visit with her and before he wrote his article. He had some difficulty in articulating just what it was about the reading that had so impressed him. He told me that during the first part of the session, Barbara made the usual assortment of general and ambiguous statements. His mind began to wander and he found himself thinking about a problem he was having with his girl friend. Suddenly his attention was brought back to what Barbara was saying. He heard her say, "I feel that you are worried about a relationship."

This conjunction in time between her statement about a relationship and his conscious concern about his relationship with a girl friend hit Mike with an emotional wallop. He had no doubt that the relationship Barbara was talking about was *the* relationship he was worried about. This emotional release converted an otherwise typical reading into something special. And Mike was now convinced that Barbara had some special powers.

From the skeptical viewpoint, Mike's account does not justify attributing special knowledge or insight to Barbara. Mike was impressed by her. To account for this impression he pointed to her accuracy in telling him he was worried about a relationship. We do not need to assume any powers to account for this sort of accuracy. We can safely assume that any individual who comes to a psychic has some sort of a problem with a relationship. Mike, of course, is sufficiently sophisticated to realize this. Yet he could not shake the conviction that Barbara's ac-

0077-8923/81/0364-0169 $01.75/0 © 1981, NYAS

curacy was more than just the use of statements that could apply to anyone. He was sure that Barbara had somehow tapped into his innermost secrets. And he had experienced, as a result, a compelling and rewarding emotional experience.

We can raise a number of questions about Mike's encounter with the psychic reader. The question that naturally occurs to the psychologist involves validity. Did the psychic's statements actually correspond to the facts of Mike's personality and situation? Did they do so in a way that would differentiate him from other clients? Notice that this same focus on the accuracy of the reading was the basis that Mike employed to justify his positive evaluation of Barbara. Another question asks what is it that the reading actually does for the client. Why do clients such as Mike experience the reading as both revealing and helpful? Notice that a reading need not be accurate to be helpful. We can ask, also, how much the client gets out of the reading in relation to how much he or she puts into it.

This last question relates the psychic reading to the topic of this conference—The Clever Hans Phenomenon. When the horse, Clever Hans, was asked a question, he would often give the correct answer by tapping an appropriate number with his hoof.[1] This would occur under circumstances that seemingly precluded the horse's actions being under the control of signals from the owner. Because the questioner knew he was not cuing the horse, he assumed that Hans's answer was a response to the verbal question and that the answer, by being correct, revealed the conscious understanding of the question and the requisite knowledge to supply the answer. In fact, Hans was responding to a simple, involuntary postural adjustment by the questioner, which was his cue to start tapping, and an unconscious, almost imperceptible head movement, which was his cue to stop. The horse was simply a channel through which the information the questioner unwittingly put into the situation was fed back to the questioner. The fallacy involved treating the horse as the source of the message rather than as a channel through which the questioner's own message is reflected back.

The psychic reading shares this fallacy with the Clever Hans situation. In most such readings the psychic is simply a channel through which information unwittingly emitted by the client is fed back to the client. The client typically assumes that the message originates from some secret or occult source to which the psychic has access.

But the psychic reading is richer and more complicated than the Clever Hans situation. Both situations deal with an individual who is unknowingly both the source and the destination of the message. In both cases, the assumption is made that actual communication is taking place because the information being received is "accurate." Hans answers the questions accurately. The psychic apparently tells the client things that he or she accepts as accurate. And this is just how it should be if actual communication were taking place. But it is just at this point in determining the accuracy of the communication that we notice an added feature that contributes to the success of the reading.

The questioner has no difficulties in deciding whether Hans has answered correctly or not. Either Hans has tapped the appropriate number of times or he has not. But the output from a psychic reading is both complex and highly ambiguous. The referent is the life history, personality, concerns, and problems of the client. And the language of personality description is notoriously difficult to interpret and apply.[2] This gives rise to what has become known as the "Barnum effect"—the phenomenon whereby people willingly accept personality interpretations comprised of vague statements with a high base rate occurence in the general population."[3]

Beginning with the 1949 publication by Forer on "the fallacy of personal validation,"[4] an area of research has emerged which follows more or less the same paradigm: (1) the subject completes a personality test or supplies information relevant to an assessment procedure; (2) the subject waits while the assessment information is processed; (3) the subject then receives a personality sketch allegedly derived from the assessment information; and (4) the subject rates the sketch for its "accuracy." In actual fact, all the subjects receive the same stock spiel composed of "Barnum" statements. The results are quite consistent and robust. When the subjects believe that the sketch was specifically meant for them they tend to rate it as a highly accurate and unique description of themselves.[3]

The appeal of these Barnum-type statements is so strong that it even overrides personality sketches especially written for the subjects. Sundberg, for example, administered the Minnesota Multiphasic Personality Inventory to 44 students.[5] Subsequently each student was presented with two personality sketches. One was written by a trained psychologist especially for the student on the basis of the student's answers to the inventory. The other was a fake sketch. Each student was asked to choose "Which interpretation describes you better." Of the 44 students, 26, or almost 60%, chose the fake sketch over the one written especially for them. Since Sundberg's study, several other investigators have demonstrated the same result.[3]

In a way this outcome should not be surprising. The fake sketches are composed of items that are true of almost everyone. The personalized sketches are composed of statements that discriminate the subject from others. When Barbara told Mike that he was worried about a relationship, we should not be surprised that Mike accepted the statement as accurate because such a statement applies with high probability to just about every young, single male. But Mike accepted the statement not because he recognized its universal applicability, but rather because he understood it to apply to his unique circumstances. And subjects in the experiments on the Barnum effect accept the sketch as not only accurate, but as uniquely descriptive of themselves as distinct from others.

Much of the research on the fallacy of personal validation tries to isolate the conditions under which the fake sketch will be read and accepted as a unique description. Subjects *can*, under appropriate circumstances, recognize that the items in the sketch apply to them just because they are universally true for everyone. If the subject is handed the fake sketch and simply told it is a general description rather than one made especially for the subject, then he or she is less likely to accept it as a unique self-description. But the subject will not only accept as accurate the very same sketch, if presented under the belief that it was prepared especially for him or her, but will fail to realize that it is just as accurate for the general population. In addition, as a result of this acceptance, the subjects also increase their faith in the assessment procedure and the skill of assessor.[3]

Almost certainly the situations in which subjects accept the fake sketch as unique and those in which they recognize its general applicability are experienced quite differently. The meaning of the sketch, like the meaning of any literary product, cannot be separated from the reader or listener. As the semioticians and structuralists keep emphasizing, there is no unique relationship between signifier and signified — nor is there a unique referent or reading for any given message.[6] The reader of the sketch that is not allegedly prepared especially for him or her is doing something different from the reader of the sketch that is allegedly prepared especially for him or her. Each reader is actually reading a different sketch.

All cognition and all sign-interpretation (the two are almost synonymous) involve a heavy contribution by the recipient in addition to that of the sign and the sender. In both the psychic reading and the Clever Hans situation, the contributions of the receiver almost totally determine the message and its interpretation. The fallacy in these situations is due to the fact that the receivers do not realize how much of the message and its meaning is their own contribution.

The Barnum effect is sufficient to account for the apparent accuracy in certain kinds of psychic readings. In some types of readings, the client is presented with a complete sketch. In such situations the psychic or "sender" need not even be physically present (getting an astrological writeup through the mail, receiving a printout of personality descriptions selected by a computer, etc.). No opportunity is afforded for the client to ask questions, clarify statements, or to agree or disagree while the message is being delivered. Nor does the psychic have the opportunity to alter or modify statements as a result of the client's reactions.

In contrast to such a static reading, most psychic readings are "dynamic" in the sense that the message is delivered to the client sequentially with the opportunity for the client to interact with the psychic during this process. It is in these dynamic readings that the Clever Hans effect combines with the Barnum effect to yield very effective results. In the typical reading, both the psychic and the client are equally victims of these two effects. Both believe that the client's acceptance of the reading and the client's conviction of having been helped stem from the psychic's access to hidden knowledge and to skills in diagnosing and advising.

But there is a class of psychic readers who are quite conscious of how they are managing the reading and the client to create the illusion that the psychic is the source of both the information and successful solutions to problems. These readers have sometimes written manuals to guide other readers. (Examples of such manuals, which I consulted, are listed in the bibliography.[7-13]) The manuals divide readings into two types: (1) the psychological reading and (2) the cold reading. The psychological reading, which corresponds to the static reading, involves delivering a stock spiel to the client. Because the sketch is usually memorized and delivered to several different clients, the studies of the Barnum effect apply directly. The cold reading is so called because the client is encountered without any prior knowledge. The cold reading employs the dynamics of the dyadic relationship between psychic and client to develop a sketch that is tailored to the client. The reader employs shrewd observation, nonverbal and verbal feedback from the client, and the client's active cooperation to create a description that the client is sure penetrates to the core of his or her psyche. The cold reader achieves this goal by feeding back to the client information the client has unwittingly revealed during the course of the reading. (Elsewhere I have written another version of how this is done and why it apparently works.)[14]

The manuals for both the psychological reading and the cold reading essentially provide guidelines for creating just those conditions that convince the client that the reading is accurate and that the accuracy is a function of the special knowledge of the psychic. These guidelines, while sometimes vague, clearly indicate that the writers operate under assumptions and theories very much in accord with modern cognitive science, literary structuralism, semiotics, and social cognition. None of these writers, of course, are familiar with such disciplines. And their assumptions and theories antedate the development of these fields of inquiry.

But before I indicate some of the ways the manuals anticipate current academic disciplines that deal with communication and signs, it is worth considering again the issue of accuracy. The success of the psychic reading depends entirely upon acceptance by the clients — upon personal validation. And, on the surface, such acceptance seems to be related to how accurate the sketch appears to the client. But why should the client care or be concerned about such accuracy? The client accepts what the psychic says about him or her only if it agrees with what the client already knows. Why should the client pay for being told what he or she already knows?

Several answers come to mind. One is that many of the things that the client is being told are recognized or accepted as "true" but were not consciously considered previously. In this sense, the client is learning new things about himself or herself — but things that are acceptable because they "ring" true.

But probably the most important reason is that to the extent the psychic can tell the client things that are true, but which the client believes could not be known to the psychic through normal means, the psychic validates the belief that he does, indeed, have special powers. Like the shaman, the psychic reader is most successful if the client attributes to him mysterious powers and occult sources of knowledge.

The client does not patronize the psychic to discover things that the client already knows. The client is typically a person with a problem who seeks help. If the helper clearly has access to hidden knowledge and magical forces, this increases the chances that he can work wonders for the client. If the psychic can convince the client that he knows things that could only come from mysterious sources, then his powers are validated. So the client has a big stake in the accuracy of the psychic. And part of the success of the reading stems from the client's need to see the psychic as omniscient.

The manuals, then, can be seen as guidelines for creating in the client the illusion of the reader's omniscience. As such, they embody a theory or a set of presuppositions about how to manage the dyadic interaction so that the client "reads" the results in the desired manner. And these presuppositions, as previously indicated, seem to anticipate the ideas emerging from the variety of overlapping contemporary disciplines that study humans as sign-using systems. To illustrate this point, I will consider the types of advice provided by the manuals under four categories: (1) setting the stage; (2) preliminary observations and categorization; (3) constructing the preliminary script; (4) delivering the message and revising the script.

Setting the Stage

The psychological and cold readers attribute much of their success to how the client is prepared prior to the reading. Such preparation consists of advertising, word-of-mouth accounts by previous clients, the preliminary introduction and statements to the client, the dress and mannerisms of the psychic, the furnishings and arrangement of the consultation room, and other ways that help to "define the situation" for the client and that specify what role the client is to play in this interaction.

Such preparations accomplish a number of important goals. The manuals specify that the psychic should make it clear from the outset that the psychic is the one who is in control of the situation. The psychic is not only experienced and an authority in this sort of interaction, but he or she has already proven that the psychic can do his or her part successfully. But, as the psychic will indicate to the client, the reading is a cooperative venture that requires the active participation of the client. This puts much of the burden for success on the client. It also emphasizes that the client has to collaborate with the reader to produce a satisfactory outcome.

It also reinforces the idea that the reader is omniscient and knows what he or she is saying and doing. If something that the reader later says does not tally with the client's beliefs or does not make sense, the client has been prepared to treat the apparent confusion as due to the client's own failure to adequately understand rather than to the psychic's lack of knowledge.

Some manuals further suggest ways to encourage this attitude in the client by having the reader state, as part of the preparation, something like, "During the course of the reading, impressions and images will come to me that will make no sense to me, but are very meaningful for your current or future situation. It will be up to you to make sense of these thoughts."

This setting of the stage recognizes what is now taken for granted in cognitive science—that the listener/reader reacts not to a message or test as such, but to the message as encoded. And this encoding must always be in terms of frameworks, concepts, expectations, and attitudes that the listener/reader brings to the situation. The role of the listener/reader in constructing the context and meaning of stories and literature is also recognized in semiotics and structuralism.[6] The concept of genre, for example, is considered to be a guide to the reader on how to interpret the text. The same set of words, when set out in the form of a poem is "read" differently than when set out in the form of a brief note. And it is recognized that no matter how concrete and specific a message appears to be, it still can bear a variety of "readings."

The client who is actively processing the message from the psychic as a meaningful account of the client's current situation and the client who is skeptically processing the same message as a set of Barnum propositions are experiencing two qualitatively different types of communication.

Another very important part of the preparation of the client is not directly under the control of the reader. This is the client's objectives in participating in the reading. Most clients have problems, concerns, worries, and feelings of inadequacy. They seek solace, advice, support, or simply a good listener. For some the problems are chronic and of long standing. For others, the difficulty is a current and acute crisis. Such clients obviously have a vested interest in extracting the greatest benefit possible from their exchange with the reader. They are going to encode and react to what the reader says much differently than the casual curiosity seeker might. Even in the latter case, such as happened with Mike, the client who is merely curious or even skeptical is often surprised and impressed by what takes place.

But, given the highly involved client and the proper setting of the stage, even the crudest psychological reading—as evidenced by the work on the Barnum effect—is almost sure to meet with "success." But the readers have further items in their bag of tricks.

PRELIMINARY OBSERVATIONS AND CATEGORIZATION

Even the manuals that supply the user with stock spiels to memorize and deliver suggest that some accommodations be made to take into consideration characteristics of the clients such as age, sex, socioeconomic status, and obvious signs of health. Sometimes such adjustments require minor alterations in statements indicating whether a problem was in the past, present, or future or adapting sexually and age-related references. In other cases, the manuals actually supply different stock spiels for different categories of clients: the young, unmarried female; the young, married female; the young, unmarried, male; the elderly woman; and so forth. Such adjustments based on preliminary observations and categorization of the client move the psychological reading towards the flexibility of the cold reading. The cold reader goes beyond this preliminary categorization and keeps refining and revising the reading to reach what is eventually a customized description.

All the manuals emphasize the surprising amount of specific information that one can pick up from a careful study of dress, physical characteristics, gestures and mannerisms, posture and attitudes, speech, jewelry, name, address, coat labels, and the like. The shrewd reader employs this information not only to make preliminary categorizations, but also to surprise the client with statements that the client believes could only have come from some occult source.

Constructing the Preliminary Script

The psychological reader, having set the stage and made some preliminary categorizations of the client, is ready to deliver the final version of the reading. This final version is a combination of facts based on the preliminary observations and a generalized or schematic script, which serves as a framework for ordering these facts. The psychological reading is based on the explicit assumption that there are universal features to all human lives as well as common characteristics and problems that face individuals in similar times and environments. Many manuals offer stock spiels for delivering a cradle-to-grave reading — with provisos for adjustments based upon sex, age, and other obvious characteristics of the client. Such a reading is based on the assumption that most of us live through the same major milestones: birth, childhood, school, work, marriage, children, and death. And during that life span we face more or less the same general problems such as career choice, love, sex, health, financial matters, and family.

The success of a book such as Gail Sheehy's *Passages* confirms what the psychological readers already know — that we share, despite our various lifestyles and unique histories, common problems and "predictable crises" at various ages such as the 20's, 30's, 40's, 50's, and beyond. The paperback edition says on its cover, "At last this is your story. You'll recognize yourself, your friends, and your lovers." It is just such recognition that the reader counts upon in his clients to make the reading succeed. Indeed, contemporary manuals strongly urge the psychic readers to study such books as *Passages* to better make their scripts realistic and convincing.

The use of a generalized, universal script around which to construct the reading contributes in many ways to its acceptability. Many of the ideas in structuralism and contemporary cognitive science testify to the value of such an underlying schema. Both the psychological and the cold reading are constructed on the basis that there are universal themes shared by all human lives. And this idea that diverse lives and unique histories are constructed from a limited set of constituent components motivates structuralist and semiotic approaches to literature, myths, and cultures.[6] In psychology, the idea that we both encode and remember our experiences in terms of underlying schematas was emphasized by Bartlett in his classic *Remembering*.[16] However, it is only recently that psychologists and cognitive scientists have begun seriously applying Bartlett's insights into how we encode, understand, and remember prose.[17-19]

The reader's generalized script, among other things, acts both as a memory probe and an organizational framework for the client. To encode and make sense of what the reader is saying, the client has to supply the flesh to the skeleton. He retrieves from memory incidents and examples to instantiate the more general things being described. Mike did not simply hear Barbara say something about a relationship; he heard her talking about his particular problem with his girlfriend. Furthermore, this particular problem was now placed within an organized setting created by the underlying script and other relevant incidents that it brought to mind. Anything that Barbara said that did not make sense in terms of this script or anything that Mike thought about that was inconsistent or contradictory to the overall unfolding story would later be forgotten or difficult to retrieve. The script not only provides organization and meaning to the client's experiences, but it also strongly guides how and what he or she will recall from the reading. Both laboratory research and what we know about actual psychic readings predict that the client will remember mainly those things the psychic said that were consistent with the overall script and will also remember them in terms of those concrete memories that the client brought to bear in order to make sense of the reading.

DELIVERING THE MESSAGE AND REVISING THE SCRIPT

The cold reader does not stop with the preliminary script. He delivers the message in the presence of the client and modifies it as it unfolds on the basis of reactions and inputs from the client. This is where the Clever Hans phenomenon works to supplement the Barnum effect. The cold reader uses Barnum type statements organized around the preliminary script as trial probes. These statements are modified, withdrawn, revised, or elaborated in terms of the reactions of the client. These reactions are typically nonverbal such as pupillary enlargement, eye movements, postural adjustments, facial changes, and the like. The reader employs these cues to quickly pin down those topics of most interest to the client and to gauge when he is or is not on the right track.

Often the client emits verbal responses as well. These vary from exclamations to questions and comments. The manuals encourage the reader to promote such verbal feedback. One technique for doing this, for example, is "fishing" — the psychic seemingly makes statements but phrases them as subtle questions. "I see two dark and tall men in you life — do you recognize them?" "I'm getting a vague image concerning a deed or some such financial document — does that make sense to you?" Once the client does begin responding with questions and comments, the reader takes pains to reward such behavior with careful and attentive listening. By listening to whatever the client says, the reader accomplishes many goals simultaneously. The client is convinced that the reader is sincerely interested. He or she probably has a strong desire for someone simply to listen. The attention encourages further such responses by the client. And, finally, the psychic is carefully storing this information in memory for later use in the reading.

When the client is not talking, the reader overwhelms him or her with a steady stream of patter. Part of the reason for this fluent outflow is to continually monitor the client's reactions and pick out topics and items that create the most reaction. Another reason is to prevent the client from realizing how much the client has already said. The successful reading, as the manuals make clear, is one in which the client ultimately is given what he or she considers a devastatingly accurate and penetrating personality analysis. The information for this analysis has come from both the verbal and nonverbal behavior of the client. The reader has merely fed back to the client information which the client has unwittingly supplied to the reader. The reading succeeds because the client is completely unaware that he or she has been the source of the description.

In effect the reader is like a ghost-writer who helps the client construct a coherent autobiography on the basis of information supplied by the client. In the typical psychic reading, both the psychic and the client falsely attribute the accuracy of the final reading to the occult sources to which the reader has access. In the cold reading, the reader is fully conscious that the client is the source of all the information and takes steps to maximize the client's contributions.

The impact and success of the psychic reading go beyond the illusion of accuracy. The client's acceptance is ultimately based on an emotional experience rather than a dispassionate assessment of the accuracy. Indeed, as the case of Mike and Barbara suggests, the attribution of uncanny accuracy to the reader probably stems from the emotional impact achieved during the reading. The attribution of accuracy is based upon information that the client has supplied to the reader. In a sense, the reading which the psychic later gives back to the client contains nothing "new." In another sense, however, the repacking of this material and putting it into a coherent order provides new insights to the client. In addition, the client is now looking at his or her memories and experiences from a new vantage point. This provides something similar to what Shklovsky attributes to the function of literacy devices. They serve not to represent familiar events, but rather to make them strange—to defamiliarize them.[6] In a way, the reading does for the client's self-concept what Poincaré claims that mathematical discoveries do for already familiar concepts—"they reveal to us unsuspected kinship between other facts, long known, but wrongly believed to be strangers to one another."[20]

It is thus conceivable, then, that despite the false attribution of secret knowledge to the reader, the client emerges from the reading with a new and more adapative model of his situation. He or she may have a new insight into the conflicts and problems that precipitated the consultation. And new alternatives for coping with the situation may have been opened up. Whether or not the new perspective and vision of the client's life and situation is ultimately beneficial, it is easy to understand how the client might experience the reading as revealing and rewarding.

Much more can be said about the psychic reading. But I hope I have said enough to convince you that it is a rich and challenging opportunity for studying the ways signs, witting and unwitting, combine to create illusions of communication in dyadic settings. Both the Barnum effect and the Clever Hans phenomenon combine to induce the overwhelming conviction that the psychic is the source rather than a mirror of an accurate appraisal of the client and his or her circumstances. In some cases it might be revealing to look upon the reader as a catalyst that aids the client in the construction of a coherent and helpful self-description. In general, however, the psychic reading, like the Clever Hans case, is but one of many illustrations of the human propensity to project meaning into situations even when the situations, themselves, have no meaning.

REFERENCES

1. PFUNGST, O. 1965. Clever Hans. Holt, Rinehart & Winston, New York, N.Y.
2. BROMLEY, D. B. 1977. Personality Description in Ordinary Language. Wiley & Sons, New York, N.Y.
3. SYNDER, C. R., R. J. SHENKEL & C. R. LOWERY. 1977. Acceptance of personality interpretations: The "Barnum Effect" and beyond. J. Consult. Clin. Psychol. 45: 104–114.
4. FORER, B. R. 1949. The fallacy of personal validation: A classroom demonstration of gullibility. J. Abnormal Soc. Psychol. 44: 118–123.
5. SUNDBERG, N. D. 1955. The acceptability of "fake" versus "bona fide" personality test interpretations. J. Abnormal Soc. Psychol. 50: 145–147.
6. HAWKES, T. 1977. Structuralism and Semiotics. University of California Press, Berkeley, Calif.

7. ANONYMOUS. 1971. Pages from a Medium's Notebook. Micky Hades, Calgary, Alberta, Canada.
8. BOARDE, C. L. 1947. Mainly Mental: Volume I: Billet Reading. Globe Service, New York, N.Y.
9. CORINDA. 1968. Thirteen Steps to Mentalism. Louis Tannen, New York, N.Y.
10. HESTER, R. & W. HUDSON. 1977. Psychic Character Analysis: The Technique of Cold Reading Updated. Magic Media Ltd., Baltimore, Md.
11. MAGNUSON, W. G. 1935. The Twentieth Century Mindreading Act or the Modern Spiritualist Medium's Act. Albino, Chicago, Ill.
12. NELSON, R. A. 1951. The Art of Cold Reading. Nelson Enterprises, Columbus, Ohio.
13. RUTHCHILD, M. 1978. Cashing in on the Psychic. Lee Jacobs Productions, Pomeroy, Ohio.
14. HYMAN, R. 1977. The Zetetic 1 (Spring/Summer): 18-37.
15. SHEEHY, G. 1977. Passages: Predictable Crises of Adult Life. Bantam Books, New York, N.Y.
16. BARTLETT, F. C. 1932. Remembering. Cambridge Univ. Press, London, England.
17. BLACK, J. B. & G. H. BOWER. 1980. Story understanding as problem-solving. Poetics. In press.
18. BOWER, G. H. 1976. Experiments on story understanding and recall. Q. J. Exp. Psychol. 28: 511-534.
19. BOWER, G. H. 1978. Experiments on story comprehension and recall. Discourse Proc. 1: 211-231.
20. POINCARÉ, H. 1955. Mathematical creation. In The Creative Process—A Symposium. B. Ghiselin, Ed. pp. 33-42. Mentor, New York, N.Y.

Pavlov's Mice, Pfungst's Horse, and Pygmalion's PONS: Some Models for the Study of Interpersonal Expectancy Effects*

ROBERT ROSENTHAL

Department of Psychology and Social Relations
Harvard University
Cambridge, Massachusetts 02138

PAVLOV was very much interested in the concept of the inheritance of acquired characteristics; so much so, indeed, that he conducted experiments to show the operation of this phenomenon. For a time it seemed that Lamarck was correct. But then, in 1929, Pavlov discovered that it was not so much a matter of improved learning by his mice but a matter of improved teaching by his research assistants. The research assistants, knowing which mice were likely to become "smarter," may have communicated their expectations unwittingly to these mice, creating what Merton has called a "self-fulfilling prophecy."[1-3]

Von Osten's horse, the Clever Hans we are gathered to celebrate, had a similar and even earlier lesson to teach us. Questioners of the horse communicated to Hans the information he needed to tap out the correct answer with his hoof, thus making him, like Pavlov's mice, the victim, or the beneficiary, of a self-fulfilling prophecy or interpersonal expectancy effect.[3-6]

Pavlov's mice and von Osten's horse have many lessons to teach us, but the lessons that become our personal favorites tell as much about ourselves as they do about the mice and the horse. The latter are only inkblots for the behavioral scientist. These famous cases in the history of science teach some of us that scientists make mistakes, are taken in by design, or by accident. They teach others of us the surprising subtlety of cues that can serve to affect the performance of those organisms with whom we come into contact. They teach still others of us that people may be prone to the operation of self-fulfilling prophecies.

Building on the work of Merton, Pavlov, and Pfungst, my own work

* This work was supported by the National Science Foundation.

has emphasized the social importance of the self-fulfilling nature of many interpersonal expectations.[1,3,4] There are now many experiments showing that when experimenters in behavioral research are led to expect certain responses from their research subjects, they are substantially more likely to obtain the results they have been led to expect.[3] In addition, there are now many experiments showing that when teachers are led to expect better intellectual performance from their pupils, they are substantially more likely to obtain such improved performance.[3,7,8]

A recent review of the literature of studies of interpersonal expectancy effects summarized quantitatively the results of 345 such studies.[9] The estimated average size of the effect of experimenters' expectations on the results of their research or the effect of teachers' expectations on the intellectual performance of their pupils was very substantial: 0.70 of a standard deviation—a result that was very close in magnitude to the average size of the effect of psychotherapy found by Glass in his review of several hundred studies of psychotherapy.[10] This effect size is equivalent to about 10 IQ points in terms of most IQ tests (e.g., mean IQ = 100; SD = 15).

Although these results have important substantive implications (e.g., for educational theory and practice) and important methodological implications (e.g., for how we conduct experiments in behavioral research), they also have important implications for our understanding of the everyday operation of various processes of nonverbal communication. A number of the studies included in the 345 summarized by Rosenthal and Rubin suggested that experimenters' expectations for their subjects' behavior and teachers' expectations for their pupils' performance were communicated nonverbally.[9] We turn first to a consideration of some of this evidence.

EVIDENCE FOR THE IMPORTANCE OF NONVERBAL CUES

Many of the experiments on the operation of experimenter expectancy effects employed a standard photo rating task in which subjects were asked by their experimenter to rate the degree of success or failure that appeared to be reflected in the 10 or more photographs of faces shown to the subject by the experimenter. One of the earliest and strongest hints that nonverbal cues were probably involved in the mediation of these expectancy effects came from the fact that all experimenters read the same standard instructions to their subjects; despite this precise standardization of the *verbal* content of experimenters' communications, subjects responded in accordance with the expectations we had experimentally induced in the minds of the experimenters. If the words did not differ as a function of experimenters' expectations, then the *non*verbal cues must be critical, we felt.[3,11]

When screens were placed between experimenters and their research subjects so that visual access was denied the participants, the size of the effect of experimenter expectations was approximately cut in half. This suggested that although expectancy effects could be mediated by tone of voice alone, visual cues could also contribute to their operation.[12]

A two-stage study conducted by Adair and Epstein further supported the idea that expectations might be communicated by an experimenter's tone of voice.[13] In stage I, experimenters were led to expect either high or low ratings of success from their research subjects. Just as in many other studies of this kind, experimenters did in fact obtain results significantly in the direction of their expectations. In stage II, there were no experimenters at all. Instead, the tape-recorded voices of the experimenters instructing their subjects in stage I were played for new groups of subjects. Thus, stage II subjects were given their instructions not by "real" experimenters but by the voices of experimenters who had been given different expectations for how their subjects should rate the photos of faces. Results of stage II showed that the effects of experimenters' expectations were just as effectively communicated by their tape-recorded voices as they had been when experimenters had interacted directly with their subjects in stage I. Audio cues, then, were sufficient to communicate to subjects the expectations of their experimenters.

The results of the studies by Rosenthal and Fode and by Adair and Epstein were supported further by the work of Zoble and Lehman who found substantial effects of experimenter expectations in a tone-length discrimination task even when subjects were restricted to auditory cues alone.[12-14] Zoble and Lehman went on to show, however, that visual cues alone could also mediate substantial effects of experimenter expectations in this task. These results taken together, suggest that when subjects are deprived of either visual or auditory information, they focus more attention on the channel that is available to them. This greater attention and perhaps greater effort may enable subjects to extract more information from the single channel than they could, or would, from that same channel if it were only one part of a two-channel information input system. Although we cannot give the details here, it should be noted that there are other studies also showing the importance of nonverbal cues in the mediation of experimenter expectancy effects,[15-17] and some of these have been summarized elsewhere.[8,11]

Many of the 345 studies of interpersonal expectancy effects summarized by Rosenthal and Rubin[9] were studies of the effects of teacher expectations on the intellectual performance of their pupils. Enough studies were available that examined the actual behavior of the teachers toward the pupils for whom they held favorable or unfavorable expectations to warrant the postulation of a preliminary four-factor "theory" of the communication of expectancy effects in teacher–pupil interactions.[3,8] This "theory" suggests that teachers (and perhaps clinicians, supervisors, and employers) who have been led to expect superior performance from some of their pupils (clients, trainees, or employees) appear to treat these "special" persons differently than they treat the remaining less-special persons in the four ways shown in TABLE 1.

Careful reading of the several dozen studies upon which the four-factor theory is based has led to the development of a preliminary model that may be useful in pointing out areas most in need of attention before we can understand (a) the variables serving to moderate or alter the size of interpersonal expectancy effects and (b) the variables serving to mediate the operation of interpersonal expectancy effects, including the role of various channels of nonverbal communication.

A PRELIMINARY MODEL FOR THE STUDY OF INTERPERSONAL EXPECTANCY EFFECTS

The model utilizes an underlying dimension of time and therefore looks path analytic in form. The model is not dependent upon the set of assumptions underlying the employment of the path analytic model, however, and does not imply any particular data analytic method (e.g., regression analysis). The model makes explicit the classes of variables that must be examined in relation to one another before we can achieve any systematic understanding of the social psychology of interpersonal

TABLE 1

SUMMARY OF A FOUR-FACTOR "THEORY" OF THE MEDIATION OF TEACHER EXPECTANCY EFFECTS

Factor	Summary of the Evidence
Climate	Teachers appear to create a warmer socio-emotional climate for their "special" students. This warmth appears to be at least partially communicated by nonverbal cues.
Feedback	Teachers appear to give their "special" students more differentiated feedback, both verbal and nonverbal, as to how these students have been performing.
Input	Teachers appear to teach more material and more difficult material to their "special" students.
Output	Teachers appear to give their "special" students greater opportunities for responding. These opportunities are offered both verbally and nonverbally (e.g., giving a student more time in which to answer a teacher's question).

expectation effects. The basic elements of the model include (a) distal and (b) proximal independent variables, (c) mediating variables, and (d) proximal and (e) distal dependent variables.

Distal independent variables refer to such more stable attributes of the expecter (e.g., teacher, therapist, experimenter) or expectee (e.g., pupil, patient, subject) as gender, status, ethnicity, ability, and personality. (It should be emphasized that distal independent variables refer to stable attributes of the expectee as well as of the expecter. That increases the power of the model by allowing us to make use of expectee attributes as moderating variables.) The proximal independent variable in this model generally refers to the variable of interpersonal expectation—especially expectations that have been varied experimentally rather than those that have been allowed to vary naturally. When expectations are merely measured rather than varied experimentally, a correlation between distal and proximal variables is introduced (e.g., teachers usually expect superior performance from brighter students); this correlation makes it virtually impossible to disentangle the effects of interpersonal expectations from the effects of attributes of the expectee, so that the effects of interpersonal expectations per se become virtually unassessable.

Mediating variables refer to the processes by which the expectation of the expecter is communicated to the expectee. These, then, are like the process variables of the psychotherapy research literature, and our focus is on the behavior of the expecter during interaction with the expectee. By constraining the nature of the verbal communication permitted between expecter and expectee, many studies (as discussed earlier) have shown that these mediating variables must, to a great extent, be nonverbal in nature.[3]

Proximal dependent variables refer to the behavior of the expectee after interaction with the expecter has occurred. A significant relationship between these variables and the experimentally varied proximal independent variables is what we mean by an interpersonal expectancy effect. We should note that the behavior of the expectee (D) including the nonverbal behavior may have important feedback effects on the behavior of the expecter (C) and the expectation of the expecter (B). Distal dependent variables refer to longer term outcome variables such as those obtained in follow-up studies (e.g., the one-year follow-up testing in the Pygmalion research by Rosenthal & Jacobson).[7] We can present the model diagrammatically:

The 10 arrows of the model summarize some of the types of relationships that are to be examined before any claim to a thorough understanding of interpersonal expectancy effects can reasonably be made. As will be shown, each of the arrows is usually of social psychological significance with the exception of arrow AB, which is often of only

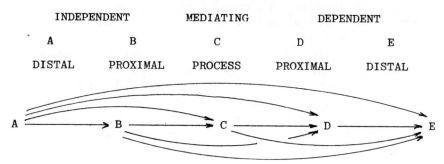

FIGURE 1. Preliminary model for the study of interpersonal expectancy effects.

methodological significance. An overview of the meaning of the 10 arrows follows.

AB. These relationships are often large in studies not manipulating interpersonal expectations. Thus, in studies in which teachers are asked to state their expectancies for pupils' intellectual performance,[18] high correlations between teacher expectations (B) and pupil IQ (A) are inevitable. These high correlations make it difficult to conclude that it is the teacher's expectancy rather than the pupil's IQ that is "responsible" for subsequent pupil performance. Covariance analysis and related procedures can be useful here, however, and have been creatively employed (e.g., by Gerard & Miller).[19] When expectancies are varied experimentally, the expected value of the AB correlations is zero since neither the attributes of the expecter nor of the expectee should be correlated with the randomly assigned experimental conditions. A nonzero correlation under these circumstances serves as a methodological warning of a "failure" of the randomization procedure.

AC. These relationships describe the "effects" on the expecter's interactional behavior of various characteristics of the expecter, expectee, or both. The joint "effects" of therapist gender and patient gender on the therapist's subsequent behavior toward the patient serve as one illustration. The ethnicity of pupils as a determinant of teacher behavior toward them serves as another illustration.

AD. These relationships describe the effects of (usually) the expecter's characteristics on the subsequent behavior of the expectee. A relationship between teacher attitude and pupil learning would be an illustration.

AE. These relationships are like those of AD except that behavior E occurs at some time in the future relative to D.

BC. These relationships describe the effects on the expecter's behavior toward the expectee of the expectation that has (usually) been induced experimentally in the expecter. An example is Rosenthal's four-factor "theory" of the mediation of teacher expectancy effects that summarizes several dozen studies of BC relationships.[3,8]

BD. These relationships define the phenomenon of interpersonal expectancy effects when the expectancy has been experimentally manipulated, and they have been summarized recently.[3,9] (In addition, these relationships may be self-moderating over time as when expectee behavior (D) affects the subsequent expectation of the expecter (B).)

BE. These relationships define the longer term effects of interpersonal expectations. There are very few studies of this type available.[7]

CD. These relationships provide suggestive clues as to the type of expecter behavior that *may* have effects on expectee behavior. It is often assumed that CD relationships tell us how teachers, for example, should behave in order to have certain desirable effects on pupil behavior. Except in those very rare cases where mediating variables are manipulated experimentally, such assumptions are unwarranted. Finding certain teacher behaviors to correlate with certain types of pupil performance does not mean that teachers changing their behavior to emulate the behavior of the more successful teachers will show the same success with their pupils. (These relationships may also be self-moderating over time as when expectee behavior (D) affects the subsequent behavior of the expecter (C).)

CE. These relationships are like the CD relationships except that the outcome variables are of the follow-up variety.

DE. These relationships may merely assess the stability of the behavior of the expectee, as when the measures employed for D and E are identical. When these measures are not very similar, the DE relationship may yield an index of predictive validity that is of substantive interest.

Of the 10 arrows of the model, three are clearly most important and should ideally be included in most studies of interpersonal expectancy effects: BC, BD, and CD.

The BD relationship, if significant, tells us that interpersonal expectancy effects occurred. The BC relationship, if significant, tells us that the (usually experimental induction of) expectancy was a determinant of a particular type of behavior of the expecter toward the expectee. The CD relationship tells us that certain behaviors of the expecter are associated with changes in the behavior of the expectee. If the BC results show an increase in behavior X due to the induced expectations, and the CD results show that increases in behavior X by the expecter are associated with changes in the performance of the expectee, behavior X becomes implicated as a candidate to be regarded as a mediating

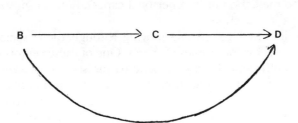

FIGURE 2. Most essential features of a model for the study of interpersonal expectancy effects.

variable. Among the most outstanding recent investigations of these relationships are the studies of Brophy and Good[18] (AC or BC), Jones and Cooper[20] (CD), Snyder, Tanke, and Berscheid[21] (BC, BD), and Word, Zanna, and Cooper[22] (BC, CD).

MEASURING SENSITIVITY TO NONVERBAL CUES

As discussed earlier, much of the research on interpersonal expectancies has suggested that mediation of these expectancies depends to some important degree on various processes of nonverbal communication. Moreover, there appear to be important differences among experimenters, teachers, and people generally, in the clarity of their communication through different channels of nonverbal communication. In addition, there appear to be important differences among research subjects, pupils, and people generally, in their sensitivity to nonverbal communications transmitted through different nonverbal channels. If we knew a great deal more about differential sending and receiving abilities we might be in a much better position to address the general question of what kind of person (in terms of sending abilities) can most effectively influence covertly what kind of other person (in terms of receiving abilities). Thus, for example, if those teachers who best communicate their expectations for children's intellectual performance in the auditory channel were assigned children whose best channels of reception were also auditory, we would predict greater effects of teacher expectation than we would if those same teachers were assigned children less sensitive to auditory nonverbal communications.[23]

Ultimately, then, what we would want would be a series of accurate measurements for each person, describing his or her relative ability to send and to receive in each of a variety of channels of nonverbal communication. It seems reasonable to suppose that if we had this information for two or more people we would be better able to predict the outcome of their interaction regardless of whether the focus of the analysis were on the mediation of interpersonal expectations or on some other interpersonal transaction.

Our model envisages people moving through their "social spaces" carrying two vectors or profiles of scores. One of these vectors describes the person's differential clarity in sending messages over various channels of nonverbal communication. The other vector describes the person's differential sensitivity in receiving messages in various channels of nonverbal communication. TABLE 2 illustrates this model.

In general, the better persons A and B are as senders and as receivers, the more effective will be the communication between persons A and B. However, two dyads may have identical overall levels of skill in sending and receiving and may yet differ in the effectiveness of their nonverbal communication because in one dyad the sender's skill "fits" the receiver's skill better than is the case in the other dyad. One way to assess the "fit" of skills in dyad members is to correlate the sending skills of A with the receiving skills of B and to correlate the sending skills of B with the receiving skills of A. Higher correlations reflect a greater potential for more accurate communication between the dyad members since the receiver is then better at receiving the channels that are the more accurately encoded channels of the sender.

The mean (arithmetic, geometric, or harmonic) of the two correlations (of A's sending with B's receiving and of B's sending with A's receiving) reflects how well the dyad members "understand" each other's communications. That mean correlation need not reflect how well the dyad members like each other, however, only that A and B should more

TABLE 2

MODEL OF VECTORS OF SKILL IN SENDING AND RECEIVING IN
VARIOUS CHANNELS OF NONVERBAL COMMUNICATION

Channels of Nonverbal Communication (e.g.)	Vectors of Person A		Vectors of Person B	
	Sending	Receiving	Sending	Receiving
1. Face				
2. Body				
3. Tone				
4. etc.				
. . .				
. . .				
. . .				
k. . . .				

quickly understand each other's intended and unintended messages, including how they feel about one another.

As a start toward the goal of more completely specifying accuracy of sending and receiving nonverbal cues in dyadic interaction, an instrument was developed that was designed to measure sensitivity to various channels of nonverbal communication: The Profile of Nonverbal Sensitivity, or PONS. The details of the test, its development, and the results of the extensive research that has been conducted employing the PONS are given elsewhere.[24] Here we can give only a brief overview.

The Profile of Nonverbal Sensitivity. The PONS is a 45-minute 16 mm sound film comprised of 220 two-second auditory and/or visual segments. The printed answer sheet employed by the viewer has 220 pairs of descriptions of real-life situations. From each pair of descriptions the viewer circles the description that best fits the segment that has just been seen and/or heard. Twenty scenarios are each represented in the eleven nonverbal "channels" shown in TABLE 3.

TABLE 3

THE 11 CHANNELS OF THE PONS

Channels	Descriptions
1. Face	Only the sender's face is on screen.
2. Body	Only the sender's body from the lower neck to the knees is on screen.
3. Face + Body	Both the sender's face and body are on the screen.
4. Randomized Spliced Voice	Only the voice can be heard, the sound track having been scrambled randomly to eliminate the content.
5. Content-Filtered Voice	Only the voice can be heard, the sound track having been treated electronically to filter out high frequencies so as to eliminate the content.
6. Face + Randomized Spliced Voice	Combined channels 1 + 4.
7. Face + Content-Filtered Voice	Combined channels 1 + 5.
8. Body + Randomized Spliced Voice	Combined channels 2 + 4.
9. Body + Content-Filtered Voice	Combined channels 2 + 5.
10. Face + Body + Randomized Spliced Voice	Combined channels 3 + 4.
11. Face + Body + Content-Filtered Voice	Combined channels 3 + 5.

TABLE 4

20 SCENARIOS OF THE PONS ARRAYED IN 4 QUADRANTS

Quadrant	Scenario
Positive-Dominant	1. Expressing motherly love 2. Talking to a lost child 3. Admiring nature 4. Talking about one's wedding 5. Leaving on a trip
Positive-Submissive	6. Expressing deep affection 7. Trying to seduce someone 8. Helping a customer 9. Expressing gratitude 10. Ordering food in a restaurant
Negative-Dominant	11. Nagging a child 12. Expressing jealous anger 13. Criticizing someone for being late 14. Expressing strong dislike 15. Threatening someone
Negative-Submissive	16. Talking about the death of a friend 17. Asking forgiveness 18. Returning faulty item to a store 19. Saying a prayer 20. Talking about one's divorce

The twenty scenarios were selected such that half were judged to communicate positive affects and half were judged to communicate negative affects. Orthogonally to this dimension, half the scenarios were judged to reflect dominant affects and half were judged to communicate submissive affects. Thus, there are five scenarios in each of the following four quadrants: Positive-Dominant, Positive-Submissive, Negative-Dominant, and Negative-Submissive. The scenarios in each quadrant are shown in TABLE 4.

The eleven channels and four quadrants can be conceptualized in terms of an analysis of variance model so that each person assessed provides five replications (scenarios) in each cell of a $2 \times 2 \times 11$ design (positive–negative $\times$ dominant–submissive $\times$ channels). An even more useful model is based on the assumption of a hypothetical 12th channel called "no video—no audio." The expected accuracy rate for this channel if subjects simply guessed is 50% but adding it to the model permits a more powerful and more compact analytic model: $2 \times 2 \times 2 \times 2 \times 3$ or

(A) Positive vs Negative
(B) Dominant vs Submissive
(C) Face shown vs Face not shown
(D) Body shown vs Body not shown
(E) No Audio vs Randomized Spliced vs Content-Filtered.

For each person or for any homogeneous group of persons, this model permits an evaluation of accuracy in nonverbal communication as a function of these five orthogonal factors and the 2-, 3-, 4-, and 5-way interactions among them. In addition, individuals and groups can be compared with one another on the relative importance to each person or to each group of all five factors taken singly or in interaction.

For some purposes a sixth factor of order or learning is added. Thus, within each combination of channel and scene type there are five scenes that can be arranged for analysis into the order in which they are shown in the PONS test. This order or learning factor with its five levels is fully crossed with the five factors listed above. The one-df contrast[24] for linear trend is an overall index of improvement over time in PONS performance. Individuals and groups can, therefore, be compared for their degree of learning as well as their level of performance. In addition, the interaction of the one-df learning contrast with other one-df contrasts provides interesting information on such questions as which channels show greater learning, which content quadrants show greater learning, and various combinations of such questions.

An Overview of Some Results

The reliability of the PONS is quite adequate ranging from 0.86 to 0.92 for internal consistency with a median retest reliability of 0.69. The voice, body, and face channels of the PONS contribute to accuracy in judging the scenes in the approximate ratios of 1:2:4, respectively. The factor analysis of the 11 channels of the PONS yielded four factors essentially equivalent to:

(1) The six channels showing the face (face present).
(2) The three channels showing only the body (body only).
(3) The randomized spliced channel taken alone (RS).
(4) The content-filtered channel taken alone (CF).

For 133 samples comprised of 2,615 subjects, sex differences favored females very consistently. Females were especially superior to males at judging cues of negative affect. In general, younger children were less accurate than older children and young adults at decoding nonverbal cues. However, younger samples showed a *relative* advantage at judging audio as opposed to video cues, a result suggesting that the ability to read vocal nonverbal cues may be developmentally prior to the ability to read visual nonverbal cues.

Some 2,000 subjects from 20 nations took the PONS and it was found that those nations performed best that were more similar linguistically to the United States and more similar in terms of general modernity (e.g., steel consumption) and development of communications (e.g., television, radio, telephone) to United States culture.

PONS performance was not highly correlated with intellectual ability though it did tend to be correlated with cognitive complexity. Further, people scoring high on the PONS tended to be better adjusted, more interpersonally democratic and encouraging, less dogmatic, more extraverted, more likely to volunteer for behavioral research, more popular, and more interpersonally sensitive as judged by acquaintances, clients, spouses, or supervisors. These results, based on dozens of studies, contribute strongly to the construct validity of the PONS.

Many special groups have been tested with the PONS and of these, the best have been actors, students of nonverbal communication, and students of visual arts. Clinical psychologists, psychiatrists, and other clinicians scored no higher than college students but clinicians rated as more effective by their supervisors scored significantly higher than did clinicians rated as less effective by their supervisors. Samples of psychiatric patients and alcoholic patients tended to score appreciably below the level of the norm group of high school students. Blind students tested on audio portions of the PONS performed about the same as the sighted comparison groups. Deaf students (age 10–15) tested on video portions of the PONS, however, scored substantially below the level of the comparison groups of hearing students. Among these students, those whose hearing was less impaired performed better than did those whose hearing was more impaired. Interestingly, skill at lip-reading was not related to PONS performance. When we tested deaf college students, we found *no* significant difference between their PONS performance and that of the hearing comparison groups.

Length of Communication Exposure

A 40-item brief exposure form of the PONS was developed to permit us to study the effects of length of exposure on accuracy in decoding nonverbal cues from the face and body. The 20 face-only and the 20 body-only scenes from the full PONS were each subdivided into four groups of scenes varying in length of exposure. Whereas in the full PONS each film clip is 2 seconds long (48 frames of film), in this test the four lengths of film clip are 1/24th, 3/24ths, 9/24ths, and 27/24ths second, corresponding to 1, 3, 9, and 27 frames of film. Thus each presentation of the face or the body is substantially shorter than in the full PONS test.

The results of 9 studies of high school and college students and U.S. adults ($N = 506$) were quite consistent in showing accuracy very much greater than chance (and large in magnitude) at even the 1 frame length of exposure. Accuracy showed a dramatic increase in going from 1 to 3 frames, but no further increase in going from 3 frames to 9 frames or to 27 frames. Accuracy rates for 1-, 3-, 9-, and 27-frame exposures were 56%, 74%, 73%, and 74% respectively. The very dramatic gain in accuracy in going from 1 to 3 frames may have been due to the introduction of motion in the longer exposure or simply to the longer visual access to the stimulus materials or perhaps both.

Decreasing exposure length to 9 or fewer frames from 27 frames actually increased accuracy for body-only but greatly decreased accuracy for face-only. Perhaps both cues are rapidly processed in high-speed exposures in an intuitive, global, nonanalytic manner with additional small increases in exposure length serving more to confuse than to facilitate decoding.

Subjects' self-ratings of their relationships with others tended to be negatively correlated with accuracy at very brief exposures. Perhaps those most accurate at the most abbreviated displays have less satisfactory interpersonal relationships because they are able to decode cues they are not intended to decode. That is, perhaps people who are especially accurate at very brief exposures "know too much" about others to be socially acceptable.

COURTESY IN NONVERBAL COMMUNICATION

Among the more recent studies of sensitivity to nonverbal communication are those suggesting that there are interesting sex differences in the degree of courtesy shown in nonverbal behavior.[25,26] The evidence suggests that, although women are superior to men in decoding nonverbal cues in general, their use of nonverbal information from others is more courteous and more interpersonally accommodating than is that of men. Women, for example, are less likely to eavesdrop on the nonverbal cues of other people. That is, they are more likely to use the nonverbal cues of others that are under greater control of the sender, i.e., are less leaky.[26]

Another series of studies showed that women rated the behavior of others as less tense, less ambivalent, and less channel-discrepant, results that fit well with the pattern of seeing only what it is polite to see. This series also showed that women were markedly superior to men in the decoding of clear, nondeceptive behavior. However, when deception cues were being emitted, women were substantially less accurate in their decoding and were more likely to interpret these cues as the deceiver wanted them to be interpreted rather than as the deceiver really felt. There was also suggestive evidence that this politeness mechanism, by which women showed decreased accuracy for more ambivalent and especially for more deceptive communications, operated relatively more strongly among women who were personally and socially more vulnerable. This profile of vulnerability on an individual level has been shown in research by Hall to be paralleled in an intriguing way on a cross-cultural level: sex differences in accommodativeness are more pronounced in those countries in which women are less liberated and more oppressed.[27] Finally, this series of studies suggested not only that women were less likely to eavesdrop on leaky nonverbal channels, but also that they were more likely to encode more honestly and to be more accurately decoded by others.

Taking together the results of our several series of studies, it seems reasonable to conclude that women are more polite in the nonverbal aspects of their social interactions than are men. They are more guarded in reading those cues that senders may be trying to hide but more open in the expression of their own affective states. Further, there are indications that women who are less accommodating in these nonverbal ways are judged by others to have less successful interpersonal outcomes. Perhaps women in our culture have been taught that there may be social hazards to knowing too much about other people's feelings. This pattern of accommodativeness by women is consistent with the standards of politeness and social smoothing-over that are part of the traditional sex role ascribed to women in our culture, a sex role that is only now beginning to change.

REFERENCES

1. MERTON, R. K. 1948. The self-fulfilling prophecy. Antioch Rev. 8: 193–210.
2. GRUENBERG, B. C. 1929. The Story of Evolution. Van Nostrand, Princeton, N.J.
3. ROSENTHAL, R. 1966. Experimenter Effects in Behavioral Research. Appleton-Century-Crofts, New York, N.Y. [Enlarged edition: 1976. Irvington Publishers, New York, N.Y.]
4. PFUNGST, O. 1911. Clever Hans. Holt, New York, N.Y.
5. ROSENTHAL, R. 1964. Experimenter outcome-orientation and the results of the psychological experiment. Psychol. Bull. 61: 405–412.
6. ROSENTHAL, R. 1965. Clever Hans: A case study of scientific method. In Clever Hans. O. Pfungst, author. Introduction. Holt, Rinehart and Winston, New York, N.Y.

7. ROSENTHAL, R. & L. JACOBSON. 1968. Pygmalion in the Classroom. Holt, Rinehart and Winston, New York, N.Y.
8. ROSENTHAL, R. 1973. On the Social Psychology of the Self-Fulfilling Prophecy: Further Evidence for Pygmalion Effects and Their Mediating Mechanisms. Module 53. MSS Modular Publication, New York. N.Y.
9. ROSENTHAL, R. & D. B. RUBIN. 1978. Interpersonal expectancy effects: The first 345 studies. Behav. Brain Sci. 3: 377–386, 410–415.
10. GLASS, G. V. 1976. Primary, secondary, and meta-analysis of research. Paper presented at a meeting of the American Educational Research Association, San Francisco, Calif.
11. ROSENTHAL, R. 1969. Interpersonal expectations: Effects of the experimenter's hypothesis. In Artifact in Behavioral Research. R. Rosenthal & R. L. Rosnow, Eds. Academic Press, New York, N.Y.
12. ROSENTHAL, R. & K. L. FODE. 1963. Three experiments in experimenter bias. Psychol. Rep. 12: 491–511.
13. ADAIR, J. G. & J. S. EPSTEIN. 1968. Verbal cues in the mediation of experimenter bias. Psychol. Rep. 22: 1045–1053.
14. ZOBLE, E. J. & R. S. LEHMAN. 1969. Interaction of subject and experimenter expectancy effects in a tone length discrimination task. Behav. Sci. 14: 357–363.
15. DUNCAN, S., M. J. ROSENBERG & J. FINKELSTEIN. 1969. The paralanguage of experimenter bias. Sociometry 32: 207–219.
16. DUNCAN, S. & R. ROSENTHAL. 1968. Vocal emphasis in experimenters' instruction reading as unintended determinant of subjects' responses. Lang. Speech. 11: 20–26.
17. SCHERER, K. R., R. ROSENTHAL & J. KOIVUMAKI. 1972. Mediating interpersonal expectancies via vocal cues: Differential speech intensity as a means of social influence. Europ. J. Soc. Psychol. 2: 163–175.
18. BROPHY, J. E. & T. L. GOOD. 1974. Teacher-Student Relationships. Holt, Rinehart & Winston, New York, N.Y.
19. GERARD, H. B. & N. MILLER. 1971. Desegregation: A Longitudinal Study. Unpublished manuscript. University of Southern California, Los Angeles, Calif.
20. JONES, R. A. & J. COOPER. 1971. Mediation of experimenter effects. J. Pers. Soc. Psychol. 20: 70–74.
21. SNYDER, M., E. D. TANKE & E. BERSCHEID. 1977. Social perception and interpersonal behavior: On the self-fulfilling nature of social stereotypes. J. Pers. Soc. Psychol. 35: 656–666.
22. WORD, C. O., M. P. ZANNA & J. COOPER. 1974. The nonverbal mediation of self-fulfilling prophecies in interracial interaction. J. Exp. Soc. Psychol. 10: 109–120.
23. CONN, L. K., C. N. EDWARDS, R. ROSENTHAL & D. P. CROWNE. 1968. Perception of emotion and response to teachers' expectancy by elementary school children. Psychol. Rep. 22: 27–34.
24. ROSENTHAL, R., J. A. HALL, M. R. DIMATTEO, P. L. ROGERS & D. ARCHER. 1979. Sensitivity to Nonverbal Communication. Johns Hopkins University Press, Baltimore, Md.

25. ROSENTHAL, R. & B. M. DePAULO. 1979. Sex differences in eavesdropping on nonverbal cues. J. Pers. Soc. Psychol. **37**: 273–285.
26. ROSENTHAL, R. & B. M. DePAULO. 1979. Sex differences in accommodation in nonverbal communication. *In* Skill in Nonverbal Communication. R. Rosenthal, Ed. Oelgeschlager, Gunn & Hain, Cambridge, Mass.
27. HALL, J. A. 1979. A cross-national study of gender differences in nonverbal sensitivity. Unpublished manuscript. Johns Hopkins University, Baltimore, Md.

The Ultimate Enigma of "Clever Hans": The Union of Nature and Culture

THOMAS A. SEBEOK*

Research Center for Language and Semiotic Studies
Indiana University
Bloomington, Indiana 47405

> Tell me, where is fancy bred,
> or in the heart or in the head?
> How begot, how nourished?
> Reply, reply.
>
> — *The Merchant of Venice*

THE SCOPE of application of the epithet "Clever Hans" turned out to be very elastic indeed. Almost from the birth of the appellation, especially after it was reinforced by the initial appearance of Pfungst's 1907 study, in German, there sprung up unregenerate skeptics, who simply refused to accept the findings of Stumpf's commission of inquiry that was charged with the examination of Mr. von Osten's eponymous horse, such as the obsessive jeweler, Krall,[1] and a 1911 Belgian Nobel Prize winner, the symbolist playwright, Maeterlinck.[2] Nearly two decades later, the late Joseph Banks Rhine, a self-styled psychic adventurer, instead of employing Ockham's razor to shave away pseudo-explanatory entities, dubbed the mare Lady Wonder "the greatest thing since radio," ascribing the filly's alleged powers to telepathy;[3,4] (later, this act was proved to be an instance of a well-known mentalist trick, called, in the business, "pencil reading"[5]). In the late 1970's, one is still distressed to read assertions, by a self-proclaimed "centaur," that horses communicate by means "more subtle than man himself: extrasensory perception and telepathy."[6]

In contemporay psychology and some adjacent disciplines, the narrow, literal, and simplistic view of the Clever Hans episode prevails, to wit, that what was involved in that uproar constituted a perhaps amusing aberration in the history of science, featuring a canny stallion that operant-conditioned the learned world of Berlin and beyond, and, as

* Address for correspondence during the 1980–81 academic year: National Humanities Center, Research Triangle Park, N.C. 27709.

199

countless other smart horses before and since (as well as pigs, dogs, a goat or two, and many geese and other birds), succeeded in duping the gullible public at large. Those who cling to this ingenuous definition of the happening are unable to grasp the cross-specific, indeed, universal pervasiveness of the effect, the general implications of which were, however, perfectly understood by the reviewer of the first American edition of the classic account, when he remarked that "Mr. Pfungst has made a lasting contribution to both human and animal psychology, which cannot be passed over lightly by any serious student in either field."[7]

According to a particularly disturbing interpretation, exemplified recently by several leading workers with the allegedly language-like behaviors of chimpanzees, the Clever Hans effect *may* apply to horses, but "oversimplifies" the problem, for, "The chimpanzee behaviors . . . even if they are meaningless . . . are far too complex to be controlled by a simple 'go' or 'no go' cue."[8] Although it has been shown, in painstaking detail, that many experimental results in this field achieved so far can be more parsimoniously explained in terms of the Clever Hans effect than otherwise,[9] there seems to remain a lingering doubt about the pertinence of the effect to primates, including particularly humans, even though M. Polanyi has correctly argued that even philosophers are prone to be influenced by the Clever Hans effect, specifying that "this is exactly how [they] make their descriptions of science, or their formalized procedures of scientific inference, come out right."[10] I would like to augment Polanyi's observation—and for this assertion I adduced evidence elsewhere[11]—that the principle is a naturally connate part of every verbal and nonverbal interaction of all human dyads, as well as an ineradicable component of each and every instance of man–animal communicative interchanges, specifically scientific experiments; as Hediger has so perceptively noted: "Every experimental method is necessarily a human method and must *per se* constitute a human influence on the animal . . . . The concept of an experiment with animals—be it psychological, physiological or pharmacological—without some direct or indirect contact between human beings and animals is basically untenable."[12] The fundamental reason for the ubiquity of the Clever Hans phenomenon in the realm of biology is best understood when viewed as a special case of the physical law imposing the ultimate limitation on the knowability of the "real" state of affairs: the quantum of action couples the observer (subject) with the system observed (object), such that, as in wave mechanics, an observation necessarily changes the observed system. (In the human/animal context, a further perturbation may frequently be caused by the intrusion of the Pathetic Fallacy.[13])

The omnipresence of the Clever Hans phenomenon in all dyads—whether only one partner is human or both are—seems to me no longer at issue, although its paradigmatic operation in the ontogenesis of infant behavior remains to be shown in full detail in its differential role across cultures.

In this paper, I wish to reopen an ancient and perhaps perennial problem area by redrawing it in the context of the Clever Hans phenomenon. This indeed constitutes a profound enigma, which can be stated in several isomorphic ways, depending on one's philosophical preconceptions and terminological preferences. As Jakob von Uexküll might have phrased it: how are semiotic strings—verbal or nonverbal—emanating from the organism's *Umwelt*, transmuted into beneficial or harmful effects in the body's *Innenwelt*?[14] Or, to adapt Lévi-Strauss's opposition, how are subjective states—let us call them "Culture"—transformed into objective states, that is, "Nature"? The model presupposed here is, of course, a dualist-interactionist mock-up (sometimes known as "the correlation hypothesis"), involving the flow of information across the interface between the brain and the conscious self, in the tradition of Sherrington's philosophical position, as clarified to an extent, collaboratively, by Popper and Eccles, in their famous discussions of World I and World 2.[15] My personal image of the brain is that of a physical network, or structure, of neurons. I define "mind" as a system of signs, or representations, of what is commonly called "the world," or, more exactly, the *Umwelt* (von Uexküll) or "ground" (Peirce).[16] This model implies confidence that there exist one–one relationship patterns between the physical fabric of the brain onto which are coded the signs of the mind. Since we are ignorant of the linking principles, the enigma is, of course, how is this coding accomplished? Or just how do signs represent?

Walter Cannon's classic paper, of 1942, reports many instances of mysterious, sudden, and apparently psychogenic, death, from all parts of the globe, traceable to voodoo: the victim is cursed by a witch doctor, or perhaps because of the violation of a powerful taboo.[17] However, if the warlock removes the spell, recovery is immediate. Faith, obviously, is a pivotal factor, just as it is in the workings of the placebo effect (although suggestibility cannot be clearly separated from spontaneous change, as Lasagna has recently underlined[18]), as well as in Christian Science, the central tenet of which was pronounced by Mary Baker Eddy as: "Mortal belief is all that enables a drug to cure material ailments [since] the so-called laws of health are simply laws of mortal belief."[19] The same may be true of hypnosis, which, Sacerdote tells us, "may be in many ways the most powerful of placebos."[20] The sudden death syndrome has been described not only in man, but also in rats (by Richter[21]), and many other animals, and has been variously ascribed to shock or to hopelessness. At any rate, what seems to be involved is overactivity primarily of the parasympathetic system. In an interesting book, *Scared to Death*, Barker[22] has shown that if terrified animals are first given chlorpromazine before being severely constrained, the proportion of dying is much reduced; this is likewise the case when they are administered atropine, which neutralizes parasympathetic effects. However, if given mecholyl, or similar parasympathetic stimulants, the "voodoo" effect is intensified.

Although Cannon's ideas are fascinating and have generated nearly 40 years of controversy, and although he disclosed a shadowy outline of a novel area of exciting research possibilities, the fundamental conundrum remains in sharp contradiction to the adage, "Sticks and stones will hurt my bones, but words can never harm me." We know that indeed they can, as they can cause—at least since 1224—spontaneous stigmatization. There are now about 50 more or less reasonably reliable cases in the literature, beginning with that of St. Francis of Assisi, for none of which has a satisfactory medical explanation been hitherto offered.[23]

What is missing, in all such cases, is the mind-to-body conversion mechanism, although it seems increasingly likely that the psychogenic activation of the secretion of certain brain chemicals may be implicated, as, for example: pain-diminishing endorphins (which are endogenous opiate-like substances, such as dynorphin), interferon (which counters viral infections), and steroids (which reduce inflammations). It is speculated that these or similar mechanisms may be operative in the whole range of so-called miracle cures, as well as their opposites. Thus, the cues, both verbal and nonverbal, we may further postulate, are transformed by the appropriate receptor into signals from nerve cell to nerve cell, releasing specific chemicals that affect the brain by coupling with action sites on yet other classes of nerve cells. These chemicals cohere with their receptors like keys in their unique locks, and, by so fitting, they switch off activities in the cell. Sometimes, in the manner of hormones, their influence may spread much more widely. The effects of peptides in the nervous system represent a chemical semiotic system that the brain uses to communicate with itself. Herbert Benson's current investigation of the survival patterns of Tibetan lamas under winter conditions at altitudes of 14,000 feet, allegedly by the practice of Tumo yoga, supplemented by the use of herbal remedies, may throw considerable light on systems such as these, especially if we consider the Tibetan world view, which emphasizes three components as essential in dyadic interactions: the patient's faith, the healer's belief, and, interestingly, the *karma* (literally, "deed" or "action" that binds men to the world) linking the two.[24]

In passing, I should also like to mention the serendipitous discovery of Robert Nerem that laboratory animals given special attention, greeting and cuddling several times daily, have significantly reduced fatty deposits (i.e., cholesterol) in contrast to the control group.[25] Since the Clever Hans effect, as I said at the outset, is narrowly, but nonetheless commonly, confined by psychologists to animals, it may be worth reminding ourselves of the fact that psychosomatic practice is a recognized component of veterinary medicine. As Brouwers, a well-known Belgian veterinarian, teaches, "Mental factors . . . play their part in veterinary medicine just as in human medicine," and supports this statement by many examples from many species.[26] Chertok and Fontaine, in a fascinating paper,[27] collected numerous further instances of psychosomatic disturbances, ranging from eczemas in dogs to pseudopregnancy in cats, mares, and heifers, as well as in bitches. A whole series of organic and functional disturbances occur in captive animals as a result of situational stress arising from alterations in interindividual relationships, embodied in both verbal and nonverbal projections from man to animal. As in the human context, these authors tell us that, while a good deal is known about such relationships on the psychological level (transference), next to nothing is known about the somatic basis of these relationships. Kalogerakis[28] has shown that smell is certainly one of the surface mediators, but it would be valuable to know why some animals react to cues by behavioral change, others by somatic or functional upsets.

Permit me to conclude with one more example, recently and rightly marveled at by Lewis Thomas[29] — the homely wart — the taking away of which, "by rubbing . . . with somewhat that afterwards is put to waste and consumed," earlier aroused the curiosity of Francis Bacon.[30] "I do apprehended the rather, because of mine own experience," he tells us. Thomas meditates over an experiment by Surman *et al.*, who designed a test for the hypothesis that warts are treatable by hypnotherapy. Their tentative findings supported their hypothesis and suggest that hypnosis has a general effect on host response to the causative virus.[31] An earlier experiment, by Sinclair-Gieben and Chalmers, went much further — this demonstrated that hypnosis can influence lesions selectively: that is, it was suggested to the patient that the warts on one side of the body (the worse affected) would disappear (the other side serving as a perfectly matched control). In nine of the ten patients, the warts on the "treated" side disappeared, while those on the control side remained unchanged.[32] The heart of the mystery lies, of course, less in how the power of cuing resulted in the power of curing than how the immunological system of the human body—which we assume is both deaf and blind—can be induced to perform its healing function by deploying its lymphocytes upon command, by turning off bilateral arterioles in a hierarchical manner, confined, that is, to but one side of the body. If a selected work is done by a chemical mediator, it would be nice to know how and which, and in response to what.

Thus the omnipresence of the Clever Hans phenomenon in human everyday life, including human/animal contact situations in a rich variety, affords an appropriate occasion to deliberate upon what Freud so picturesquely dubbed "the mysterious leap" from the mind (psychology) to the body (neurophysiology[33]). It may also constitute a convenient investigative tool with the aid of which the Cartesian gap can, to a degree, be further narrowed. For a final resolution to the correlation version of the identity hypothesis, and the attendant coding problems, we must look to further assiduous researches in the field of brain electrochemistry, for, plainly, the Clever Hans effect is too important to be left to psychologists.

REFERENCES

1. KRALL, K. 1912. Denkende Tiere: Beiträge zur Tierseelekunde auf Grund eigener Versuche. Der kluge Hans und meine Pferde Muhamed und Zarif. 2nd edit. Friedrich Engelmann. Leipzig.
2. MAETERLINCK, M. 1914. The Unknown Guest. A. Teixeira de Mattos, Translator. Chap. IV, The Elberfeld horses. Dodd, Mead and Company, New York, N.Y.
3. RHINE, J. B. & L. E. RHINE. 1928-29. An investigation of a "mind-reading" horse. J. Abnormal Soc. Psychol. 23:449-466.
4. RHINE, J. B. & L. E. RHINE. 1929-30. Second report on Lady, the "mind-reading" horse. J. Abnormal Soc. Psychol. 24:287-292.
5. GARDNER, M. 1957. Fads and Fallacies in the Name of Science. Dover, New York, N.Y.
6. BLAKE, H. 1977. Thinking with Horses. Souvenir Press, London.
7. JOHNSON, H. M. 1911. Review of Oskar Pfungst's "Clever Hans (the Horse of Mr. von Osten)": A contribution to experimental, animal, and human psychology. J. Phil. 8: 663-666.
8. SAVAGE-RUMBAUGH, E. S., D. M. RUMBAUGH & S. BOYSEN. 1980. Do apes use language? Am. Sci. 68: 49-61.
9. UMIKER-SEBEOK, J. & T. A. SEBEOK. 1981. Clever Hans and Smart Simians: The self-fulfilling prophecy and kindred methodological pitfalls. Anthropos 76 (in press).
10. POLANYI, M. 1958. Personal Knowledge: Towards a Post-Critical Philosophy. Univ. Chicago Press, Chicago, Ill.
11. SEBEOK, T. A. 1979. The Sign & Its Masters. Chap. 5. Univ. Texas Press, Austin, Texas.

12. HEDIGER, H. 1974. Communication between man and animal. Image Roche **62**: 27–40.

13. SEBEOK. T. A. 1972. Perspectives in Zoosemiotics. Mouton, The Hague, The Netherlands.

14. VON UEXKÜLL, J. 1980. *In* Kompositionslehre der natur: Biologie als undogmatische Naturwissenschaft. T. von Uexküll, Ed. Vol. 3 (No. 4): 291–388. Ullstein, Frankfurt-am-Main, West Germany.

15. ECCLES, J. C. 1979. The Human Mystery. Lecture 10, The mind-brain problem: Experimental evidence and hypothesis. Springer International, Heidelberg, West Germany.

16. PEIRCE, C. S. 1935–66. Collected Papers of Charles Sanders Peirce. C. Hartshorne, P. Weiss, and A. W. Burks, Eds. Vol. 1:292. Harvard Univ. Press, Cambridge, Mass.

17. CANNON, W. B. 1942. "Voodoo" death. Am. Anthropol. **44**: 169–181.

18. LASAGNA, L. 1980. The powerful cipher. The Sciences **20**(1): 31.

19. EDDY, M. B. 1934 [1st edit., 1875]. Science and Health With a Key to the Scriptures. 2nd edit. pp. 174, 184. Trustees under the will of Mary Baker G. Eddy. Boston, Mass.

20. SACERDOTE, P. 1977. Quoted from C. Holden. Pain control and hypnosis. Science **198**: 808.

21. RICHTER, C. P. 1957. On the phenomenon of sudden death in animals and man. Psychosom. Med. **19**: 191–198.

22. BARKER, J. C. 1968. Scared to Death. F. Muller, London.

23. THURSTON, H. 1952. The Physical Phenomena of Mysticism. Burns Oates, London.

24. SOBEL, D. 1980. Placebo studies are not just "all in your mind." The New York Times (Sunday, January 6).

25. ANONYMOUS. 1980. Try a little TLC. Science 80 **1**(2, Jan/Feb): 15.

26. BROUWERS, J. 1956. Le role du système nerveux en pathologie générale. Ann. Med. Vet. **44**: 245–270.

27. CHERTOK, L. & M. FONTAINE. 1963. Psychosomatics in veterinary medicine. J. Psychosom. Res. **7**: 229–235.

28. KALOGERAKIS, M. G. 1963. The role of olfaction in sexual development. Psychosom. Med. **25**(5): 420–432.

29. THOMAS, L. 1979. The Medusa and the Snail: More Notes of a Biology Watcher. Chap. 13. Viking Press, New York, N.Y.

30. BACON, F. 1900. The Works of Francis Bacon. J. Spedding, R. L. Ellis, and D. D. Heath, Eds. Houghton, Mifflin and Company, Boston, Mass.

31. SURMAN, O. S., S. K. GOTTLIEB, T. P. HACKETT & E. L. SILVERBERG. 1973. Hypnosis in the treatment of warts. Arch. Gen. Psychol. **28**: 439–441.

32. SINCLAIR-GIEBEN, A. H. C. & D. CHALMERS. 1959. The evaluation of treatment of warts by hypnosis. Lancet (2): 480–482.

33. DEUTSCH, F., Ed. 1959. On the Mysterious Leap From the Mind to the Body: A Workshop Study on the Theory of Conversion. International Universities Press, New York, N.Y.

The Semiotic Function Underlying the Referential Communication Paradigm*

BARBARA R. FOORMAN

Department of Educational Psychology
College of Education
University of Houston
Houston, Texas 77004

RESEARCH into the development of communication skills in humans and nonhumans has suffered from crises of paradigms. What does it mean "to communicate"? and how do we as experimenters decide upon criteria for assessing whether our human and nonhuman subjects are, in fact, communicating? Are our criteria flexible enough to accommodate novel messages from speakers? And, perhaps more importantly, are our experimenters even aware of the influence of subtle cuing on our subjects' responses? Cuing varies. Cuing may be as simple as in the case of Clever Hans in which the horse responds to two basic cues: (1) to start tapping, and (2) to stop tapping. In the cases which Herbert Terrace has brought to our attention, the matter is surely more complex.

In this paper I will not be approaching the issue of deception through experimenters' unknowing cues and assumptions. Rather, I will consider an accidental deception of the experimenter by primed, 4-year-old children in a referential communication paradigm. Referential communication involves describing an object out of an array of similar objects.[1] In the present study the stimulus sets were photographs of everyday objects — pictures of human emotional expressions[2] and pictures of dogs. The priming procedure consisted of asking general probes about how people's faces differ from one another and how dogs differ from one another and specific probes about how one photograph differs from another. It was anticipated that the priming experience would elicit knowledge in the child's head that was relevant to the task at hand. Such knowledge was indeed elicited but it was not expressed verbally in the expected manner. What did we, the adult experimenters, the con-

* This work was partially supported by the Department of Applied Behavioral Sciences, University of California, Davis as part of the author's postdoctoral research. Some of these data were presented at the American Psychological Association Convention in New York, September, 1979.

0077-8923/81/0364-0206 $01.75/0 © 1981, NYAS

structers of referential communication paradigms expect? We expected conventional, feature-analytic descriptions. What did we get? From the primed 4-year-olds we got an increased production of unconventional, holistic descriptions that were metaphorical in nature. The point is that we had deceived ourselves into thinking that a paradigm that has referential language as its focus would not seem conducive to metaphoric production. Happily, a series of probes into how the 4-year-old constructs the world cued us in to our deception. This paper, then, examine 4-, 5-, and 7-year-olds' descriptions on a referential communication task with an eye towards their metaphorical content and their informativeness.

A definitional classification is in order regarding the notion of metaphorical production. Traditionally, metaphors involve some tension or lack of shared features between the topic and vehicle. For example, in the metaphor, "American society is a fruit salad, not a melting pot," there is a certain amount of tension between the topic "American society" and the vehicles "fruit salad" and "melting pot" if these terms are taken literally. But metaphor is often described as involving the nonliteral use of language, in which the intended proposition is mediated by a comparison of two distinct conceptual domains.[3] According to the "semantic feature spaces" notion of metaphor,[4] the distances between these conceptual spaces will affect the aesthetic quality and the comprehensibility of the metaphor.

In contrast to an emphasis on semantics in our definition of metaphor, there is the emphasis on pragmatics, the study of speech acts and the contexts in which they occur. From a pragmatic point of view, metaphors are recognized as such only through the contexts in which they occur. But at an even more fundamental level than the distinctions between literal and nonliteral and pragmatic and semantic is the issue of the underlying psychological processes involved. Rumelhart[5] takes a constructivist point of view when minimizing the need for extraordinary explanations for metaphor. He points out that children naturally produce metaphor as they acquire a language, and, therefore, metaphor is a normal linguistic phenomenon. Verbrugge goes a step further and suggests that "metaphoricity" is a normal part of children's prelinguistic development and that "these playful metaphors of perception and action are the *prototype* of genuine metaphoric processes."[6] My perspective is constructivist as well as pragmatic. To the extent that we as speakers are attempting to construct or reconstruct our view of reality in communicating an idea to a listener, we are seeking linguistic representations that, given contextual cues, will be interpretable. Put into a broader theory of reference, in our attempt as speakers of a language to communicate our intentions, we create extensions or referents, which may or may not be shared by our listeners. The definition of metaphor lies in the external judge's view of appropriate tension between the expressed intension (i.e., the topic) and the expressed extension. But, ultimately the listener is the judge by the very nature of his or her response.

In this study the mode of presentation and construction of the referent–nonreferent array maximized a constructivist approach to metaphor. The referent photographs invited a certain amount of tension between "topic" and "vehicle" because they were selected so that no one-label message would readily apply. The referent dog photographs were pictures of mutts, while the referent photographs of human emotional expressions[2] were comprised of multiple emotions involving happiness, sadness, fear, anger, surprise, disgust, and contempt. Each referent was presented to the speaker in isolation, while the listener viewed a nonreferent array comprised of the referent and three "high agreement" photographs (i.e., pure-bred dogs or clear human emotional expressions). The primary rationale for presenting the referent in isolation instead of in the context of the entire array was an appeal to ecological validity: in the real world the pool of nonreferents available to a listener is rarely evident to the speaker. Part of the task of a competent speaker is figuring out what features are potentially relevant to the communication. In short, in attempting to image the potential nonreferent comparisons, the speaker must construct linguistic representations that take these potentially relevant comparisons into account.

One way for the speaker to encode these comparisons liguistically, while dealing with the lack of a single label to apply to the referent, is to give a feature-analytic description. An example of a feature-analytic description for one of the facial expressions might be, "It's a man with dark hair and dark eyes"; for one of the dog photographs it might be, "It's big and brown and has a long tail." Non-feature-analytic descriptions are coded as holistic or part-inferential.[7] Holistic descriptions are encodings that attempt to capture the gestalt of an object. A holistic encoding for one of the facial expressions might refer to a relatively conventional emotional expression, such as, "she's real mad," or to a more novel encoding, such as, "he looks like Frankenstein." For the dog photographs a relatively conventional holistic encoding might be, "It looks like a wolf," while an unusual one would be, "It looks like a dinosaur dog." In addition to these holistic encodings are part-inferential descriptions, which encode a feature with an image. Examples are: "his eyes look worried," for one of the facial expressions; and "a tail like a 'c'" for one of the dog photographs. Taken together, these holistic and part-inferential encodings were considered, from a constructivist perspective, as metaphorical productions.

METHOD

Subjects

The study involved 96 children, 4, 5, and 7 years old, from middle-class families where English was the dominant language. There were 24 5-and 7-year-olds from an elementary school in Piedmont, Calif., and 48 4-year-olds from preschools in Davis, Calif. The 4-year-olds were divided into experimental and control groups, with the experimental group receiving a preexperimental treatment in which features relevant to the communication task were primed. The average age for the preschoolers during the time of the study was 4.5, while for the kindergarteners it was 5.6, and for the second graders it was 7.7.

There were two experimenters. In the case of the 5- and 7-year-old subjects, the author served as the listener in the communication task and administered the pre- and post-test procedures; another adult female was present during the communication task to give directions and take note of nonverbal behavior. In the case of the 4-year-olds, a female graduate student of human development served as a listener and collected the posttest data and one of the pretest measures, while a male human development undergraduate helped collect pretest data and served as the observer in the communication task.

Task Description and Procedure

Pretest

All subjects received a mental attentional energy measure called the Cucui test.[8] The Cucui test consisted of 26 line drawings of a potato-shaped man called Mr. Cucui. The first six pictures were for practice. In each picture different parts of Mr. Cucui's body were colored. The subject was shown each picture for five seconds and then asked to place on an outline of Mr. Cucui an "X" on the places where Mr. Cucui had been wearing colors. The more spots of colors a subject could remember, the greater the subject's short-term memory and, hence, mental processing capacity. The number of colors Mr. Cucui was wearing in the test items varied randomly from one to four. There were five items at each of these four levels. The test was used to refine the factor of grade level by excluding from the preschool, kindergarten, and second grade sample any subject whose performance fell outside a specified range for the average number of schemes capable of being coordinated at ages 4, 5–6, and 7–8.[9]

In addition to the Cucui test, half of the 4-year-olds received a preexperimental priming treatment consisting of a series of probes asked in the context of the practice trial for each stimulus set. General probes asked, "How are people's faces different from one another?" or, "How are dogs different from one another?" When general probes no longer elicited information from the subject, specific probes were asked about the difference between pairs of pictures: "How do these two faces differ?" or, "how do these two dogs differ?" Specific probes were asked until a subject failed to respond. The experimenter completed the priming procedure with the general probe, "Can you think of any other way that people's faces (dogs) are different from one another?"

Main Test: Communication Task

The goal of the referential communication task was to describe in isolation a black-and-white photograph of a dog or a person's facial expression well enough that the adult listener could pick out the referent from among an array of nonreferent photographs of the same stimulus set. The task instructions were nearly identical for both sets of stimulus materials. For example, for dogs the instructions were: "Pretend that you lost your dog and you have to go to the dog pound to see if you can get it back. You must tell the dog catcher what it looks like and she will look at the dogs that are in the pound to see if she can find it." If the speaker's spontaneous initial encoding was not adequate for distinguishing the referent from the nonreferent array, two forms of feedback were provided by the listener: (1) nonspecific verbal feedback of the form, "I still don't know which one it is. Is there anything else you can tell me about it?"; and, if an adequate description was still not given, (2) specific, visual contrasts (i.e., nonreferents) that fit the description already provided by the subject but were inconsistent with respect to those relevant features not yet mentioned.

In sum, the communication task consisted of two stimulus sets presented in the two possible orderings to groups of subjects who were counterbalanced for grade level. Each stimulus set consisted of six items, the first one being a practice item. The influence of item order was taken into consideration by having two random orders within each stimulus order. Each items involved four photographs in the referential array, the referent and three nonreferents.

Posttest

The posttest revealed the extent of a subject's repertoire of values for specific features by posing questions about the referent photographs of the communication task (e.g., "How does the tail look?"). All subjects were tested on all features listed in the feature matrices (see *Scoring and Treatment of Data*) so that the spontaneously produced vocabulary from the communication task could be correlated with that produced on demand in the posttest. There was approximately a four-week time lapse between administration of the main test and posttest for a subject.

Scoring and Treatment of Data

Pretest

Scores for the Cucui test were generated in the following manner: One point was assigned if 4 out 5 items in a level were "passed"; 1/5 of a point was added for any other item scored perfectly.

Scoring procedures for the priming treatment consisted of dividing responses to general and specific probes into *categories* and *values*. A *category* was a response that named a feature, such as *hair color* or *length of tail*. A *value* gave the specific attribute of a feature, such as "black" or "short tail." Subjects' scores were then reported in terms of the proportion of the total number of general and specific probes that were categories and values.

Communication Task

Audio- tapes of the communication task were transcribed and scored according to several dependent variables. Consideration was given to the adequacy of the speaker's initial description for distinguishing the referents from the nonreferents, the relevancy of the features mentioned, and the number of steps of listener feedback required before an adequate description was produced. If a speaker's initial description enabled the listener to distinguish the referents from the nonreferents, the description was scored as one adequate initial encoding. For any one subject, the potential range of adequate initial encodings within stimulus sets was 0 to 5. If a speaker's initial encoding did not differentiate the referent from the nonreferents, four steps of feedback were available: the first step of feedback corresponded to nonspecific feedback (i.e., "I still don't know which one it is. Is there anything else you can tell me about it?"); the second, third, and fourth steps of feedback corresponded to specific, visual contrasts. Thus, on any one trial a subject's feedback score could range from 0 to 4. Obviously, if 0 steps of feedback were required, then the subject had an adequate initial encoding.

The scoring criterion for relevant features was determined by a score of 1 for every feature mentioned that was in the relevant matrix for each stimulus set. The matrices contained such conventional features as: *hair color, size, orientation,* and *length/position of ears* and *tail* for dogs. These matrices also noted the nonanalytic categories called *holistic* and *part-inferential,* the two categories of metaphorical productions. As noted earlier, holistic encodings included such conventional gestalts as, "She looks like she's about to hit her kid," and, "It looks like a pig." Part-inferentials described a specific feature with an image, such as, "His ears look like they had an operation," and, "His tail is like a snake going down a tree." Holistic and part-inferential encoding were totaled for each subject and represented as a proportion of the total number of relevant features produced per stimulus set. If a feature was so universal or so personalized as not to help distinguish the referent from the nonreferents (e.g., "She has a nose," or, "This dog looks like my dog.") it was coded as an irrelevant feature. On any one trial the scores for relevant features ranged from 0 on up.

Posttest

The scoring procedure for relevant features mentioned in the posttest utilized the same relevant feature matrices as were used in the main communication task.

When the raters of the tests were compared with each other, interrater reliability coefficients of 0.90–0.99 were obtained for the coding of relevant features and holistic and part-inferential encodings.

TABLE 1

TABLE OF GROUP MEANS FOR COMMUNICATION MEASURES
FOR EACH STIMULUS SET

Dependent Variable	Group*			
	4nP	4P	5	7
Adequate initial encoding				
Dogs	0.25	0.34	0.27	0.47
Faces	0.28	0.39	0.36	0.44
Relevant features				
Dogs	1.84	2.27	3.27	4.23
Faces	1.61	2.11	3.21	4.12
Steps of feedback				
Dogs	1.67	1.51	1.61	1.07
Faces	1.90	1.45	1.58	1.17
Relevant features in the posttest				
Dogs	3.92	3.92	4.62	5.48
Faces	3.47	3.47	4.90	5.81

	Trials				
	1	2	3	4	5
Adequate initial encodings					
Dogs	0.28	0.46	0.27	0.38	0.28
Faces	0.21	0.29	0.58	0.38	0.37
Relevant features					
Dogs	2.69	3.07	3.03	2.99	3.02
Faces	2.30	2.70	2.88	2.86	2.87
Steps of feedback					
Dogs	1.83	1.09	1.58	1.19	1.64
Faces	2.11	1.54	.98	1.48	1.50

* Groups of 4-year-old nonprimed, 4-year-old primed, 5-year-old, and 7-year-old children, respectively.

Results

Communicative Informativeness

Analyses of variance were performed on the following dependent variables: the number of adequate initial encodings, the number of (potentially) relevant features, the number of steps of feedback required, and the number of relevant features given in the posttest. TABLE 1 presents the table of means and TABLE 2 presents the significant results for each dependent variable for the stimulus sets dogs and faces. These stimulus sets were analyzed separately since previous work[10,11] indicated different encoding styles for descriptions of dogs. Facial expressions tended to be described holistically (e.g., "He looks sad"), while pictures of dogs tended to be described analytically (e.g., "It's big and brown, with a long tail").

Examination of TABLE 2 reveals significant group differences in the following variables: adequate initial encodings for dogs ($F_{3,88} = 3.46, p < 0.02$), relevant features for faces ($F_{3,88} = 20.42, p < 0.01$) and dogs ($F_{3,88} = 12.10, p < 0.01$), and relevant features in the posttest for faces ($F_{3,88} = 22.35, p < 0.01$) and dogs ($F_{3,88} = 15.91, p < 0.01$). Post-hoc analysis, utilizing Scheffe's statistics, revealed relatively little difference between the groups of 7-year-olds, 5-year-olds, 4-year-old primed subjects (4P), and 4-year-old nonprimed subjects (4nP), except in the posttest measure of vocabulary. Significant group differences may be summarized in the following manner, where ">" indicates "significantly greater than" and "≯" indicates "not significantly greater than."

TABLE 2

ANOVA RESULTS FOR FOUR DEPENDENT VARIABLES AND
TWO STIMULUS SETS

	Stimulus Sets	
Dependent Variable	Faces	Dogs
Adequate initial encodings		
$F_{3,88}$ (Groups)		3.46 ($p < 0.02$)
$F_{4,352}$ (Trials)	11.15 ($p < 0.01$)	3.76 ($p < 0.005$)
Relevant features		
$F_{3,88}$ (Groups)	20.42 ($p < 0.01$)	12.1 ($p < 0.01$)
$F_{4,352}$ (Trials)	5.55 ($p < 0.01$)	
Steps of feedback		
$F_{4,352}$ (Trials)	13.44 ($p < 0.01$)	9.23 ($p < 0.001$)
Posttest		
$F_{3,88}$ (Groups)	22.35 ($p < 0.01$)	
$F_{9,396}$ (Sets)	30.11 ($p < 0.01$)	

* Because analyses were conducted separately for dogs and faces, the significance level was divided in half: $a/2 = 0.05/2 = 0.025$. Scheffe's statistic at $a = 0.05$ is 2.87.

	Faces	Dogs
Adequate initial encodings		7≯5≯4P≯4nP, 7>4nP
Relevant features	7≯5>4P≯4nP	7≯5, 5≯4P, 5>5nP
Posttest	7>5>4P≯4nP*	7>5>4P≯4nP*

We can see from the results of the posttest that all subjects had more vocabulary terms for the dog photographs than the facial expressions. In addition, we see that 7-year-olds had a significantly greater vocabulary repertoire than 5-year-olds, that the 5-year-olds' repertoire was significantly greater than the 4-year-olds', and that the primed and nonprimed 4-year-olds did not differ significantly. Given these vocabulary differences, the relative lack of age level differences in the communication task is striking. There were no group differences in regard to steps of feedback, nor in regard to adequate initial encodings for faces. In terms of adequate initial encodings for dogs, the only significant group difference lay between 7-year-olds and 4-year-old nonprimed subjects. In regard to relevant features the significant group differences for faces lay between the 7- and 5-year-olds versus both groups of 4-year-olds, while for dogs the significance lay between the 5- and 7-year-olds and the nonprimed 4-year-olds.

In sum, age seemed to be a good predictor of vocabulary effectiveness, accounting for 42% of the explained variation in posttest performance. However, in terms of the informativeness of the descriptions given prior to feedback and with feedback, age was not very indicative of performance. In the case of significant group differences in adequate initial encodings for dogs, age accounted for only 5% of explained variance. However, in the case of group differences in relevant features, age was more indicative of performance, accounting for 29% of explained variance for dogs and 40% for faces.

Significant trials effects depicted in TABLE 2 will not be presented since trend analysis revealed the patterns to be typical learning curves (i.e., quadratic and quartic) and there were no group differences in these patterns.

METAPHORICAL PRODUCTIONS

Group differences in metaphorical productions were examined by representing each subject's total number of holistic and part-inferential encodings for each stimulus set as a proportion of the total number of relevant features produced. The resulting distributions were highly

---

* Values obtained with pictures of dogs were greater than those obtained with pictures of faces.

skewed, thereby suggesting a nonparametric analysis of the data. The proportion of metaphorical productions for the descriptions of dogs was so small for each group that a chi-square analysis was performed, contrasting "no metaphors" to "some metaphors" (i.e., one or more metaphor produced per subject). The result was no significant difference between groups (χ_3^2 = 6.75, p < 0.08).

However, there were significant group differences in the number of metaphors produced in describing facial expressions. A one-way analysis of variance, utilizing the Kruskal-Wallis statistic, revealed significant group differences between mean ranks of 61.40 for the 4-year-old primed subjects, 44.19 for the nonprimed 4-year-olds, 47.90 for the 5-year-olds, and 40.52 for the 7-year-olds (χ_3^2 = 7.70, p ≤ 0.05). Post-hoc contrasts indicated that the 4-year-old primed group was producing significantly more metaphors than the nonprimed 4- year-olds or the 7-year-olds. These primed subjects were not, however, producing significantly more metaphors than the 5-year-olds. TABLE 3 breaks the number of metaphors produced by each age group for facial expressions into categories of holistic and part-inferential and, within that, into *conventional* and *novel*. Holistic-conventional metaphors were metaphors that applied a single, relevant label to the facial expressions, even though the referent photographs were not easily described by a single label. The labels most frequently chosen were "worried," "sad," "mean," "frightened," "scared," "glad," and "smiling." These same labels, when applied to specific features, were called part-inferential-conventional encodings (e.g., "scary mouth," "happy teeth," and "sad eyes").

Metaphors categorized as holistic-novel were either unusual but relevant one-label descriptions, such as "Frankenstein," "gorilla," "witch," or "Indian," or were phrases describing imminent action. Examples of such

TABLE 3

AGE GROUP TOTALS FOR CATEGORIES OF METAPHORS FOR FACIAL EXPRESSIONS

	Holistic		Part-Inferential		Totals		
Age*	Conventional	Novel	Conventional	Novel	Metaphors	Relevant Features	%†
4P	51	8	15	2	76	522	15
4nP	28	2	4	2	36	416	9
5	37	17	1	1	56	787	7
7	43	8	2	1	54	1003	5

* 4P, 4-year-old primed; 4nP, 4-year-old nonprimed.
† (Total metaphors/Total relevant features) × 100.

phrases were: "starting to cry;" "about to scare someone;" "he's not so sure what they're doing;" and "she looks like she's going to bite you." Novel descriptions for part-inferential metaphors included: "teeth going 'eee,'" "eyebrows like a clown," and "ears like he had an operation." Inter-rater reliability coefficients for these classifications ranged from 0.93 for the coding of conventional and novel categories within holistic encodings and 0.98 for the coding within part-inferentials.

Examination of TABLE 3 shows that 15% of the 4-year-old primed subjects' relevant encodings were metaphorical in nature, compared to 9% for the nonprimed 4-year-olds, 7% for the 5-year-olds, and 5% for the 7-year-olds. These percentages reflect the significant difference between the 4-year-old primed subjects and 7-year-olds revealed in the analysis of variance ratios but do not suggest the significant difference between the two groups of 4-year-olds. The reason for this discrepancy is that the data were ranked for the analysis of variance so that a ratio of 0 metaphors to 1 relevant feature was less than 0 metaphors to 2 relevant features. Consequently, the 4-year-old nonprimed subjects received relatively low ranks because of their lower number of relevant features and metaphors produced, but proportionately their total metaphorical production to total relevant feature production is relatively higher.

The other striking part about TABLE 3 is the "15" in the part-inferential-conventional column for the 4-year-old primed subjects and the "17" in the holistic-novel column for the 5-year-olds. The primed 4-year-olds' higher production of metaphors lies in the conventional category of holistic and part-inferential encodings. Therefore, priming appears to enhance metaphorical production in the area of conventional metaphors. The relatively high number of holistic-novel encodings is not readily explainable except to say that the one-third of the 5-year-old subjects who were responsible for these productions were each producing several metaphors.

DISCUSSION

The 4-year-old primed subjects' impressive performance in the area of metaphoric production is tempered by their relatively less impressive performance in the area of informativeness of the message. 15% of the relative features produced by the 4-year-old primed subjects, compared to 9% for the nonprimed 4-year-olds, 7% for the 5-year-olds, and 5% for the 7-year-olds were metaphoric. Yet the percentage of the descriptions that were sufficiently informative to make listener feedback unnecessary was 37% for the 4-year-old primed subjects, compared to

26% for the 4-year-old nonprimed subjects, 25% for the 5-year-olds, and 44% for the 7-year-olds.

These percentages, based on relevant features, are further clarified by considering the percentage of features produced that were irrelevant to the task: 13% for the primed 4-year-olds, 14% for the nonprimed 4-year-olds, 10% for the 5-year-olds, and 4% for the 7-year-olds.[10,11] The irrelevant features produced were either so universal in nature as not to be discriminating (e.g., "This person has a nose") or so personalized as not to be informative (e.g., "This person looks like my mother"). These personalized encodings were, in fact, metaphorical in the holistic sense decribed above. But these personalized encodings constitute such a small percentage of total encodings that their importance is minimal: 1% for the primed 4-year-olds; 6% for the nonprimed 4-year-olds; 2% for the 5-year-olds; and 1% for the 7-year-olds.

The conclusion to be drawn from these percentages is that although 56% to 75% of the initial descriptions produced by subjects required feedback, 86% to 96% of all features produced in these initial encodings were relevant to the task. Of these relevant encodings, 85% to 95% were feature analytic in nature. But the fact that metaphors were produced at all in a task that would seem to value analytic encodings is worthy of note. And the fact that the 4-year-old primed subjects were producing significantly more metaphors for encodings of facial expressions than both the nonprimed 4-year-olds and the 7-year-olds is particularly interesting. True, the majority of the holistic and part-inferential encodings produced by this primed group were categorized as conventional. But, given the real-world nature of the stimulus materials, it is not surprising that conventional rather than novel metaphors were chosen. The price for novelty was clearly displayed by the reaction to one 7-year-old's description of, "This looks like a gorilla." When the adult listener repeated this encoding with a surprised intonation, the child replied, "Because . . . I don't know. Ok, I'll decribe it." The child then proceeded to analyze the holistic description "gorilla" into specific features. But there was a slight tone of disappointment in her voice, as if to say "I gave you the best description, but now, if you insist, I will break it down," This subject's interpretation of task instructions was clear: "to describe" meant to give a feature-analytic description. As experimenters we must not be distracted by such potentially constraining task instructions that influence the nature of our subjects' encodings.

The ability of the priming procedure to enhance metaphoric production in the communication task has important implications for the field of child language. Without the priming procedure we might have assumed that the younger subjects' relative lack of informative messages

in referential communication tasks resulted form a mediation deficiency as well as a production deficiency. In short, we might have
assumed that not only were younger children not producing informative messages, but perhaps they were not even mediating their perceptions about discriminating features. But the priming procedure revealed
that younger children were aware of discriminating features and would
encode these features linguistically so as to communicate relevant information to a listener. Moreover, the metaphorical nature of these linguistic encodings produced by the primed subjects is important. One
reason why younger subjects may not be producing very many relevant
features without priming is that, as their posttest results suggest, they
are relatively limited in task-relevant, conventional, feature-analytic
vocabulary. In addition, younger children had less attentional energy (as
measured by the Cucui test) than older children for overcoming misleading field effects or affective factors. Given such limitations of
vocabulary and processing capacity, we must appreciate the richness of
expression tapped by the general probes, "How are people's faces different from one another?" and the specific probes, "How are these two
faces different from one another?"

Child language researchers have used the "rich interpretation" rationale to explain how young children's productions can be com-
prehended by adults in spite of seeming anomaly. Such "over-
generalizations" as *doggie* for all four-legged creatures are classic examples. We understand the child. We give the child credit for emerging
cognitive structures and an emerging representational system for map-
ping these constructed cognitions.

I was recently viewing a video-tape of a 22-month-old boy talking to
his mother. The mother asked him whom he took a bath with. He answered, "Mommy." Then she asked him, "Where do you take a shower?"
The little boy paused, repeated "shower," and while touching his toes
said, "My hair takes it with my toes." A little while later he produced a
well-formed "wh-" question, asking where the dog had gone. How are we
to interpret, "My hair takes it with my toes"? We laugh. Children are so
cute—and witty. Then we pause to analyze why we understand, yet
smile. It seems that the child is responding not to the "where" question
because the "where" question is obvious: "Where do you take a shower?"
In the shower, of course. Instead, the child refers back to the "who"
question and responds with *parts* of himself—his hair and toes— rather
than with his *whole* self. Is this child giving an acceptable "communicative" response to "wh- cues" that signal responses of person, location, or things. However you answer this question, the point is that,
because we share in events with our children, we must aid in the fram-
ing of an emergent semiotic function that seeks common extensions in
the midst of developing intensions.

ACKNOWLEDGMENTS

The author would like to thank Kathy Gaustad for her help in collecting, coding, and analyzing part of this data and Kathy Denkowski for her help in scoring and key punching. Gratitude is also extended to Andy McCurrin for her help in preparing this manuscript.

REFERENCES

1. GLUCKSBERG, S., R. M. KRAUSS & E. T. HIGGINS. 1975. The development of referential communication skills. *In* Review of Child Development Research. F. Horowitz, E. Hetherington, S. Scarr-Salapatek & G. Siegel, Eds. Vol. 4. University of Chicago Press, Chicago, Ill.
2. EKMAN, P. & W. V. FRIESEN, 1975. Unmasking the Face. Prentice-Hall, Englewood Cliffs, N.J.
3. FRASER, B. 1979. The interpretation of novel metaphors. *In* Metaphor and Thought. A. Ortony, Ed. Cambridge University Press, Cambridge, England.
4. STERNBERG, R. J., T. TOURANYEAU & G. NIGRO. 1979. Metaphor, induction, and social policy: The convergence of macroscopic and microscopic views. *In* Metaphor and Thought. A. Ortony, Ed. Cambridge University Press, Cambridge, England.
5. RUMELHART, D. E. 1979. Some problems with the notion of literal meanings. *In* Metaphor and Thought. A. Ortony, Ed. Cambridge University Press, Cambridge, England.
6. VERBRUGGE, R. R. 1979. The primacy of metaphor in development. New Directions Child Devel. 6:78.
7. HEIDER, E. R. 1971. Style and accuracy of verbal communication within and between social classes. J Personal Soc Psychol 18:33–47.
8. DIAZ, S. 1974. Cucui Scale: Technical Manual (Multilingual Assessment Program). Stockton Unified School District, Stockton, Calif.
9. PASCUAL-LEONE, J. 1970. A mathematical model for transition rule in Piaget's developmental stages. Acta Psycholog. 32: 301–345.
10. FOORMAN, B. R. 1977. A neo-Piagetian analysis of communication performance in young children. Ph.D. dissertation. University of California, Berkley. 1979. J. Psycholing. Res. In press.
11. FOORMAN, B. R. 1979. The effect of priming on referential communication in 4 year olds. Unpublished manuscript.

On Self-Deception

THEODORE R. SARBIN

Board of Studies in Psychology
University of California, Santa Cruz
Santa Cruz, California 95064

THE POINT OF DEPARTURE for this confer-
ence is the Clever Hans story. Because the story was written as a scientific treatise, the horse's trainer, Herr von Osten, appears primarily as a
stimulus object. Scientists observed and recorded Herr von Osten's
postural movements and concluded that such movements were signals
to the horse to tap his hoof. Although not entirely neglected in the
scientific write-up, Herr von Osten's unshakable belief in the horse's
human-like rationality received only brief notice. Both Stumpf[1] and
Pfungst[2] declared that von Osten was not a trickster nor a swindler;
rather they settled on a diagnosis of self-deception.

Although in its naked form, the term "self-deception" contains a contradiction, it continues to be widely used to denote states of affairs
similar to that of Herr von Osten in the Clever Hans story. Herr von
Osten held to his anthropomorphic belief in the horse's rationality in
the face of adverse evidence presented by respectable and responsible
scientists. Parenthetically, one should note that anthropomorphism is a
rather common human belief, and Herr von Osten's apparent espousal
of anthropomorphic doctrine was not taken by his contemporaries as a
sign of senile dementia (nor should it have been).

My aim in this paper is to illuminate the conduct frequently labeled
self-deception. Von Osten is an appropriate example. To show that the
phenomenon is not unique, I will identify several additional examples
taken from experimental and field studies. These examples, taken
together, serve both as a working definition of self-deception and a
framework for a sketch of previous attempts to come to grips with the
problem. At the end of my paper, I offer a theoretical statement about
self-deception based on the narrative as the root-metaphor for knowing.

OPTON'S STUDY

In 1969, intensive interviews were conducted with a representative
sample of San Francisco Bay area residents, the subject being the
recently published photographs and stories of the My Lai massacre in
Vietnam. Under the direction of Edward E. Opton, open-ended inter-

0077-8923/81/0364-0220 $01.75/0 ©1981, NYAS

views were conducted with 42 respondents. These data were later supplemented by surveys conducted by the *Wall Street Journal,* the *Minneapolis Tribune, Time,* and the Harris Poll. About two-thirds of the
respondents expressed attitudes that Opton epitomized in the title of
his report: "It Didn't Happen and Besides They Deserved It."[3]

The respondents were shown the photographs and asked to comment. Typical of the majority of the responses were the following:

> I don't believe it actually happened. The story was planted by Viet Cong sym-
> pathizers and people inside this country who are trying to get us out of Viet-
> nam.

> I can't believe that a massacre was committed by our boys. It's contrary to
> everything I've learned about America.

> I can't believe anyone from this country would do that sort of thing.

The refusal to believe that American soldiers would engage in
atrocities in the face of the type of evidence that is ordinarily granted
credibility (newspaper and news magazine stories and photographs) is
similar to the behavior of Herr von Osten. If self-deception is a valid
category, then Opton's respondents were engaging in self-deception.

Psychiatric Cases

The most common examples of self-deception are to be found in the
psychiatric literature. Abundant clinical material supports the view that
persons who are singled out as deviant, abnormal, disordered, disturbed,
abberant, and so on fit the criteria of self-deception. When they hold
beliefs that are contrary to consensus or to socially approved "facts,"
they are said to experience delusions. When they report socially disap-
proved imaginings with unconventional metaphors, they are said to be
hallucinating. The epistemic activities of people diagnosed as delusional
or hallucinatory fit the criteria of self-deception.

A capsule summary of a case report will illustrate. Dorothy, age 19,
brought to a mental health center by her parents, refused to believe that
her older brother had been killed in an automobile collision three months
before. Tenaciously, she argued that her brother was still alive. Conventional evidence supported the contradictory conclusion: medical certifi-
cate of death, parents' identification of the body in the morgue, and
police and newspaper reports of the fatal accident. In short, Dorothy
expressed belief in "belief-adverse" circumstances.[4] Like Herr von
Osten and Opton's respondents, Dorothy held to her belief in the face
of contradictory evidence.

In this case, the therapist working with Dorothy uncovered additional information that made the strongly held belief understandable. The accident had occurred a half-hour after she and her brother had engaged in a shrill and bitter quarrel. As he was leaving the scene, she shouted after him: "I hope you get smashed up in your car!" It required no hermeneutical expert to uncover Dorothy's unvoiced belief that wishes are equivalent to deeds; and the implication that death wishes are equivalent to murder.

MILITARY STRATEGISTS

The field of military strategy provides many examples of conduct which, upon analysis, fit the criteria of self-deception. Ben-Zvi[5] has assembled accounts of military engagements, the outcomes of which were determined by conduct on the part of the strategists that in principle was the same as that of Herr von Osten. The cases studied were: the Nazi attack on the Soviet Union in 1941; the Chinese intervention in the Korean War in 1950; the Sino–Indian Border War in 1962; and the Arab–Israeli (Yom Kippur) War of 1973. In all these cases, strategies were held as unshakable beliefs in the face of available intelligence that could have influenced the strategists to modify their beliefs about relative strengths, mobility of troops, armor, and so on. In the typical cases, strategists paid little heed to information that would support beliefs contrary to those embedded in the strategy. The most dramatic case is the Barbarossa operation. Stalin steadfastly held the belief that Hitler would not attack the Eastern front without first declaring an ultimatum. Whaley[6] has documented no less than 84 separate pieces of contrary information available to Stalin, all of which were rejected or distorted in the interest of keeping whole the belief. (Parenthetically, Hitler encouraged the self-deception. He planned his actions and his communications so as to reinforce Stalin's belief.) Contemporary analysts find "self-deception" an appropriate label for Stalin's conduct.

HYPNOSIS

The conduct of an identifiable class of hypnotic subjects fits the criteria for self-deception, i.e., convincingly expressing belief in a counterfactual proposition in the presence of contradictory evidence. Of special interest is the considerable body of work on hypnotic amnesia. It is a commonplace observation that responsive hypnotic subjects under amnesia instructions do not recall events occurring during the hypnotic performance. Memory for very recent events is tested during the period that the subjects are engaged in the hypnotic role. Under special demand conditions, such as a demand for honesty, about half the subjects "breach" the amnesia — these are subjects who also report being in control of their actions. During postexperimental inquiry, they report having used various tricks or stratagems not to remember, such as concentrating on some distracting stimulus. It is important to note that when the amnesia is "lifted," the experiment concluded, the "forgotten" material is quickly recalled. As a first step in interpreting the conduct of this class of subjects, we might make use of the dramatistic idiom: the subjects were enacting their roles in the manner of actors in the theater, and like actors were monitoring their performances. If we could remove the moralistic taint from the term, we could say they were engaged in performing deceptively, in creating the illusion of amnesia, and so forth.[7]

The remaining half of the subjects engage in conduct that fits the criteria of self-deception. These are subjects who do not breach the amnesia, even under the special demand conditions of the laboratory. They describe their actions as not being under their own control during the hypnotic performance. They say, "I don't remember," and the critical observer makes the judgment that they believe their counterexpectational utterances.[7] The self-deception label is an appropriate one. They show the paradox of "knowing" and "not knowing" at the same time.

The foregoing reference cases share the following description: A belief, serving as a guide to action, is expressed that is unwarranted by the evidence — unwarranted, that is, from the perspective of a critical observer. The belief is supported by the claim of "ignorance" of the contradictory belief that is supported by the evidence. Further, information that would challenge the claim of ignorance is rejected, distorted, or reinterpreted. Stated in another way, the person says, "I don't know," when the context calls for the contradictory statement, "I know."

LABELING SELF-DECEPTION

What are the conditions that permit the assignment of the label "self-deception"? One would first rule out such possible influences on unconventional reasoning as cognitive immaturity, lack of syllogistic skill, and cerebral incompetence. The label "self-deception" is employed when a person utters a conclusion that is counterexpectational, given the premises and the person's assignment of credibility to the classes of evidence from which the premises are adduced. The respondents in Opton's study ordinarily read newspapers and news magazines and assigned credibility to reports of tennis matches, stock market fluctuations, local elections, international trade agreements, and romantic attachments of Hollywood celebrities. When confronted with exemplars drawn from the same class of evidence (news reports of the My Lai massacre), they withdrew their credibility. To "believe" say, a photograph of the finish of a horse race, and "not believe" a photograph of the My Lai incident in the same newspaper would influence a critical observer to consider self-deception as a proper description.

Brief mention should be made of the problem generated in the recognition that self-deception as a term descriptive of ongoing behavior cannot be assigned to oneself. The sentence "I am deceiving myself" is self-contradictory, like the sentences "I am lying," "I am sleeping," and "I am hallucinating." Only when the subject of the sentence is in the second or third person are the sentences freed from internal contradiction. "You are (or he is) lying," "You are (or he is) sleeping," "You are (or he is) hallucinating," "You are (or he is) self-deceiving" are acceptable sentences. Like the other illustrative predicates, "self-deception" can be employed only in a social context. Two actors are required: one whose epistemic activities are under scrutiny, and another who is ready and willing to construct inferences about the epistemic activities of the first actor.

ATTEMPTS TO EXPLAIN SELF-DECEPTION

Self-deception, then, becomes the label of choice when observations lead to the inference that the person "knows" and "does not know" at the same time. A popular, if futile, way of expressing this contradictory state of affairs is to invoke knowing at different levels of consciousness. Because of the opacity of "consciousness" and cognate terms, writers fall back on the use of metaphors that have been chartered by poetic usage. Fingarette,[8] for example, argues that the self-deceiver "knows" the whole story "in his heart." Such metaphorical knowings, which Stephen Crites[9] has dubbed "coronary knowledge," does not take us very far in understanding how people organize their conduct and their epistemic activities so that others assign the label self-deception.

Influenced by psychiatric theories, self-deception is most often used in connection with practices that call out descriptions such as rationalization, denial, repression, selective inattention, subception, hysteria, paranoia, and so on. The usual formulation is that these processes are mechanically activated by unconscious forces. Behind this postulate is the remote assumption that unconscious forces are provided in ontogenetic development as a device for defending against psychic (metaphorical) pain.

FINGARETTE'S THEORY

Fingarette attempted to move out of the mechanistic paradigm. He began from the assertion that the contradictory state of affairs is the product of willed action, not a passive reaction to mechanical forces. With the help of existentialist language, he offered a portrayal of the intentional acts that lead to self-deception:

> . . . the self-deceiver is one whose life-situation is such that, on the basis of his tacit assessment of the situation, he finds there is overriding reason for adopting a policy of not spelling out some engagement of his in the world.

> He does not stop at refusing to spell out what is so. He is forced to fabricate stories in order to keep his explicit account of things and the way things really are in some kind of harmony such as will make his account plausible. However, he does not spell-out that he is doing this. That is, . . . the fabrication he tells us he also tells himself. [Reference 8, page 62.]

"Spelling-out" is the central metaphor in Fingarette's analysis of self-deception. In the context of his analysis, spelling-out is an apt metaphor—it conveys to the reader the imagery of the person articulating, elaborating, detailing, describing causal relations, pointing to intentions and motives, and so on. However, Fingarette's account of the conditions that lead to the refusal or inability to spell-out engagements is less than helpful. He goes on to say:

> ". . . in the particular case of self-deception it is spelling-out, a skill aspect of consciousness, which lies at the heart of the matter. Therefore, if we wish to be able to put matters directly and non-paradoxically . . . we must turn to this largely unexplored way of characterizing consciousness instead of to the familiar knowledge–belief–perception approach. [Reference 8, page 64.]

Consistent with his intentionalist framework, Fingarette discussed spelling-out as a skill that makes it possible for a person to become "explicitly conscious" of something. Modeled after language skill, the skill of spelling-out is universal, and like all skills, is subject to individual variation. The person who does not spell-out or reflect upon his engagements in the world, then, is deficient in this skill. To the skill of spelling-out modeled after a language skill, Fingarette joined an unexpected proposition: that a person's not spelling-out his engagements represents the actions of a morally-flawed person. The conclusion is similar to axiological proposals based on Sartre's notion of "bad faith" (*Mauvaise foi*).[10]

The thrust of my paper differs from Fingarette's analysis. Whether or not the practice of self-deception is consistent with moral rules will depend upon considerations other than the observation that at a given time and place a person did not spell-out his engagement in the world. Cervantes raised the same issue in the 17th century: Is it better to side with Sancho Panza and see the world "as it is," or to join up with Don Quixote and see the world as it ought to be? The ultimate answer is still forthcoming.

SELF-DECEPTION AS THE OUTCOME OF SKILLED ACTIONS

I agree with Fingarette that human conduct is influenced by the skill in spelling-out. I propose an additional skill: the skill in not-spelling-out. The operation of this skill is consistent with the world view that human beings are active participants in their worlds, not passive information processors. Consider Herr von Osten's refusal to repudiate his belief in equine rationality. Rather than attributing this refusal exclusively to his inability to spell-out his engagement with the world (as a form of cognitive or moral default), I would argue that Herr von Osten demonstrated a high degree of skill in not-spelling-out certain features of his world. Similarly, it seems more continuous with observations to attribute to Opton's subjects a skill in not-spelling-out their engagements with reports of the My Lai atrocities rather than a defect in the skill in spelling-out. The same conclusion could be applied to the conduct of the self-deceiving military strategists, to the so-called delusional patient, and to the conduct of hypnosis subjects who cannot remember (about which more will be said presently).

Most previous systematic attempts to explain self-deception, as Fingarette correctly points out, have begun with premises that are at home in analyzing cognition and perception. These premises originated in mentalism, a paradigm that in the 19th and 20th centuries leaned heavily on the machine metaphor. Through logical analysis, answers were sought to the question: What causes the mental machinery to operate in such an irrational way? The outcome of mentalistic analyses are generally schematic or abstract restatements of the phenomenon under scrutiny. Freud, perhaps unwittingly, provided a bridge from prevailing mechanistic models to models that emphasized intentionality. To describe the workings of the mental machinery, he employed the battlefield as a grand metaphor, with allegorical figures at war with one another, deceiving one another, and so on. The rise of phenomenology influenced the unmasking of the allegorical figures and the recognition that instances of self-deception, like instances of deception, were actions of human beings trying to make their way in problematic and ever-changing worlds.

Fingarette's concept, "the skill in spelling-out," and my addition, "skill in not-spelling-out," belong to such an action framework. The task remains: to construct a general statement the aim of which is to identify the conditions under which such epistemic skills are activated. Fingarette employed the philosophy of morality as a fruitful source for his categories; so did Sartre. I propose to use the narrative as the root metaphor.

THE NARRATIVE AS A ROOT-METAPHOR FOR THOUGHT

The use of a metaphor drawn from the humanities rather than from traditional scientific sources requires an apologia. This is a time of "blurred genres" as Clifford Geertz has identified the current refiguration of social thought. The study of human affairs can no longer be regarded as undeveloped natural science awaiting a breakthrough in methodology. Geertz shows how contemporary scholars draw their inspiration and their metaphors from the humanities. Rather than machines or organisms as instruments for reasoning about human affairs, contemporaries are finding the categories of games, dramas, and texts more and more useful.[11] My use of narrative occurs in the same context. The narrative is a vehicle for refiguring actions that appear to be self-deceiving.

This audacious departure from tradition requires the abandonment of such cherished abstractions as libido, instinct, drive, reinforcement, mentality, and so on. Almost 60 years ago, John Dewey put the matter succinctly:

> . . . the novelist and the dramatist are so much more illuminating as well as more interesting commentators on conduct than the schematizing psychologist. The artist makes perceptible individual responses and thus displays a new phase of human nature evoked in new situations. In putting the case visibly and dramatically he reveals vital actualities. The scientific systematizer treats each act as merely another sample of some old principle, or as a mechanical combination of elements drawn from a ready-made inventory.[12]

In the remainder of my paper, I shall try to show that the epistemic skills of spelling-out and not-spelling-out serve a more general purpose than that suggested by those who adopt an axiological framework. The more general purpose served by these skills is story-telling. In a word, the successful narrative takes the flow of experience and makes it intelligible through the use of plot and story line.

I shall make no further argument to support the proposition that the study of self-deception is coterminous with the study of knowing, of epistemic activity. (Here and elsewhere I have elected to use epistemic activity as synonymous with knowing. I avoid the more common word, cognition, because it has become contaminated with the residues of mentalistic psychology.) To enter into the further study of knowing, I adopt the narrative as the root-metaphor.

Etymological analysis is often supportive of a proposition in the absence of refined empirical observation. Such is the case with the connection between narration and knowing. Skeat,[13] among others, traces "know" and "narrate" to a common origin from the Indo-

European *gna*. To narrate and to know are intimately related actions, and current usage would locate the two words at opposite ends of a public–private semantic dimension. In my formulation, narrating and knowing are interdependent concepts, as suggested by the etymology.

To entertain the proposal that the narratory principle is central to human conduct, we can consider almost any slice of life. Our dreams, for example, are experienced as stories, as dramatic encounters, often with mythic overtones. Our fantasies and daydreams are stories. The rituals of daily life and the pageantry of special occasions are organized as if to tell stories. Our rememberings, our plannings, our loving and hating, are guided by narrative. Even films of geometrical figures in random motion are spontaneously emplotted into coherent stories containing common human sentiments and actions.[14,15] The claim that the narratory principle facilitates survival is not to be dismissed as hyperbole. Survival in a world of meanings is problematic without the talent to make up and to interpret stories about interweaving lives.

The action of organizing bits and pieces of experience into a coherent story with a beginning, a middle, and an ending may be called emplotment.[16,17] Unorganized, chaotic, and unsettled "facts," images, recollections, fantasies, and records are ordered into a coherent story, whether in the genre of history, fiction, biography, or autobiography. For the moment, let us consider the historical narrative as a familiar example of emplotment.

EMPLOTMENT

The uninterpreted chronicle of a historical record is not a story. The chronicle must be emplotted according to some governing theme in order to answer questions of order, sequence and connexity. Emplotment is an activity that depends on the use of fictions, i.e., imaginings or as–if constructions that give meaning to the raw materials of the chronicle. Hayden White has convincingly argued that the chronicle of the 19th century could be (and was) emplotted differently by different historians. The style of emplotment expressed by each historian was not dictated by logic but by aesthetic taste. White identified four styles of emplotment that correspond to four common narrative forms: romance, comedy, tragedy, and satire.[16] Similarly, in emplotting a biography the aesthetic preference of the writer intrudes upon the "facts" and he emplots a narrative that is a blend of fact and fancy. The fourfold classification of narrative plots may be used for biography and autobiography as well.

It is only a short distance from biography to autobiography and an even shorter distance from autobiography to the imaginings that a person constructs sequentially to give order to his own life. The silent story, the muted personal narrative, the sequence of episodic imaginings are the unwritten versions of history, biography, and autobiography. As in the historical narrative, the bits and pieces of happenings, doings, and imaginings do not tell a story. To make sense of the flux of events, the person fashions a narrative. The concept of emplotment may be profitably employed here no less than in the discussion of the historical narrative. And the choice of plot or story line does not follow from logic or argument but from aesthetic requirements. For the theory proposed here, the grammatical "I" — the self — is a figure in the constantly expanding silent narrative.

The listener or reader will better grasp the notion of the silent narrative if he or she will recall the organized fantasy life of Thurber's character, Walter Mitty, or any stream of consciousness novel, or his/her own ordered reflections when planning a future event or remembering things past. These are common examples of story-telling where the self is the central narrative figure.

The emplotment of a narrative — whether a novel, a history, or an unvoiced fantasy — is a creative act. From the bits and pieces of fact and fancy, the author selects some items for elaboration and ignores items that would render the plot overly cumbersome, absurd, unconvincing, or lacking in charm and grace. The author in any genre has authorial (i.e., poetic) license. The extrapolation of authorial license from literature to the self-narrative is not gratuitious. If we accept the postulate that human beings are all poets, authors, and creators of tales, then the differential use of spelling-out and not-spelling-out is expected and allowed. We do not charge the poet or other creator of fiction with deception, but rather assign a positively valued label: creative. When the made-up story is about the self, the mere exclusion of a particular fact or even a particular fancy need not carry the implication of neurosis or dishonesty, but of artistry and creativity.

Before we return to Herr von Osten and the other exemplars of self-deception, a further elaboration of the narratory principle is in order. It is necessary to posit a feature of humankind: the need to see things in some order, sequence, and connexity. The plot, or story line, provides the means of creating order out of chaos, linear relations out of randomly occurring events, and a system to say what goes with what. The origins of storylines are obscure. In the history (and probably in the prehistory) of humanity, story-telling has been a pervasive human activity supported by oral traditions. At least as remote as the Homeric epics, the narrative has been recited and heard by eager participants in the world of imagery and imagination. The ancient and still extant practice of guiding moral behavior through the telling of parables and fables, and the continued use of proverbs (which are condensed fables), also suggests the pervasiveness of story-telling. The universality of the story and the poem to entertain and to enlighten, and the omnipresence of special kinds of stories, sacred myths, to illuminate cosmological questions, may also be cited as evidence to support the assertion that the narratory principle is indeed pervasive. The myths contained in political ideologies no less than cosmological myths provide plot models. Contemporary novels and histories promote plots for self-narratives. Goethe's novel, *The Sorrows of Werther*, provided a plot for the suicidal resolution of unrequited love for 18th century romantics. It is said that General George Patton constructed his self-narrative from reading military history. The story of Charlemagne served as the basis for Napoleon's self-narrative.

It is not necessary to invoke some postulated "deep structure" within the nervous system to account for the pervasiveness of the narrative. The skill in using symbols, in talking about absent things as if they were present and present things as if they were absent, and the survival value of sharing the meanings of happenings, are enough to account for the universality and pervasiveness of story-telling. The narratory principle is so much a part of our daily actions that it is unrecognized by most of us. A moment's reflection should convince even the most skeptical that his or her epistemic activities are guided by the search for a coherent story. Whether assembling facts and imaginings for a theory of communication, for predicting the outcome of a personal strategy, or for understanding the jumble of international events, he or she strives for a coherent story that will answer questions of order, sequence, and connexity.

THE CRITERIA FOR COHERENCE

Individual variation in the criteria for coherence is a factor in understanding self-narratives, no less than the narratives of novelists and historians. When the self-narrative remains unvoiced and private, the story-teller must forego the potential benefits (and costs) of public literary criticism. If the story-teller gives voice or pen to his self-story, an observer can play the role of literary critic, or, if the self-narrative is enacted, of drama critic. For example, the critic could declare the self-story unconvincing because the author failed to invent a connection between apparently unconnected episodes. The same critic might call up the label "self-deception" if the self-narrative violated the law of non-contradiction in fashioning a story that had an Alice-in-Wonderland quality. Another critic, however, might express satisfaction, even pleasure, with a self-story that transcended conventional rules of causality.

The narrator, as I said before, cannot employ the self-deception label about his own story. His task is no different from the task of other poets: to experience the world and to tell the story. The criteria for constructing a self-story, then, cannot be the criteria of truth or falsity. By definition, the self-story, like any narrative, is emplotted through the use of fictions. The criteria for coherence, then, must be sought in the vocabulary of aesthetics. Taking some hints from Berlyne,[18] we can characterize the achievement of the aesthetic response as pleasing. Pleasant feelings are associated with the self-story when some problem has been solved and the strain of uncertainty dissolved. As in other art forms, a transformation of elements takes place: in the narrative, isolated fragments of empirical facts and bits of imagery and imagination are transformed into a scenario, a story, a tale. To the extent that the transformation is patterned and shows organization or plot structure, to that extent is the aesthetic goal achieved. Another criterion is that of complexity, necessarily a relative criterion, given such constraining variables as age, education, language facility, role-taking skill, degree of skill in spelling-out, and degree of skill in not-spelling-out. If one were to imagine a scale of simplicity–complexity, aesthetic experience would be achieved when the degree of complexity was appropriate to the comprehension of the story teller.

Another criterion is convincingness; that is, the self-narrative must be credible to the narrator, a condition that is achieved when the narrator prepares his plot so that a new fact or fiction is assimilated into the story—it does not create the conditions for incoherence. To take an example from the genre of the novel: Over and over again Cervantes violates the law of noncontradiction. Yet the reader has no difficulty in following the Quixotic adventures. The violation of conventional principles of causality creates no unresolvable strain in the reader. At the beginning of Part II of Cervantes' book, the two protagonists hold a prolonged conversation with the Bachelor Carrasco. Don Quixote and Sancho express curiosity about the responses of the reading public to the adventures recorded in Part I. The Bachelor is a narrative figure no less than Don Quixote and Sancho, yet the author's artistry compels the contemporary reader to perceive the Bachelor as coming into the story from the world outside the covers of the book for the purpose of telling the two characters how their adventures affected the readers. The mixing of the fictions of the story with fictions about the responses of readers of the story require breaking out of one frame and entering another and then making the return trip. The artistry of the novelist had deftly prepared the contemporary reader who, instead of rejecting the mixture as paradoxical nonsense, is amused by the portrayal of vanity of the leading characters.

The narrator of the self-story, like Cervantes, may fashion his tale by mixing perspectives, or by adding new fictions, or by ignoring old facts or fictions. Either in remembering things past or in forming plans for the future, the self-narrative will influence actions of the story-teller *vis-à-vis* the world of empirical objects and people. It is when the story spills over into action that the observer may notice that the actor has violated conventional expectations. At that point, he may regard the actions as consistent with a diagnosis of self-deception.

REPRISE

In the first part of my paper, I identified the conduct that sometimes earns the label "self-deception." I selected five reference cases to show that the phenomenon is pervasive. From the descriptions offered by other analysts, I concluded that the actions of so-called self-deceivers could be referred to the exercise of two complementary skills: the spelling-out skill and the not-spelling-out skill. In the second part, I undertook to locate these skills in the theory of the narrative. My objective was to show that epistemic activities could be understood in the idiom of story-telling, especially, the silent story-telling in which the self is the principal narrative figure. I drew the conclusion that the appropriate criteria for story-telling had to be drawn from the study of aesthetics.

In this final section, I apply the conclusions to the reference cases, elaborating on the hypnosis studies. Herr von Osten constructed a self-narrative that influenced his conduct not only with Hans but also with the scientists and others who participated in the experiments. The plot from which von Osten constructed his story was a romantic one, a plot that attributed human qualities to animals, including not only rationality, but stubbornness, whimsy, and rascality. His story was kept viable by the exercise of skill in not-spelling-out such features of his world as the findings of the scientists. Opton's subjects were also skillful in not-spelling-out certain features of their worlds. Their refusal to believe the evidence of My Lai was consistent with the myth of the decency of American soldiers and other heroes. The so-called delusional patient had no choice but to believe that her dead brother was alive: to believe otherwise would have made her a murderess, given a story line in which the myth of the interchangeability of wish and deed is emphasized. The stories—called strategies—of military planners were created under highly involving conditions. It is a truism that under high emotional involvement, metaphors become transformed to myths, and fictive stories are transformed to truths. Incoming evidence contrary to the requirements of the plot is rejected or distorted. In rejecting contrary information, the strategist, like von Osten, must exercise his epistemic skills to keep the story whole.

The hypnosis studies have special relevance for our story. It is now an everyday observation among researchers that amnesia for recent events—a counterexpectational occurrence—is not sustained in about half the subjects. Although at first they assert that they cannot remember, under prodding, incentive conditions, and demands for honesty, they "breach." The remainder of the subjects continue to express their inability to remember, even under the usual demand characteristics of the laboratory. It is important to recognize that the hypnosis experiment on amnesia is an almost-ideal laboratory model for testing the implications of theories of self-deception. There is no question about the inputs into the stories for both the experimenter and the subject; they both share the same events. The experimenter remembers, the subject forgets.

The conduct of the two classes of subjects is amenable to analysis through the use of the ideas presented before. Both sets of subjects construct self-narratives, taking into account the actions of the experimenter, the setting, and so on. For the subjects who breach amnesia, the self as narrative figure is spectator. In the silent story, the plan for action, the self is narrator and controls the action. In fact, in the studies reported by Coe and associates,[19,20] these subjects report being in control of their epistemic behavior. Dramaturgy is an apt metaphor for the creative actions of these subjects. When pressed to remember, the form of self-narrative leads to actions related to the self as spectator. For the nonbreaching subjects, the subjects whose behavior keeps alive romantic stories about the effects of states of entrancement or enchantment, the self-narrative is different. The self as narrative figure is in the vortex of action. The silent story requires the self to be highly involved in the actor role. Under such conditions, the narrator must be very skillful in not-spelling-out certain engagements with his world. In the same way that a reader can become involved in a published story, or a listener can become involved in attending to an oral story, so can the hypnotic subject (or anyone else) become engrossed and absorbed in the self-narrative. Kermode, interested in similar phenomena, wrote: ". . . to have the capacity to subvert manifest senses is the mark of good enough readers and good enough texts."[21] I would add "and good enough self-narratives." In the same way that a reader becomes deeply involved in the actions of fictive characters, so can he or she become deeply involved in the self-narrative.

For both classes of subjects, the self-narrative provides a guide to action when the experimenter presses the subject to remember. The differential responses can be attributed to the location of the self in the respective silent plans. The typical nonbreaching subject, in saying, "I don't remember," engages in communicative acts to assimilate the ongoing social action to the plot of the self-narrative. The performance vis-à-vis the experimenter allows the subject publicly to ratify his role as central figure, as hero, in the silent drama; the attribution of heroism is supported by prowess in not-remembering.

A FINAL WORD

The heterogeneity of my examples—Herr von Osten, the psychiatric patient, the respondents in the My Lai study, the planners of military strategy, and the hypnosis subjects—suggests that no one is immune from performing in ways that lead others to apply the label "self-deception." If space permitted, I could increase the heterogeneity by including the epistemic activities of eminent scientists whose theories were challenged, not to mention the actions of ordinary people committed to a particular religious belief or to a political ideology. The universality and the pervasiveness of the phenomenon challenges the claim that persons who engage in certain kinds of epistemic activities are morally tainted or psychologically flawed.[22]

On the assumption that silent story-telling is a common feature of the human condition and on the argument that such story-telling influences belief and action, can anyone of us avoid the risk of being diagnosed as self-deceived? My analysis raises an additional query: What credentials must a person acquire to warrant his/her declaring another person's story as inept or incoherent, or, in the scientific arena, lacking in truth-value?

ACKNOWLEDGMENTS

I am grateful to Professors William C. Coe, Philip Hallie, Joseph B. Juhasz, John I. Kitsuse, Karl E. Scheibe, Joseph Silverman, and Hayden White. Each at one time or another listened patiently to my halting attempts and gave me encouragement as well as advice and criticism. Professor Don Mixon directed me to the apt quotation from John Dewey. Professor Stephen Crites supplied me with a number of his published and unpublished writings which helped me develop my line of argument.

REFERENCES

1. PFUNGST, O. 1965. Clever Hans (The Horse of Mr. von Osten). R. Rosen-

thal, Ed. New York: Holt, Rinehart and Winston, New York, N.Y. [Translated from the German by Carl L. Rahn and published by Henry Holt and Co., 1911.]

2. STUMPF, C. *In* Reference 1.
3. OPTON, E. E. 1971, *In* Sanctions for Evil. N. Sanford & C. Comstock, Eds. Jossey Bass, San Francisco, Calif.
4. CANFIELD, J. V. & G. F. GUSTAFSON. 1962. Self-Deception. Analysis **23**: 32–36.
5. BEN-ZVI, A. 1976, Hindsight and foresight: A conceptual framework for the analysis of surprise attacks. World Politics **28**: 381–395.
6. WHALEY, B. 1973. Codeword Barbarossa. MIT Press, Cambridge, Mass.
7. SARBIN, T. R. & W. C. COE. 1979. Hypnosis and psychopathology: Replacing old myths with fresh metaphors. J. Abnormal Psychol. **88**: 506–526.
8. FINGARETTE, H. 1971. Self-Deception. Routledge and Kegan Paul, London.
9. CRITES, S. 1979. The aesthetics of self-deception. Soundings **42** (2): 197–129.
10. SARTRE, J.-P. 1956. Being and Nothingness. [Transl. Hazel Barnes.] Philosophical Library, New York, N.Y.
11. GEERTZ, C. 1980. Blurred Genres: The refiguration of social thought. Am. Scholar **80**: 165–179.
12. DEWEY, J. 1922. Human Nature and Conduct. Henry Holt and Co., New York, N.Y.
13. SKEAT, W. W. 1963. A Concise Etymological Dictionary of the English Language. Capricorn Books, New York, N.Y.
14. MICHOTTE, A. E. 1963. The Perception of Causality. Methuen and Co., London.Translated by T. R. Miles and E. Miles, from La Perception de la Causalité. 1946. L'Institut Supérior de Philosophie, Louvain, France.]
15. HEIDER, F. & E. SIMMEL. 1944. A study of apparent behavior. Am. J. Psychol. **57**: 243–259.
16. WHITE, H. 1973. Metahistory. Johns Hopkins Univ. Press, Baltimore, Md.
17. SARBIN, T. R. 1977. Contextualism: A World View for Modern Psychology; 1976. Nebraska Symp. on Motivation. A. Landfield, Ed. Univ. Nebraska Press, Lincoln, Neb.
18. BERLYNE, D. 1971. Aesthetics and Psychobiology. Appleton-Century-Crofts, New York, N.Y.
19. HOWARD, M. L. & W. C. COE. 1980. The effects of context and subjects' perceived control in breaching posthypnotic amnesia, J. Personal. (in press).
20. SCHUYLER, B. A. & W. C. COE. 1980. A physiological investigation of volitional & non-volitional experience during posthypnotic amnesia. J. Abnormal Psychol. (in press).
21. KERMODE, F. 1979. The Genesis of Secrecy: On the Interpretation of Narrative. Harvard Univ. Press, Cambridge, Mass.
22. GUR, R. C. & H. A. SACKHEIM. 1979. Self-deception: A concept in search of a phenomenon. J. Personal. Soc. Psychol. **37**: 147–169.

Magical Thinking in the Analysis of Scientific Data*

PERSI DIACONIS

Department of Statistics
Stanford University
Stanford, California 94305

INTRODUCTION

THE CLEVER HANS phenomenon points to the inter-action between an experimenter and the subject being studied. An experimenter's biases, views, and (even) presence can affect the outcome of a study in complex ways. Much of the work on the Clever Hans phenomenon (as surveyed by Rosenthal,[1] for example) focuses on the interaction between experimenters and animal or human subjects. In the examples discussed here, the experimenter is interacting with himself, often through a nonhuman interface: computer output.

I will describe a part of modern statistical practice: exploratory data analysis—the art of finding structure, or simple descriptions, in a set of data. Data analytic goals often clash with more classical statistical goals. This clash is currently being resolved in ways that I hope are instructive to participants at this conference.

The practice of data analysis as described below seems a clear example of magical thinking in the analysis of scientific data. It provides approximate formulas (the simple descriptions) with little or no attempt to explain the formulas from more basic principles. One looks at numbers and tries to find patterns. One is encouraged to follow up leads suggested by background information, imagination, and the patterns perceived. A discussion of magical thinking, relevant to the issues discussed here, is in Schweder.[2]

Exploratory data analysis is described, by example, in the next section. Some problems associated with subjective analysis are then described. After that, available remedies are discussed. The final section argues in defense of magical thinking, when properly labeled.

* This work was partially supported by the National Science Foundation (Grant MCS77-16974).

Statistics and Exploratory Data Analysis.

Classical statistics is, in many ways, an antidote for magical thinking. In principle, models and hypotheses are formulated *before* seeing the data. Then estimates or tests of these assumptions are carried out. Over the past 20 years, statisticians have been widening their role in the analysis of scientific data. A great deal of useful science gets done before more formal evaluation. Also, after formal testing has been carried out there is often much potential juice to be squeezed from a data set. The statistics community has been working on better ways of exploring data sets. The main tools are:

> *Graphical methods:* Visual displays of numbers, scatter plots, and their higher dimensional cousins are examples.
> *Robust/resistant methods:* Statistical techniques that are not sensitive to a few "wild" or outlying observations.

Both sets of tools are aided by interactive computing. The idea is to find:

> Simple structures that describe the bulk of the data, and
> Isolated values — outliers — which are apart from the bulk of the data.

The ideas are best illustrated by example. At the Stanford Linear Accelerator we have a device called PRIM-9. This allows us to view three-dimensional scatter plots. The idea is simple. A scatter plot is shown on a television screen. Scatter plots are normally used to picture two-dimensional data, such as the height and weight of 150 subjects. A third dimension, such as age, can be pictured by rotating the two-dimensional projection in such a way that points that would be closest to the viewer move faster. Parallax then fools the eye into seeing in three dimensions. The machine is set up so that the viewer can interactively rotate the three-dimensional view around in up to nine-dimensional data to try and find "interesting" or "structured" views. Interesting parts of a data set can be masked off and isolated for separate viewing. The name, PRIM-9, is for Picturing, Rotation, Isolation, and Masking in up to nine dimensions. Further details of PRIM-9 can be found in Fisherkeller *et al.*[3] or in a 20-minute color sound film available from the computation research group at SLAC. An example of the successful use of PRIM-9 on a diabetes data set is in Reaven and Miller.[4]

The above is just the tip of the iceberg. But I must refrain from giving further details of data analytic tools and be content with the following references: The leading expositor of innovative data analytic techniques is John Tukey. His book, *Exploratory Data Analysis*,[5] is a fascinating introduction to the subject. It can be read by people who do not have

238 ANNALS NEW YORK ACADEMY OF SCIENCES

strong statistical backgrounds. A center of modern exploratory techniques is Bell Labs at Murray Hill, New Jersey. A recent survey of some of the Bell Labs techniques at work on large, real problems is in Mallows.[6] For those with some statistical training, I recommend Mosteller and Tukey[7] and Gnanadesikan[8] as useful surveys.

SOME PROBLEMS

The skills, background information, and biases of a data analyst will clearly affect the course and final outcome of an investigation. People can imagine patterns in data when there is nothing there. Without a standard ritual to follow, different researchers may come to different conclusions based on the same evidence. A vigorous documentation of the pitfalls associated with subjectivity in the analysis of data is in the interesting book by Barber.[9] A fascinating summary of modern psychological research on the failings of man as an intuitive statistician is in Nisbitt and Ross. Both books describe numerous cases of mistaken inferences based on experimenter bias, preconception, or inadequacy.

I want to report an interesting example here. In "Deathday and Birthday," D. Phillips[11] describes a purely data analytic finding. Phillips found that people's birthdays and time of death are associated, more people dying just after their birthday than just before. The effect seemed more pronounced among famous people. The evidence consists of some graphs and averages.

I put Phillips' findings to a test during a course on sample surveys at Stanford, Winter Quarter, 1980. Thirteen students in the course tested Phillips' claim on new data as part of their final project. The students read Phillips' article, designed a test statistic, and then each took a sample from a book such as *Who's Who*. Without exception, each student's formal test rejected Phillips' hypothesis. I want to mention one student's analysis in particular.

This student worked with *Baker's Biographical Dictionary of Musicians*, 3rd edit. (Schirmer, 1919). She first took a preliminary sample to determine an appropriate sample size. She wound up taking a cluster sample of all names on 100 pages. For each person she recorded (among other things) a "1" if the person died in the six months following their birthday, and "0" otherwise. She was then in a position to estimate the proportion, p, of people in the book who died in the six months following their birthday. If Phillips' claim is right, p should be larger than 0.5. Her estimate was $\hat{p} = 0.5125$ with a standard deviation of 0.0283. Standard tests do not permit rejecting the null hypothesis, $p \leqslant 0.5$, at any reasonable significance level.

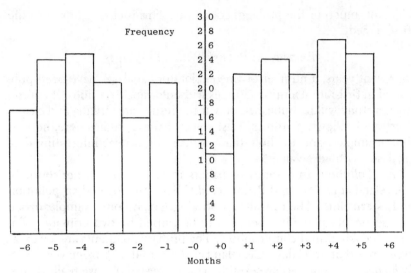

FIGURE 1. The number of months separating birthday from deathday for a large group.

In addition to formal testing, this student also did some clever exploratory analysis of the data she collected. One of her findings is summarized by FIGURE 1.

This graph reports the frequency of people in the sample who died k months from their birthday, $-6 \leqslant k \leqslant 5$. People who died within a month of their birthday are divided into the two groups labeled -0 and $+0$. There seems to be a decline in the 6-month period prior to the birth month. The graph is a fascinating discovery, reflecting hard work and imagination. It highlights a familiar problem. A formal test rejects an hypothesis; an informal, innovative analysis finds a pattern. Which should be believed? In this case we may summarize what we know as follows: The effect is not very strong, if it is there at all. A more sensitive statistic which weights deaths close to birthdays more heavily should be used in the next stage of analysis. Further analysis of this problem is presented in Schulz and Bozerman.[12]

SOME REMEDIES

Statistical practice has developed a number of useful ways of dealing with the problems posed by exploratory analysis. A first step is to carefully distinguish between confirmatory and exploratory analysis. By giving the problem a name and calling attention to the potential subjective element, much of the problem goes away. This is closely linked to the first remedy.

Remedy A: Publish Without p-Values

In recent years, a number of fine exploratory analyses have been published in first-rate scientific journals without a single p-value. Of course, such a study will be published only if the results are striking and clearly worthy of follow up studies. I list some of these studies here, both as fine examples, and to show that journals will accept interesting data analyses without p-values.

Air Pollution. In a series of papers in *Science* and other journals, Cleveland et al. [13-16] and Bruntz et al.[17] have investigated air pollution in Eastern cities. Their results are striking. To give one example: ozone is a secondary pollutant, believed to be caused by two primary pollutants "cooking" in the atmosphere. The primary pollutants are lower on weekends than weekdays. Cleveland et al.[13] found that ozone was slightly *higher*, on average, on weekends. This suggested that we really do not understand how ozone is produced. The papers cited above are particularly noteworthy because of the notorious difficulty of working with air pollution data. The available data base is huge, and not very reliable. There are many groups around the country trying to fit more or less standard statistical models to this data base. I think it is fair to say that the Bell Labs group, using exploratory techniques, have triumphed while classical techniques have faltered.

Psychology. Carlsmith and Anderson[18] contains a nice example of exploratory techniques (and common sense) overturning a finding bolstered by p-values. Much of the work on clustering and scaling can be considered here too.

Economics. Slater[19,20] contain various examples.

Statistics. Hoaglin[21] attacks the problem of finding a simple approximation for chi-squared percentage points by using Tukey's median polish method of fitting two-way tables. The idea of using exploratory techniques to find a theoretical approximation is refreshing.

History. Singer's work[22] is a friendly tutorial of exploratory techniques.

Other nice examples are in Reaven and Miller,[4] Gabb et al.,[23] and Mallows.[6]

Remedy B: Try to Quantify the Damage

There have been several attempts to try to quantify how much standard p-values change in the face of widely used data analytic procedures.

Multiple Comparisons. When many contrasts are considered, some will appear significant by chance. There is a large body of work on adjusting for multiple comparisons. This is eloquently surveyed in Miller.[24,25] Some more specialized problems have been studied. For instance, Bickel and Doksum[26] have studied the effect of allowing nonlinear transformations on the usual two-sample t-test. Freedman[27] has studied the effects of first fitting a big model and then investigating the smaller model dictated by significant coefficients. Both studies show that naive use of p-values can be very misleading.

Bayesian Quantification. A far reaching effort to deal systematically with the problems considered here is presented by Leamer.[28] A widely discussed example is Bode's law. This says that the mean distances of planets from the sun are approximately proportional to $4 + 3 \cdot 2^n$, $n = -\infty, 0, 1, 2, \ldots$. The law sort of fits the facts (there are deviations). A fascinating attempt to quantify a degree of validity for the law, taking into account its failures and the richness of the fitting procedures, is in Good[29] and Efron.[30] Both authors offer an answer to the question: suppose a new solar system is observed, will the distances follow Bode's law? Good concludes the odds are $30+$ to 1 in favor of Bode's law. Efron concludes there are roughly even odds. One problem I have in thinking about Bode's law relates to "the selection effect." We do not know that Bode decided to look at the distance of planets to the sun. Rather, he might well have begun by looking, more generally, at numbers connected to the solar system: weights, separation between planets, density, and much else. Extracting some order for some part of this larger set of numbers might be far easier than suggested by Good and Efron. Good (private communication) told me he did take the selection effect into account. The next rememdy is useful in combating the selection effect.

Remedy C: Try It Out on Fresh Data

This is, of course, a mainstay of the scientific method. Replication on fresh data by another group of experimenters. Even this precaution may fail to insure a valid result. Blondlot's N rays (see Klotz[31] for a modern account) is a well-know case. Some further examples are described in Chapter 5 of Vogt and Hyman.[32] In some problems, such as Bode's law, replication is not a practical possibility. One compromise, which works for large data sets, is what statisticians call cross validation. The idea is that take a (random) sample of a data set, use exploratory techniques on the sample, and then try the result out on the rest of the data. Mosteller and Tukey[7] or Efron[33] contain further discussion.

Remedy D: Remedies to Come

As data analysis becomes more widely recognized, and used, we can look forward to research aimed at aiding the analyst in avoiding self-deception. Here are two suggestions: In the middle of an interactive data analysis session it might be useful to have a display of random, unstructured noise available. This should come in a form close to the data being examined. It is easy to imagine structure, dividing lines, and outliers in uniformly distributed scatter plots. A second suggestion for research is to have an adaptation or specialization of the heuristics and biases described by Tversky and Kanamann[34] along with Nisbet and Ross[10] to the situations encountered in routine data analysis.

In Defense of Magical Thinking

Many of us believe that the new exploratory techniques are a mandatory supplement to more classical statistical procedures tied to normal error models and linearity. The argument is two-pronged:

> First, exploratory techniques work in finding useful structure where classical techniques fall flat. This is shown in the examples cited above in Remedy A.
>
> Second, in many ordinary applications of statistics the "usual assumptions" are not even approximately valid. Therefore, the resulting p-values or "levels of significance" are meaningless. Let us give up the pretense of p-values along with the limitations of classical procedures.

It seems clear that magical thinking is here to stay. Modern data analysis provides a nice example, showing how an acceptance of subjectivity of this sort can lead to rich rewards via an active pursuit of the goals of magical thinking. At the same time, an open labeling of subjective, frankly speculative analysis seems a useful antidote to the problems described above.

Acknowledgments

I thank Ellen Silberg for allowing me to report her study of the birthday/deathday phenomenon. Bill Cleveland, Jerry Friedman, and I. J. Good provided helpful comments.

References

1. Rosenthal, R. 1966. Experimenter Effects in Behavioral Research. Appleton-Century-Crofts, New York, N.Y.
2. Shweder, R. 1977. Likeness and likelihood in everyday thought. Curr. Anthropol. 18: 637–658.
3. Fisherkeller, M. A., J. Friedman, & J. Tukey. 1974. Prim-9, An interactive multidimensional data display system. Stanford Linear Accelerator Publication, 1408.

4. REAVEN, G. & R. MILLER, R. 1979. An attempt to define the nature of chemical diabetes using multidimensional analyses. Diabetologia 16: 17–24.
5. TUKEY, J. 1977. Exploratory Data Analysis. Addision-Wesley, Reading, Mass.
6. MALLOWS, C. 1979. Robust methods — Some examples of their use. Am. Stat. 33: 179–184.
7. MOSTELLER, F. & J. TUKEY. 1977. Data Analysis and Regression. Addison-Wesley, Reading, Mass.
8. GNANADESIKAN, R. 1977. Methods for Statistical Data Analysis of Multivariate Observations. Wiley, New York, N.Y.
9. BARBER, T. 1976. Pitfalls in Human Research. Pergamon Press, New York, N.Y.
10. NISBETT, R. & L. ROSS. 1980. Human Inference. Prentice Hall, Englewood Cliffs, N.J.
11. PHILLIPS, D. 1972. Deathday and Birthday: An unexpected connection. In Statistics: A Guide to the Unknown. J. M. Tanner et al., Eds. pp. 52–65. Holden Day, San Francisco, Calif.
12. SCHULZ, R. & M. BOZERMAN. 1980. Ceremonial occasions and mortality: A second look. Am Psychol. 35: 253–261.
13. CLEVELAND, W., T. GRAEDEL, B. KLEINER & J. L. WARNER. 1974. Sunday and workday variations in photochemical air pollutants. Science 186: 1037–1038.
14. CLEVELAND, W., B. KLEINER, J. McRAE & J. WARNER. 1976. Photochemical air pollution: Transport from the New York City area into Connecticut and Massachusetts. Science 191: 179–181.
15. CLEVELAND, W., B. KLEINER & J. WARNER. 1976. Robust statistical methods and photochemical air polluation data. J. Air Poll. Control Assoc. 26: 36–38.
16. CLEVELAND, W. & T. GRAEDEL. 1979. Photochemical air pollution in the Northeast United States. Science 204: 1273–1278.
17. BRUNTZ, S., W. CLEVELAND, T. GRAEDEL & B. KLEINER. 1974. Ozone concentration in New Jersey and New York. Science 189: 257–259.
18. CARLSMITH, M. & C. ANDERSON. 1979. Ambient temperature and the occurrence of collective violence: A new analysis. J. Personal. Soc. Psychol. 37: 337–344.
19. SLATER, P. 1974. Exploratory analyses of trip distribution data. J. Regional Sci. 14: 377.
20. SLATER, P. 1975. Petroleum trade in 1970: An exploratory analysis. IEEE Trans. Systems–Man and Cyber. (March): 278.
21. HOAGLIN, D. 1977. Direct approximation for chi-squared percentage points. J. Am. Stat. Assoc. 72: 508.
22. SINGER, B. 1976. Exploratory strategies and graphical displays. J. Interdiscip. Hist. 7: 57–70.
23. GABB, S. D. 1967. Statistical analysis and modeling of the high-energy proton data from the Telestar satellite. Bell Syst. Technic. J. 46: 130.
24. MILLER, R. 1977. Developments in multiple comparisons 1966–1976. J. Am. Stat. Assoc. 72: 799–788.
25. MILLER, R. 1966. Simultaneous Statistical Inference. McGraw-Hill, New York, N.Y.

26. BICKEL, P. & K. DOKSUM. 1980. An analysis of transformations revisited. J. Am Stat. Assoc. In print.

27. FREEDMAN, D. 1980. A remark on multiple regression equations. Mimeographed report. Dep. Statistics, Univ. California.

28. LEAMER, E. E. 1978. Specification Searches: Ad Hoc Inference with Non-Experimental Data. Wiley, New York, N.Y.

29. GOOD, I. J. 1968. A subjective evaluation of Bode's Law and an objective test for approximate numerical rationality. J. Am. Stat. Assoc. 64: 23-66.

30. EFRON, B. 1971. Does an observed sequence of numbers follow a simple rule? (Another look at Bode's Law). J. Am. Stat. Assoc. 66: 552-559.

31. KLOTZ, I. M. 1980. The N-ray Affair. Sci. Amer. 242: 168-175.

32. VOGT, E. & R. HYMAN. 1968. Water Witching U.S.A. Univ. Chicago Press, Chicago, Ill.

33. EFRON, B. 1981. Lecture Notes on Bootstrap and Jackknife Techniques. SIAM, Philadelphia Pa.

34. TVERSKY, D. & D. KANAMANN. 1974. Judgment under uncertainty: Heuristics and biases. Science 185: 1124-1131.

Success at Detecting Deception: Liability or Skill?

BELLA M. DePAULO
Department of Psychology
University of Virginia
Charlottesville, Virginia 22901

THE STUDY OF DECEPTION, in my research, has been an investigation of humans as lie-detectors. I have been interested in people's ability to detect the lies told by other people, in the ways that they usually detect lies in their everyday lives—namely, without the aid of polygraphs, stress-detectors, or any other mechanical gadgetry. One might think that the importance of this research endeavor depends in part on the pervasiveness of deceit; if people rarely ever tell lies, then the study of human lie-detection skills might seem like a frivolous indulgence. I have no idea how often people tell outright, bald-faced lies, but if we include under the rubric of deception little white lies, and also what we might call lies of self-presentation — that is, attempts to present ourselves as a little more kind, a little more sensitive, a little more intelligent, and a little more altruistic than in fact we really are—then I suspect that the phenomenon of deception is quite pervasive indeed.

But I think that our perceptions of whether or not deception is occurring are important regardless of their accuracy. If we believe that another person is lying to us, and if we are confident enough about that inference to base our subsequent behavior on it, then in some ways it is almost irrelevant whether or not that the other person actually is lying. So accuracy at detecting lies is interesting and important, but so is inaccuracy. Our perception of the honesty, sincerity, or genuineness of another person's verbal or nonverbal behavior might be construed as a kind of meta-communication judgment: we are trying to determine not so much what the communication means, what the semantic content is, but whether or not we should believe it, whatever its content. So in this way, our perceptions of deceptiveness might moderate many of the other kinds of judgments we make about other people, and influence the way we interact with them.

0077-8923/81/0364-0245 $01.75/0 © 1981, NYAS

Numerous paradigms have been used to study the effectiveness of the human lie-detector, but all share one common constraint: there must be a criterion for accuracy. If researchers are to determine when people are accurate in their judgments of deceptiveness, they must know when the people who are being judged are in fact lying and when they are telling the truth. One common approach is to ask people various questions, usually questions about autobiographical material, or inquiries about their opinions on various issues, while surreptitiously signalling them to tell a lie in response to a certain subset of those questions. These people are either observed right then and there by live human lie-detectors, or they are videotaped, and those tapes are later shown to judges who then try to determine when the people on the tapes were lying and when they were telling the truth. In one of the more realistic studies,[1] subjects were induced to cheat on an ESP task, and were subsequently asked what they thought might account for their amazing performance. Most often, the subjects in these studies have been college students, but other populations that have been sampled include children,[2-5] nurses,[6-8] mental patients,[9] and official U.S. customs inspectors.[10]

Since there are currently more than two dozen studies of human lie-detectors,[11] we are now in a position to make an educated estimate of just how effective people really are at this judgmental task. The first point that should be noted is that there is considerable variability in the outcomes of these studies: sometimes people are quite successful at detecting lies, and other times they are simply miserable. The second point is that they are never perfect and are usually far from it. The third and final point about overall accuracy is that, if we average over all of these studies, accuracy at detecting lies is quite substantially better than it would be if subjects were just guessing. The studies in which subjects were usually fooled by the liars are far outnumbered by the studies in which subjects were correct in their judgments more often than would result by chance alone.

There are several ways in which the existing studies of lie-detection might overestimate people's abilities at this task. For one thing, in real life, people probably have to have some reason to suspect that another person might be telling a lie; otherwise, it may never occur to them that this is even a possibility. In psychological research, subjects are necessarily already given this edge: to find out whether people can detect lies, we have to ask them whether or not they think someone else is lying. It is much harder for us to find out if they would have realized that there was deception going on, had we not explicitly asked them.

Also, even though people are often right about their judgments of deceptiveness, they do not always know when they are right. So some of the time when their judgment is correct, they may not be confident enough about their judgment to actually act on the basis of it. Other times, they might be very confident, but very wrong. Effectively, then, people's useful lie-detection capacities might fall short of their actual lie-detection skills.

In other respects, though, our paradigms for studying the detection of deception might underestimate people's actual abilities. The samplings of behavior upon which subjects are asked to base their judgments are often quite short—mere snippets from the stream of behavior. A 20-second clip, for example, would not be at all unrepresentative. Also, the subjects in our studies are usually deprived of all sorts of situational and contextual cues that might usually aid them in their real-life lie-detection attempts. Certain commonly available cues about the liars are absent, too, since the liars in psychological research are usually complete strangers to the lie-detectors. Furthermore, we sometimes hold constant just those factors that really do covary with deception. Let us suppose, for example, that men lie more than women; we do not really know if this is true, but let us suppose that it might be. Well, a subject who tried to use that knowledge in his or her inferences would probably be led astray, since researchers are usually very careful to make sure that their male and female subjects are instructed to lie equally often. Finally, most lie-detection paradigms are noninteractive. Subjects are constrained merely to observe the liar, and are not allowed to unleash their whole repertoire of probes, traps, and leading questions that they might ordinarily set up to try to catch a liar when left to their own devices.

Given all of these qualifications, we can not really say with certainty just how effective people are in their real-life lie-detection attempts, but we do have some informed ideas about the kinds of conditions that affect people's rate of success at detecting lies. Most of the conditions that we know the most about have to do with the sources of cues or sources of information that are available to the human lie-detector. Usually these sources of information are particular channels or modalities, such as the face, the body, the tone of voice, or words. One methodology for studying the effects of different kinds of information on lie-detection success is becoming increasingly popular. Usually, the researchers start with a videotape of people who were lying or telling the truth in response to various questions. Then, the subjects who are to detect the lies are given access to different parts of the available information. Some observe the full videotape, with the accompanying sound track. Others see only the visual portions of the tape, with all sound turned off. Within the visual-only conditions, sometimes some subjects are shown only the faces of the liars and others are shown only the liars' bodies. Still other groups of subjects hear only the sound track, without any visual information whatsoever. Other subjects might hear an audiotape that has been distorted in various ways so that the tone of voice is intelligible but the actual words are not. And finally other subjects might simply read typed transcripts made from the audiotapes. Any given study usually includes only a subset of these conditions, but enough different studies sampling enough different combinations of conditions have been conducted to suggest a number of conclusions.

But first let me say a word about the theoretical rationale behind this particular methodology. I want to keep this part brief because the theory is Ekman and Friesen's,[9] and Paul Ekman might want to tell you more of the details later this afternoon. Basically, Ekman and Friesen argue that, for a variety of social, cultural, and anatomical reasons, people are better able to control and monitor what they express through certain channels than through others, and they are also more likely to try to control certain channels than others. So, for example, people are very likely to try to control their facial displays when they are telling lies, and they are also quite skillful at doing so. In effect, when people lie to you, they try to fake you out with their faces.

It should follow from that, then, that access to facial cues might actually be misleading, rather than useful or informative, to the person who is trying to detect lies. The available data are quite consistent with this formulation. In the five studies I know of that measured subjects' accuracy at detecting deception when they had access either to speech and to facial cues or when they only heard speech without seeing any facial cues, subjects never did significantly better at detecting lies when they did have access to facial cues and sometimes actually did worse.[12-16]

But an even stronger effect that has emerged from all of these "channels" or "modalities" studies is a somewhat unexpected one: namely, that people are much better at detecting lies when they have access to any source of information that includes words (whether in the form of an undistorted audiotape, a videotape with sound, or just a transcript) than they are when they have access only to purely nonverbal cues (such as the face alone or the body alone). This conclusion is based on the results of more than 20 studies.[11] The reason it came as a surprise to some is that one might think that the verbal channel might be the most readily controlled of all channels, and hence the one that would be most likely to fool the human lie-detector. Adult liars, particularly college student liars, probably have quite a lot of practice and a good deal of skill at carefully picking and choosing their words. But this skill is countered by the skill of the detector, who has had just as much, if not more, practice at deciperhing verbal messages as the liar has had at constructing them.

Probably other factors are important, too. For instance, although we can perhaps quite readily pick and choose the meanings that we want to convey, we may be somewhat less adept at selecting just the right way to convey that meaning, so as not to be caught in our lies. There are many different ways that we could use words to convey the impression that we really like a person whom we actually detest, but not all of those ways would be equally convincing.

Although verbal communications of deception, on the whole, are more likely to be detected than nonverbal ones, there are some indications that not all verbal messages are alike in this respect. It seems to be the case that lies are most readily detected when judges have access to words and tone of voice, as they do when listening to an undistorted audiotape or when communicating with someone by intercom or over the telephone. Again, this conclusion was derived from studies that restricted judges' access to only the audiotape, or only transcripts, or only face or body cues, or some combination of these.

In real life, however, these kinds of restrictions are probably the exception rather than the rule; usually, we have full audiovisual access to the people that we are observing or addressing. What the study of isolated modalities suggests is that we perhaps do not apportion our attention in the optimal way when attempting to detect deception. If we did use the most effective strategy in all of our lie-detection attempts, then we should do just as well when we have words and tone and facial cues available to us as we do when we have access only to words and tone. As I noted earlier this is not the case. The fact that people do best at detecting lies when they have access only to an audiotape suggested to me that people might be helped in their lie-detection attempts by a little hint—the hint would be to pay particular attention to the tone of voice of the potential liar. To test this, Dan Lassiter, Julie Stone, and I did a very simple study[17] in which we gave all of our lie-detecting subjects access to a full videotape with sound. We told a quarter of them to pay particular attention to the tone of voice of the speakers, we told another quarter to pay particular attention to the words, another quarter to pay particular attention to the visual cues, and we left the rest to their own devices. Consistent with our hypothesis, the strongest effect in the data was that the subjects instructed to pay particular attention to the tone of voice did significantly better at detecting lies than the subjects who were given no special attentional instructions. We think this result is especially intriguing because the manipulation was in a sense a rather open-ended one: we never told our subjects what particular aspects of the voice they should be attending to. My hunch is that they did a much better job than they would have if we did try to specify particular cues.

Our manipulation in our attentional study was an asymmetrical one:

we told our judges to pay particular attention to the tone of voice cues, but we never told the original people who were telling the lies that observers might be paying particular attention to their tone of voice. And in fact, one of the main reasons we think that people do so well at detecting deception either when restricted to speech cues or when told to pay particular attention to tone of voice cues is that ordinarily, people are either not very adept at controlling their voices or else are not very likely to come up with the idea that they should try to control their voice tones.

What might give people the idea that they should try to control their voice tones? One way might be to make the vocal channel particularly salient to them, such as by having them communicate by using an intercom. In one of his studies of deception, Bob Krauss and his students did just that:[14] they had half of their subjects communicate by intercom, and the others face-to-face. All subjects were secretly videotaped. They then showed these tapes of the intercom subjects and the face-to-face subjects in three different ways: the full videotape with sound, only the videotape, or only the audiotape. The highest rate of lie-detection was not earned by subjects in the audio-only condition; instead, the subjects who were best at detecting lies were those who saw only the visual (mostly facial) cues of those liars who were communicating by intercom. Remember that the intercom subjects had no idea that anyone would ever be observing their facial expressions. Thus the facial channel, which is ordinarily very carefully controlled, was probably relatively uncontrolled in this condition, and the subjects attempting to detect deception on the basis of these uncontrolled visual cues were quite successful.

Let me briefly summarize what I have said so far. Humans attempting to detect deception, though far from perfect at the task, are usually substantially better than chance. They are especially good when they have access only to the words and the tone of voice of the liar and not to the liar's facial expressions or when they have access to all cues but are told to pay particular attention to the tone of voice cues. All of this is true, when the liars are unaware that their vocal cues are going to be under special scrutiny; in these usual, unaware conditions, liars are busily managing their facial expressions, leaving their vocal cues relatively unguarded. But if they are somehow tipped off to the fact that their speech might be especially salient to observers, then they can do a much better job of controlling it than they do ordinarily.

I could go on and talk about other conditions that seem to increase or decrease people's success at reading deception, but instead I am going to stop right here and consider a different question—the question of whether skill at detecting deception is a good thing or a bad thing.

When we study nonverbal and verbal communications that are not deceptive, this question is almost never an issue. If another person feels happiness or sadness or anger of fear, and is intentionally and deliberately trying to communicate that feeling, then it should be to the observer's benefit to be skillful at understanding that feeling. And in fact, numberous studies have shown that people who are especially adept at reading these kinds of intentionally communicated messages (we sometimes call them "pure" messages to distinguish them from mixed messages or deceptive messages) are better adjusted psychologically and more effective in their interpersonal relationships than people who are less sensitive to these kinds of cues.[18]

Deceptive messages, though, are not so pure. People who are lying are expressing something different from what they really feel, and so as Ekman & Friesen have pointed out,[9] there are several different kinds of cues that observers might pick up. First, they might simply notice that something is not quite right, that outward appearances somehow do not seem very genuine or real. These kinds of impressions might lead observers to suspect that there is some deception taking place. That is the metacommunicational judgment—the judgment that the overt message, whatever it is, may not be true. That judgment is the easy part.[5,19] The much more difficult part is figuring out what the person really does feel. In the example of the person who is pretending to like someone she or he actually detests, observers might suspect that the communication is a deceptive one, and they might even identify liking as the dissembled affect—the affect that the person is trying to convey. Still, that does not automatically tell them that the person's true feeling is dislike. Instead, the person might actually feel ambivalently or indifferent toward the person being described, or the speaker might simply feel somewhat less positively than she or she is pretending to feel.

The question, again, is: which of these affects should the observer discern? Are we better off seeing right through to a person's true, underlying feelings or might we sometimes do better not to see what another person does not want us to know?

If other people are going to act toward us on the basis of their true feelings—especially if they are going to act in a way that is going to be harmful or insulting or damaging, and if our knowledge of these true feelings can help us to prevent the harmful actions—then it should be the case that skill at detecting underlying, true feelings is an important asset. Also in certain professions, such as medicine and psychiatry, sensitivity to these true but covered-up feelings might be especially beneficial.

Despite these somewhat plausible arguments that at least under some conditions, people should be good at reading true, underlying affects, most of the available evidence indicates that they are not at all skilled at this.[19,20] And further, in what I think is an even more dramatic finding, there is evidence to suggest that with age, people actually get worse and worse at picking up these true affects that are masked by feigned ones.[5] Let me tell you a little bit about how we discovered this.

We started by making a videotape of eight speakers—four males and four females—who were describing people they knew. Each speaker described someone they liked, someone they disliked, someone they felt ambivalently about, and someone they felt indifferent toward. They also gave two deceptive descriptions: they described the person they really liked, pretending to dislike him or her, and they described the person they really disliked, pretending to like him or her. We then showed this videotape to sixth graders, eighth graders, tenth graders, twelfth graders, and college students, and asked them to tell us three things about each description: first, how much the speaker liked the person she or he was describing; second, how deceptive the description was; and third, to what degree the speaker seemed to have mixed feelings about the person being described. We stressed that they were to report what they thought the speakers really felt, and not what they were pretending to feel. Thus, if a speaker was pretending to like a person that she or he actually detested, subjects were told to rate that speaker as feeling dislike.

Our subjects did get better with age at calling the deceptive descriptions deceptive and the "pure" like and dislike descriptions honest. However, in their judgments of affects—that is, in their ratings of how much the speakers liked each of the persons they were describing, subjects reported what the speakers were trying to convey, rather than what they actually felt. So if a speaker was pretending to dislike someone she or he really did like, that speaker was seen as truly disliking the person being described. Moreover this tendency to see the feigned affect rather than the true affect, which occurred at every age level, also increased markedly with age. It seems that what children are learning as they grow older, probably through socialization, is politely to read what other people want you to read, and not what they really feel.

Consistent with this "politeness" interpretation, we also found a sex difference in this judgment: females, even more so than males, tended to read the overt, intended message rather than the covert, true affect, and this sex effect was essentially stable across all five age levels.

Though females were worse than males at seeing the speakers' true, underlying affects when the speakers were lying, they were quite substantially better than males at distinguishing liking from disliking when the speakers were telling the truth. Thus, when the speakers really did like the persons they were saying that they liked, and had no reason to try to hide these feelings, females were very much better than males in their understanding of those cues. Females were not as much better than males at recognizing the mixed feelings in an ambivalent message, and they were not at all better than males at determining which descriptions were deceptive and which were not. Ambivalent messages, we think, are more covert and unintended than are pure, honest messages of liking and disliking; deceptive messages are the most covert and unintended of them all. (By "unintended" I mean that speakers usually do not want these messages to be read by other people.) This tendency for women to lose more and more of their advantage over men in understanding verbal and nonverbal cues as these cues become more and more covert and unintended has been termed the "eavesdropping" phenomenon:[21] females appear to be polite and accommodating in their judgments, as they refrain from eavesdropping on those cues that other people do not want them to perceive.

It is possible to classify people according to the degree to which they show this tendency to be especially skilled at reading overt and intended cues but less skilled at reading covert and unintended cues. These scores measuring politeness or nonverbal accommodation can then be related to other outcome measures, such as measures of people's interpersonal effectiveness or personal traits and styles. We have these kinds of measures for some of the high school and college samples that we have tested. For example, we asked the teachers of our high school samples to rate each of their students on scales measuring social understanding and popularity with the same and opposite sex. Students who were rated as more understanding and more popular tended to be just those students who showed the pattern of nonverbal politeness that I described: they were much better at reading overt and intended cues than covert and unintended ones.[21] When we asked one of our college student samples to give us their own impressions of the quality of their interpersonal relationships we found the same result: people who felt better about their relationships with other people were people who tended to see the cues that others would want them to see, but to overlook the cues that others probably would prefer that they miss.[21,22]

I think what all of this suggests is that at least in some ways, in some situations, it may be better for us to see only what other people want us to see and not what they really feel. This is what children seem to be learning as they grow up — our data suggest that they do get better at understanding the conveyed affect but do not get better at understanding the experienced affect, when that affect is different from what is being overtly expressed. Moreover, for people who do not obey this politeness formulation — a formulation which describes women especially well — there seem to be personal and interpersonal costs. They are seen by others as less understanding and less popular, and they themselves feel less satisfied with their interpersonal relationships.

By calling this a "politeness" mechanism, I might seem to be suggesting that the strategies of the polite decoder are very other-oriented ones: the decoder is attempting to accommodate to the other person's wishes, to help maintain the other person's intended appearances and in other ways facilitate the other person's goal in the interaction. And I do think that this is a large part of what is going on. But it is not only the other person who is benefiting from it. The accommodating person, as I noted before, is liked more and feels better about him or herself. Also, the polite mode of decoding is probably an easier way of dealing with interpersonal information than a more probing and skeptical style would be. The accommodating decoder seems to take things at their face value, which at least under some circumstances might be the easiest, quickest, and safest way to deal with the many complex and multileveled affective messages that people sometimes seem to be conveying.

Also, seeing only what you are supposed to see might be a more comfortable interpersonal style. That is, this style might be simpler not only cognitively but also emotionally. People who begin to doubt external appearnces are first of all going to experience more uncertainty; they may also feel guilt about their suspiciousness and lack of trust; and finally, they might find out something about the other person's feelings toward them that they might be much happier not to know.

When we move into the realm of deception, then, the rules and regulations and reward systems that usually govern our verbal and nonverbal worlds get turned inside-out and upside-down. Sources of information such as the face, which are ordinarily extremely informative, can instead be downright misleading, and the kinds of skills that we usually get rewarded for—like the ability to understand what other people are really feeling—can instead function more like liabilities. The person who knows when deception is occurring and who knows what other people are really feeling has a more accurate grasp of what the interpersonal world is really like. But in some ways, under some circumstances, maybe being this good at understanding social and interpersonal cues is just no good at all.

REFERENCES

1. MEHRABIAN, A. 1971. Nonverbal betrayal of feeling. J. Exp. Res. Pers. 5:64–73.
2. FELDMAN, R. S. 1979. Nonverbal disclosure of deception in urban Korean adults and children. J. Cross-Cultural Psychol. 10: 73–83.
3. FELDMAN, R. S., L. DEVIN-SHEEHAN & V. L. ALLEN. 1978. Nonverbal cues as indicators of verbal dissembling. Am. Educ. Res. J. 15: 217–231.

4. FELDMAN, R. S., L. JENKINS & O. POPOOLA. 1979. Detection of deception in adults and children via facial expressions. Child Devel. **50**: 350–355.
5. DePAULO, B. M., A. IRVINE, A. JORDAN & P. S. LASER. 1980. Age changes in detection. *In* Nonverbal Behavioral Skill in Children. R. Buck, Chair. Symp. Psychol. Assoc., Montreal.
6. EKMAN, P. & W. V. FRIESEN. 1974. Detecting deception from the body or face. J. Pers. Soc. Psychol. **29**: 288–298.
7. EKMAN, P., W. V. FRIESEN & K. R. SCHERER. 1976. Body movements and voice pitch in deception interaction. Semiotica **16**: 23–27.
8. EKMAN, P., W. V. FRIESEN, M. O'SULLIVAN & K. SCHERER. 1980. Relative importance of face, body, and speech in judgments of personality and affect. J. Pers. Soc. Psychol. **38**: 270–277.
9. EKMAN, P. & W. V. FRIESEN. 1969. Nonverbal leakage and clues to deception. Psychiatry **32**: 88–106.
10. KRAULT, R. E. & D. POE. 1980. Behavioral roots of person perception: The deception judgments of customs inspectors and laymen. J. Pers. Soc. Psychol. **39**: 784–798.
11. DePAULO, B. M., M. ZUCKERMAN & R. ROSENTHAL. 1980. Detecting deception: Modality effects. *In* Review of Personality and Social Psychology. L. Wheeler, Ed. Sage, Beverly Hills, Calif..
12. DePAULO, B. M., R. ROSENTHAL, C. R. GREEN & J. ROSENKRANTZ. In submission. Verbal and nonverbal revealingness in deceptive and nondeceptive communications.
13. HARRISON, A. A., M. HWALEK, D. F. RANEY & J. G. FRITZ. 1978. Cues to deception in an interview situation. Soc. Psychol. **41**: 156–161.
14. KRAUSS, R. M., V. GELLER & C. OLSON. 1976. Modalities and cues in the detection of deception. Meet. Am. Psychol. Assoc., Washington, D. C.
15. LITTLEPAGE, G. & T. PINEAULT. 1978. Verbal, facial, and paralinguistic cues to the detection of truth and lying. Pers. Soc. Psychol. Bull. **4**: 461–464.
16. MAIER, N. R. F. & J. A. THURBER. 1968. Accuracy of judgments of deception when an interview is watched, heard, and read. Personnel Psychol. **21**: 23–30.
17. DePAULO, B. M., G. D. LASSITER & J. I. STONE. In submission. Attentional determinants of success at detecting deception.
18. ROSENTHAL, R., J. A. HALL, M. R. DiMATTEO, P. L. ROGERS & D. ARCHER. 1979. Sensitivity to nonverbal communication: The PONS test. Johns Hopkins Univ. Press, Baltimore, Md.
19. DePAULO, B. M., & R. ROSENTHAL. 1979. Telling lies. J. Pers. Soc. Psychol. **37**: 1713–1722.
20. FELDMAN, R. S. 1976. Nonverbal disclosure of teacher deception and interpersonal affect. J. Educ. Psychol. **68**: 807–816.
21. ROSENTHAL, R. & B. M. DePAULO. 1979. Sex differences in eavesdropping on nonverbal cues. J. Pers. Soc. Psychol. **37**: 273–285.
22. ROSENTHAL, R. & B. M. DePAULO. 1979. Sex differences in accommodation in nonverbal communication. *In* Skill in Nonverbal Communication. R. Rosenthal, Ed. Oelgeschlager, Gunn & Hain, Cambridge, Mass.

Edible Symbols:
The Effectiveness of Placebos

DANIEL E. MOERMAN

Department of Anthropology
University of Michigan—Dearborn
Dearborn, Michigan 48128

PERHAPS THE GREATEST single work that can be said to belong to the genre of medical anthropology is Levi-Strauss's essay "The Sorcerer and his Magic" published in *Structural Anthropology* in 1963. The most interesting part of that essay is Levi-Strauss's rendering of Boas's story of Quesalid, a Kwakiutl Indian shaman. Quesalid, a skeptic of existential proportions, learns the *ars magna* of one of the great shamanistic schools of the Northwest coast: he learns how to hide the little ball of down in a corner of his mouth, how to bite his tongue in the proper place, how to suck at his patient, how to throw up the bloody down in such a way as to make it seem he has extracted the "pathological foreign body" from his stricken patient. Houdini-like, Quesalid will expose these fraudulent jugglers. And we watch him, in four short pages, learn the awesome truth: his fraudulent technique is magnificently successful. As Levi-Strauss tells us, by the end of Boas's narrative, Quesalid "takes pride in his achievements, and warmly defends the technique of the bloody down against all rival schools. He seems to have completely lost sight of the fallaciousness of the technique which he has so disparaged at the beginning."[1]

Levi-Strauss poses, but, in my opinion, does not solve, the problem of how fallacious technique can be successful. Given Western notions of causality, the problem is a terrible enigma.

One of the most interesting problems for the medical anthropologist is trying to understand how placebos work. When physicians or other healers prescribe inert medications, by design or otherwise, we find that roughly 35% or 40% of the time, the patient experiences a reduction in the severity of his illness, and in about the same proportion of cases (though not necessarily the same cases), the physician can detect an amelioration of disease. What kind of act is the prescription of a placebo? What does it mean that such an act can influence human physiological processes? How are we to think about effective but false technique?

0077-8923/81/0364-0256 $01.75/0 © 1981, NYAS

Let me note than an interest in placebo healing does not spring specifically from an interest in non-Western, primitive, or folk medicine. Modern biomedicine has as large a component of placebo healing as any. The fact that it has notably more specific effectiveness than primitive medicine in some areas probably enhances the general effectiveness of the physician as this specific success enhances the enthusiasm with which he treats others. While clearly not the only factor involved, physician enthusiasm is one dimension of nonspecific effectiveness that has been demonstrated by careful research.[2-4] This is a curious situation. Physicians and medical historians consistently overrate placebo effectiveness in primitive medicine, and underrate it in modern medicine. That a very tender nerve is involved seems clear to me from the fact that, when I suggest the contrary, I find myself accused of inflating the pharmacological effectiveness of primitive medicine, and devaluing the same in Western medicine ("Well, let's see what happens when *you* get pneumonia!").

It is doubtless the case that there is real medical power, both pharmacological and surgical, in primitive medicine. Quinine, picrotoxine, strophantine, emetine, coca, various salicylate preparations, ephedrine, strychnine, curare, chaulmoogra oil (an anti-leprosy drug), and *Rauwolfia serpentaria* are among the better known drugs derived from tribal practice. Among the most interesting recent work in this area has been Etkin's demonstration of the antioxidant qualities of several Hausa antimalarial drugs, notably *Acacia arabica, Azadiracta indica,* and *Guiera senegalensis.*[5] The pages of *Lloydia,* the *Journal of Economic Botany,* and the *Journal of Ethnopharmacology* are ample testimony to the great empirical experiment in medicine in which mankind has been involved since the Middle Paleolithic.[6]

It is also doubtless the case that there is a substantial placebo component in modern biomedicine. The healing of any sickness involves three components, which are only analytically distinct. The first and most important is autonomous healing. Most sickness is "self-limiting" for two reasons: first, the human organism is remarkably resistant and adaptable with a battery of immunological devices at its disposal, and second, it is generally in the evolutionary interest of disease pathogens to minimize damage done to their hosts that they might continue to have hosts. The second factor is specific medical treatment, be it a physician's injection of penicillin or streptomycin for an infection in one culture, or a Montagnais healer's prescription of the boiled bark of red willow (*Salix lucida*) for a headache in another.[7] The third, and the most interesting component of healing, is that curious intersection of the first two, wherein the act though not the content of medication somehow triggers the body to autonomous healing. Here we are dealing with a decisively human act, where the meaning of the injection transcends the content of the syringe, where the act of taking a pill transcends its chemical constitution. The Iroquois observation that wild cranesbill (*Geranium maculatum*) has a "hook like ensnaring quality" transcends its 10% to 20% tannin constitution as a reason for prescribing it for healing sores in the mouth.[8]

A small number of published case studies of addiction to placebos attest to the profound influence inert drugs can have on people.[9-13] Controlled studies of several sorts indicate the magnitude of this third force in medicine. It is, of course, the fact of placebo effectiveness that necessitates double-blind trials of drugs to establish effectiveness. Since few studies have been carried out specifically to determine the characteristics of placebo medication itself, most of what we know about the subject comes from such drug trials.

Since 1955, legions of investigation have cited Beecher's seminal paper to the effect that, for a variety of conditions, placebos provided satisfactory relief in 35.2 ($\pm$2.2)% of cases.[14] And, subsequently, in many studies of placebo relief of *pain*, the figure 35% (plus or minus a few) recurs. Many investigators have concluded from this that placebos are effective 35% of the time. This may in part account for the long and largely unsuccessful quest for means for identifying the "placebo reactor." As an aside I might note that I am aware of no studies designed to identify placebo "nonreactors," a hypothetical group of patients deprived, for whatever reason, of the great benefit of placebo healing who, thereby, deserve extra clinical attention. But of course the search for placebo reactors has, by and large, not been carried out to learn about placebo reactions themselves, but rather to decide how to exclude reactors from drug trials.

It seems to me that Beecher's finding has been misinterpreted. What is particularly interesting about placebo effectiveness rates is less mean effectiveness than variability in effectiveness. Thus, placebo effectiveness rates reported by Beecher ranged from 15% to 58%. And a further search of the literature indicates the following end points on this range of variation. Diligent (seemingly interminable) search has finally disclosed a substantial, though not very well controlled, study in which a placebo effectiveness rate of zero was established: late in the Second World War, 260 placebo-treated sailors and 303 untreated sailors showed, respectively, 34% and 35% seasickness rates in landing barges in waters off California.[15] Unfortunately, this finding has not been replicated; others studies of seasickness indicate substantial placebo effects. At the other end of the scale, a number of recent of controlled studies of cimetidine therapy for ulcers have demonstrated placebo healing rates ranging from 8% to 83%.[16-18] An earlier controlled study of antacid therapy showed a placebo improvement rate for ulcer and several other gastrointestinal disorders of 92%.[19] What accounts for this extraordinarily variability in the outcome of inert treatment?

There are four variables in the placebo process, which I will consider in turn: the patient, the sickness, the situation, and the physician.

Most research aimed directly at clarifying the placebo process has involved patients; this research has been, by and large, fruitless. A standard design has been to divide a group of patients into placebo responders and nonresponders. Then the groups are compared on a variety of personality measures. These studies have yielded a variety of results, indicating that reactors were more neurotic and extroverted than nonreactors in one study,[20] more acquiescent in another,[21] "outgoing, verbally and socially skilled, and generally well adjusted" as opposed to "belligerent, aggressive, and antagonistic to authority" in another,[22] tending to exhibit "higher anxiety and lower ego strength and self-sufficiency" in another,[23] and so on. So placebo reactors are neurotic yet well-adjusted, extroverted yet with low ego strength, acquiescent yet antagonistic to authority. Generally it seems that the least significant variable in the equation is the personality of the patient. Attempts to screen out placebo reactors fail: a group of nonreactors on the first trial usually produce a percentage of reactors on the second trial more or less the same size as on the first.[24-26]

Perhaps the most interesting aspect of patient response to placebos is that there is some evidence that patients can respond favorably to them knowing what they are. A nonblind placebo trial carried out among a small series of 15 outpatients psychiatric patients at Johns Hopkins was remarkably successful. Patients were told that while their cases were being fully worked up, they would be prescribed "sugar pills" which had helped many other people with similar conditions. Of 15 patients, 14 completed the trial; only one was not substantially better according to both patient and physician assessment, and her husband had attempted suicide during the week-long trial. Six of the patients found the placebos to be so effective that they concluded the physicians had lied to them and had prescribed active drugs. Others were more inventive in accounting for their improved condition: "Every time I took a pill I thought of my doctor and how I'm doing. It just reminds you that you're trying to change yourself" said one patient.[27]

There is then, little consistent evidence regarding the qualities of patients to allow us to make clear statements about placebo effectiveness. There is recent equivocal evidence suggesting that "acquiescence" is a patient characteristic which predicts placebo responders;[28] I will return to this argument later.

One might imagine that the underlying disease process in a given sickness might be of some consequence for placebo effectiveness. Substantial placebo effectiveness is demonstrable in syndromes as diverse as pain, hypertension, wound healing, depression, anxiety, rheumatoid arthritis, warts, and acne. There is evidence from Levine's important study (where naloxone, an opiate antagonist, both inhibited and stopped placebo pain relief in dental postoperative patients) indicating that endorphin release mediates placebo analgesia.[29] It seems possible that the same mechanism is partly involved in placebo relief of angina and arthritis. But it is unlikely to be involved in depression, anxiety, or warts! Several sicknesses that show substantial placebo effectiveness involve immune mechanisms (warts, arthritis, asthma). A large body of research indicates substantial correlations between various psychological states and immune mechanisms;[30] while the mechanisms involved remain unclear, this "openness" of the immune system increases one's confidence in the possibility that this system can be manipulated symbolically. Again, recent work in the etiology of coronary artery disease suggests that the classic thrombosis model (which says that myocardial infarction is caused by blood clots blocking coronary arteries) is wrong; that instead, angina and infarction may be caused by coronary artery "spasms,"[31] which in turn are caused by neurological impulses or emotional stress; heart attacks then are not random mechanical events, but are rather under the control of mental or neurological processes — this system may also be an "open" one, increasing our confidence that symbolic manipulations of it are possible.

That pain, immune mechanisms, and coronary vasospasm are all *equally* accessible to symbolic manipulation seems unlikely; in any case, published data exhibit such wide variations in placebo effect rates *within* syndromes as to prevent useful comparisons *between* them, although I remain convinced this is an important avenue for study.

A third consideration is what is referred to as "situation factors." Several studies have demonstrated a distinct influence of the form or color of medication. In one study, British medical students were told they were testing either stimulants or sedatives; pink placebo capsules tended to act as stimulants while blue placebo tablets tended to act as sedatives.[32] In an Italian study of placebo influence on sleeping, women tended to prefer blue placebo sleeping tablets while men preferred orange.[33] In a British study, the tranquilizer oxazepam was shown to improve symptoms of anxiety more effectively when presented in a green tablet, and to improve symptoms of depression more effectively in a yellow tablet.[34] Several studies have shown that two placebo capsules are more effective than one.[32,35] Another study showed that a group of patients receiving different placebos in three consecutive two-week periods improved more than a group receiving one placebo for a six-week period.[36] The effectiveness of a placebo tends to correlate closely with the reputation of effectiveness of the drug for which the placebo is substituted: in a general psychiatric clinic double-blind study, a 24% placebo response rate was found when placebo was compared to a mild tranquilizer, a 35% rate when compared to modest doses of stronger tranquilizers, and 76% placebo response rate when compared to modest doses of stronger tranquilizers prescribed along with the psychological testing.[37]

But as we move from the color of the tablet to its "strength," we are no longer only considering a situational factor but something directly and indirectly a function of the attitude of the nurse or physician, for it is he or she who professes to know about the strength of the medication. And there is strong evidence, direct and indirect, to indicate that physician attitude has a substantial effect on the outcome of treatment.

Consider a recent report from France describing the outcome of two related studies of ulcer treatment.[38] Neither double blind study showed that the drugs tested had any influence on ulcer healing; both showed substantial placebo effectiveness compared with an untreated control group. So far, these are common enough results. But, in this case, there were substantially and significantly different outcomes in placebo effectiveness in the groups treated by different physicians. The measure of treatment effectiveness was the number of days of ulcer pain reported following the first consultation (FIGURE 1). For 30 untreated patients, the mean days of pain following first consultation was 19.5 days. For several groups of patients treated with placebos, outcome varied as a function of the physician: 12 days of pain for one physician's patients, 7 days for two others, and 3.5 days for a fourth. In all cases, placebo treatment was effective, but one physician was three times more effective than another in alleviating pain using inert treatment.

These kinds of differences have been noted in studies carried out in different hospitals. One author has noted that the same antacid treatment for ulcer in one hospital study had a 79% effectiveness rate while in another hospital it had a 17% effectiveness rate; similarly in another study, placebo effectiveness rates for ulcer pain relief ranged from 25% to 45% in different hospitals.[39]

At yet another level, several authors have recently suggested that placebo healing rates for ulcer are as much as twice as high in the United States as they are in Europe and Great Britain.[40,41]

Another recent study provides strong direct evidence of the effect of physician attitude on placebo response.[42] This complex study compared the effects of status of communication of drug effects (dentist vs. technician), attitude of both dentist and technician ("warm" vs

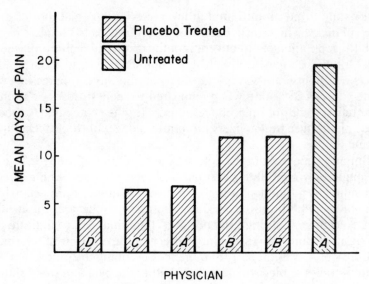

FIGURE 1. Mean days of pain in six patient groups treated by four physicians. The placebo effect varies with physician. (After Sarles *et al.*[38])

"neutral"), and the message of drug effects ("oversell" vs. "undersell") on placebo outcome. Dental patients were given placebo pills, then asked to rate the amount of pain associated with a mandibular-block injection. "The most salient of four variables [in accounting for the differences] was the type of information contained in the message of drug effects." The difference, in this study, between messages of "oversell" and "undersell" of drug effects was, in fact, quite modest:

> Oversell: This is a recently developed pill that I've found to be very effective in reducing tension, anxiety, and sensitivity to pain. It cannot harm you in any way. The pill becomes effective almost immediately.

> Undersell: This is a recently developed pill that reduces tension, anxiety and sensitivity to pain in some people. Other people receive no benefit from it. I personally have not found it to be very effective. It cannot harm you in any way. The pill becomes effective almost immediately if it's going to have an effect.

This "undersell" message seems to me more pessimistic than the "oversell" message seems optimistic. Yet this modest difference had more influence on placebo effectiveness than the source of the message, or the attitudes of the individuals providing treatment.

This study pinpoints the most important factor behind the notion that drug effectiveness is historically transitory: the 19th century French physician Armand Trousseau is alleged to have said, "You should treat as many patients as possible with new drugs while they still have the power to heal."[43]* Consider the recent paper by Benson and McCallie on the placebo effect in angina pectoris.[43] Reviewing the history of several now discredited treatments for angina, the authors note a dramatic difference in effectiveness rates depending on whether the physicians were enthusiastic or skeptical regarding the value of the medication: "The initial 70 to 90 per cent effectiveness in the enthusiasts' reports decreases to 30 to 40 per cent 'base-line' placebo effectiveness in the skeptics reports."[43] Of course, most of the initial clinical reports were open trials; but once it has been demonstrated that placebo and drug effectiveness is the same, we can infer that the enthusiasts' would have gotten the same high effectiveness rates for placebos, had their studies been genuinely double-blind. It seems possible that this may account for some of the very high placebo effectiveness rates reported in the studies of cimetidine for ulcer mentioned earlier. Recall as well that the disease Benson and McCallie are discussing is a grave one; while it may be easy for a dermatologist to get away with writing a frivolous article about "exorcising warts,"[45] the same is less easily done regarding such a fundamental organic disease as angina pectoris.

Available evidence indicates that the most important variable in nonspecific effectiveness is physicians. Twenty years ago Whitehorn and Betz were able to predict how effective psychiatrists would be in treating schizophrenics by examining their answers to 10 of 400 questions on the Strong Vocational Interest Inventory.[46] We know little more today about this problem that we did twenty years ago.

It is in this context that I will consider the one finding on "placebo reactors," mentioned earlier, that has been replicated. Fisher and Fisher in 1963, and McNair and colleagues in 1967 and again in 1979 have successfully predicted placebo response in patients who score high on the Bass social acquiescence scale.[47] On this scale, one obtains a high score by agreeing with a number of homilies and cliches: "Obedience is the mother of success; Seeing is believing; One false friend can do more harm than 100 enemies." I do not wish to characterize the personalities of medical center physicians who tend to carry out this kind of research. But in the context of the ample research that indicates strong "physician effect" and weak "patient effects," these "acquiescence" findings suggest to me a strong hypothesis.

It is clear enough that the personality of the physician, and the enthusiasm with which he embraces his procedures can create a symbolic field which, perceived by the patient, can influence him to trigger

*This quote is also attributed to William Osler.[44]

autonomous healing mechanisms. This is a rich and even visceral system of communication where the healer and the patient exchange concept and response, wittingly or otherwise, in a powerful healing ritual.

What kind of communication is this? It is not simply deceptive communication, in the style of the juggler—it *might* be, but it need not be, and in most cases it is not. Quesalid learned that lesson quickly enough.

In George Sanders Peirce's terms, placebos are signs in a multitude of ways. Inert pills, sharing sensible formal characteristics with active ones, are *icons*; insofar as they "represent . . . a parallelism in something else," they are metaphors.[48] In addition, as metonymic representations of the whole medical experience, the summation and conclusion of an entire medical conversation, they are *indexes*, or pronouns—indicating without naming or describing. As such, they share with others of this type the quality of startling us, focusing our attention.[48] (" 'It just reminds you that you're trying to change yourself,' said one patient."[27])

And finally, placebos are *symbols*, arbitrarily linked by language to their various meanings: the same inert pill might, with appropriate explanation serve to reduce pain, heal ulcers, prevent seasickness, remove warts, induce either sleep or wakefulness.

And, to add to the complexity, active pills, sharing formal sensible characteristics with inert ones, are iconic representations of them. Edible icons, edible index, edible symbol; a pill is the sign of the highest order.

Against this perspective, it is paradoxical in the extreme that the most common explanation of placebo effectiveness is that the phenomenon is a "conditioned response."[49] The source of this explanation seems to be a small number of papers published in the early 1960's asserting evidence for a placebo effect in laboratory animals. Saline injections in rats were shown to induce responses similar to those induced by scopolamine,[50] alcohol,[51] and *d*-amphetamine[52,53] in properly conditioned rats. Such responses could not be evoked when animals were treated with Dexedrine, LSD-25,[51] or chlorpromazine.[53] Since there is evidence that physiological saline is not inert when injected in rats, but rather acts as a depressant,[54,55] only the cases of the placebo amphetamine reaction, and the absence of a placebo chlorpromazine reaction seem to require explanation. Moreover, there is evidence that indicates that rat strains differing in temperament have different responses to both saline and amobarbital (another depressant).[56]

These technical criticisms aside, it seems possible that one might condition rats to evoke drug responses with inert substances. Leafing through any introductory psychology text suggests that one might train rats to do most anything. It also seems reasonable to suggest that one might similarly condition human beings. But to suggest that such animal responses are the same as the human placebo effect is the rankest form of eliminative thinking; the fact that both mice and elephants have tails does not make elephants into mice.

In general anthropological terms, specific medical treatment in some sort of ideal university-medical-center form is highly univocal, structural, context-independent, ordered, causal, deliberate, and direct. Payment is by check, preferably through a third party. General medical treatment occurs in ideal form in a farm town in Iowa, during a house call, by old Doc Jones, who stays for tea, and checks out an ailing hog while he's there; it is multivocal, antistructural, context-dependent, egalitarian, humane, and mutual. Payment is in kind: the kids shine up his Buick, and he takes home a chicken. These are both ideal types, which rarely occur in pure form. But medicine, no doubt, admits varying degrees of structure; and as such it is, perhaps, the most truly liminal profession. The physician straddles nature and culture, physiological life and symbolic life, structure and antistructure. Following Victor Turner, we cannot help but see the creative possibilities in this vital stew.[57] The physician can expand knowledge from his experience with the body; he can extend life from his experience with knowledge. Perhaps we can account for medicine's seemingly inevitable drift to monism (spiritual or statistical) by imagining that such creativity, living dialectically, is exhausting.

Exhaustion does not relieve the burden of explanation. There is yet a plethora of experiments to do, complex conceptual problems to be thought through, and ethical dilemmas to be considered. A more complex and useful semiotic study would be hard to imagine.

REFERENCES

1. LEVI-STRAUSS, CLAUDE. 1967. The sorcerer and his magic. *In* Structural Anthropology. Doubleday, Garden City, N. Y.
2. SHAPIRO, A. P., T. MYERS, M. F. REISER & E. B. FERRIS. 1954. Comparison of blood pressure response to veriloid and to the doctor. Psychosom. Med. **16**: 478–488.
3. ULENHUTH, E. H., K. RICKELS, S. FISHER, *et al.* 1966. Drug, doctor's verbal attitude and clinic setting in symptomatic response to pharmacotherapy. Pharmocologia **9**: 392–418.
4. ENGELHARDT, D. M. & R. MARGOLIS. 1967. Drug identity, doctor conviction and outcome. Neuropsychopharmacology, Vol. **5**: 543–544. Excerpta Medica Foundation, Amsterdam.
5. ETKIN, NINA. 1979. Indigenous medicines among the Hausa of Northern Nigeria: Laboratory evaluations for potential therapeutic efficacy of antimalarial plant materials. Med. Anthropol. **3**: 401–429.

6. SOLECKI, RALPH. 1975. Shanidar IV, a Neanderthal flower bural in northern Iraq. Science **190**: 880–881.
7. SPECK, FRANK G. 1917. Medicine practices of the Eastern Algonkians. Proc. 19th Internat. Congr. Americanists. pp. 303–321.
8. HERRICK, JAMES W. 1977. Iroquois Medical Botany. Ph.D. dissertation, SUNY-Albany. University Microfilms, Ann Arbor, Mich.
9. LESLIE, A. 1954. Ethics and practice of placebo therapy. Am. J. Med. **16**: 854–862.
10. VINAR, O. 1969. Dependence on placebo: A case report. Br. J. Psychiatr. **115**: 1189–1190.
11. BOLELOUCKY, Z. 1971. A contribution to the problem of placebo dependence: A case report. Activ. Nerv. Super. **13**: 190–91.
12. HONZAK, R., E. HORACKOVA & A. CULIK. 1972. Our experience with the effects of placebo in some functional and psychosomatic disorders. Activ. Nerv. Super. **14**: 184–85.
13. MINTZ, IRA. 1977. A note on the addictive personality: Addiction to placebo. Am. J. Psychiatr. **134**: 327.
14. BEECHER, H. K. 1955. The powerful placebo. J. Am. Med. Assoc. **159**: 1602–1606.
15. TYLER, D. B. 1946. The influence of placebo, body position and medication on motion sickness. Am. J. Physiol. **146**: 458–466.
16. BODEMAR, G. & A. WALAN. 1976. Cimetidine in the treatment of active duodenal and prepyloric ulcers. Lancet (2): 161–164.
17. BLACKWOOD, W. S., D. P. MAUDGAL, R. G. PICKARD, et al. 1976. Cimetidine in duodonal ulcer: Controlled trial. Lancet (2): 174–176.
18. SCHEURER, U., L. WITZEL, F. HALTER, et al. 1977. Gastric and duodonal ulcer healing under placebo treatment. Gastroenterology **72**: 838–41.
19. BACKMAN, H., H. KALLIOLA & G. OSTLING. 1960. Placebo effect in peptic ulcer and other gastrointestinal disorders. Gastroenterologia **94**: 11–20.
20. GARTNER, M. A. JR. 1961. Selected personality differences between placebo reactors and nonreactors. J. Am. Osteopath. Assoc. **60**: 377–378.
21. FISHER, S. & R. L. FISHER. 1963. Placebo response and acquiescence. Psychopharmacologia **4**: 298–301.
22. MULLER, B. P. 1965. Personality of placebo reactors and nonreactors. Dis. Nerv. Sys. **26**: 58–61.
23. WALIKE, B. C. & B. MEYER. 1966. Relation between placebo reactivity and selected personality factors. Nursing Res. **15**: 119–123.
24. BATTERMAN, R. C. 1957. Placebo and non-reactors to analgesics. Fed. Proc. **16**: 280.
25. LIBERMAN, R. 1967. The elusive placebo reactor. Neuropsychopharmacology, Vol. 5: 557–566. Excerpta Medica Foundation, Amsterdam.
26. SHAPIRO, A. K. n.d. The placebo effect. In Principles of Pharmacology. W. G. Clark & J. DelGuidice, Eds. 2nd edit. Academic Press, London.
27. PARK, L. C. & L. COVI. 1965. Non-blind placebo trial. Arch. Gen. Psychiatr. **12**: 336–345.
28. McNAIR, D. M., G. GARDOS, D. S. HASKELL & S. FISHER. 1979. Placebo

response, placebo effect and two attributes. Psychopharmacology **63**: 245-50.

29. LEVINE, J. D., N. C. GORDON & H. L. FIELDS. 1978. The mechanisms of placebo analgesia. Lancet (2): 654-657.
30. SOLOMON, G. F. 1969. Emotions, stress, the central nervous system and immunity. Ann. N.Y. Acad. Sci. **164**: 335-342.
31. MASERI, A., A. L'ABBATE, G. BAROLDI, *et al.* 1978. Coronary vasospasm as a possible cause of myocardial infarction. N. Engl. J. Med. **299**: 1271-1277.
32. BLACKWELL, B., S. S. BLOONFIELD & C. R. BUNCHER. 1972. Demonstration to medical students of placebo responses and non-drug factors. Lancet (1): 1279.
33. CATTANEO, A. D., P. E. LUCCHELLI & G. FILIPPUCCI. 1970. Sedative effects of placebo treatment. Eur. J. Pharmacol. **3**: 43-45.
34. SHAPIRA, K., H. A. MCCLELLAND, N. R. GRIFFITHS & J. D. NEWELL. 1970. Study of effects of tablet color in treatment of anxiety states. Br. Med. J. (2): 446.
35. RICKELS, K., P. T. HESBACKER, C. C. WEISE, *et al.* 1970. Pills and improvement: A study of placebo response in psychoneurotic outpatients. Psychopharmacologia **16**: 318-328.
36. RICKELS, K., C. BAUM & K. FALES. 1963. Evaluation of placebo responses in psychiatric outpatients under two experimental conditions. *In* Neuropsychopharmacology. Vol. **3**: 80-84. Excerpta Medica Foundation, Amsterdam.
37. LOWINGER, P. & S. DOBIE. 1969. What makes the placebo work? A study of placebo response rates. Arch. Gen. Psychiatr. **20**: 84-88.
38. SARLES, H., R. CAMATTE & J. SAHEL. 1977. A study of the variations on the response regarding duodenal ulcer when treated with placebo by different investigators. Digestion **16**: 289-92.
39. LITTMA, A., R. WELCH, R. C. FRUIN, *et al.* 1977. Controlled trials of aluminum hydroxide gels for peptic ulcer. Gastroenterology **73**: 6.
40. HIRSCHOWITZ, B. I. 1977. Histamine H-2 receptor antagonists. Ann. Int. Med. **87**: 373-375.
41. GUDJONSSON, B. & H. M. SHAPRIO. 1978. Response to placebos in ulcer disease. Am. J. Med. **65**: 399-402.
42. GRYLL, S. L. & M. KATAHN. 1978. Situational factors contributing to the placebo effect. Psychopharmacology (Berlin) **57**: 253-61.
43. BENSON, H. & D. P. MCCALLIE. 1979. Angina pectoris and the placebo effect. N. Engl. J. Med. **300**: 1424-1429.
44. TAYLOR, GORDON R. 1979. The Natural History of the Mind. E. P. Dutton & Co., New York, N.Y.
45. LITT, J. Z. 1978. Don't excise—exorcise: Treatment for subungual and periungual warts. Cutis **22**: 673-676.
46. WHITEHORN, J. C. & B. J. BETZ. 1960. Further studies of the doctor as a crucial variable in the outcome of treatment with schizophrenic patients. Am. J. Psychiatr **117**: 215-223.
47. MCNAIR, D. M., R. J. KAHN, L. F. DROPLEMAN & S. FISHER. 1967. Compatibili-

ty, acquiescence and drug effects. *In* Neuropsychopharmacology. Vol. 5: 536–542. Excerpta Medica Foundation, Amsterdam.

48. PEIRCE, CHARLES S. 1965. Collected Papers Vol. II: Elements of Logic. Harvard Univ. Press, Cambridge, Mass.

49. JOSPE, MICHAEL. 1978. The Placebo Effect in Healing. D.C. Heath Co., Lexington, Mass.

50. HERRNSTEIN, R. J. 1962. Placebo effect on the rat. Science 138: 677–678.

51. BLACK, C. J., JR. 1963. An attempt to establish a placebo effect in rats with psychopharmacological agents ethyl alcohol, Dexedrine, lysergic acid diethylamide. Dissertation Abstracts 24(4): 1692.

52. ROSS, S. & S. B. SCHNITZER. 1963. Further support for a placebo effect in the rat. Psychol. Rep. 13: 461–462.

53. PIHL, R. D. & J. ALTMAN. 1971. An experimental analysis of the placebo effect. J. Clin. Pharmacol. 11(2): 91–95.

54. SCHNITZER, S. & S. ROSS. 1960. Effects of physiological saline injection on locomotion activity in C57 BL/6 mice. Psychol. Rep. 6: 351–354.

55. MEIER, G. W. 1962. Suppression of activity following physiological saline injection: Still more variables. Psychol. Rep. 11(2): 333–334.

56. POWELL, B. J. 1967. Prediction of drug action: Elimination of error through emotionality. Proc. 75th Ann. Conv. Am. Psychol. Assoc. Vol. 2: 69–70. American Psychological Association, Washington, D.C.

57. TURNER, VICTOR. 1967. Forest of Symbols. Cornell Univ. Press, Ithaca, N.Y.

Mistakes When Deceiving

PAUL EKMAN

Human Interaction Laboratory
Langley Porter Institute
Psychiatry Department School of Medicine
University of California, San Francisco
San Francisco, California 94143

INTRODUCTION

WHEN A PERSON purposefully withholds certain information and presents false information in a credible fashion, two types of mistake may occur. An expression or gesture may suggest that the person is engaged in deception without revealing just what is being concealed. Wallace Friesen and I have used the phrase *deception clue* to distinguish this from those instances in which the mistake provides what we called *leakage* of the concealed information.[1] Consider the dinner party guest who leaves much earlier than the rest of the guests, telling the host how much he regrets that a very early business meeting the next morning requires that he depart this most enjoyable gathering. Suppose that as the guest says this, he engages in a prolonged hand-to-hand manipulation, using one hand to scratch and pick the other. If noticed this could be a deception clue, tipping off the host that the guest is at least quite uncomfortable, and perhaps therefore his excuse should be regarded as false and his enjoyment but a mask. Unless there was leakage, the host would not, however, know the information concealed — why the guest actually was leaving. Was he bored, irritated by the partner the host had assigned to sit next to him, off for a rendevous, or did he not want to miss the next installment of *Masterpiece Theater?*

Confession is not the same as leakage, although confession often follows or is compelled by a very noticeable leakage incident. In confession the betrayal of the concealed information is not unintended or unwitting. Instead, the confessor deliberately gives up the deceptive pretense, volunteering the true information.

Much of the current research on facial expression and body movement has been motivated by the possibility that these behaviors, more

* The research that was the basis for this report was supported by a grant (MH 11976) and a Research Scientist Award (MH 06092) from the National Institute of Mental Health and a grant from the Harry F. Guggenheim Foundation. This report is drawn from a book in preparation, *Uncovering Deceit.*

0077-8923/81/0364-0269 $01.75/0 © 1981, NYAS

than the words, provide leakage and deception clues. Freud expressed this hope in an often cited quote, ". . . if his lips are silent, he chatters with his finger-tips, betrayal oozes out of him at every pore."[2] This seductive possibility has fascinated the popular press,[3] and been attested to by clinicians.[4-7] Yet there has been little research directly examining this matter and a paucity of theory explaining why nonverbal behavior† might be more reliable than words. Even those investigators who have studied nonverbal behavior in a deceptive interaction have failed to provide much explanation of why nonverbal behaviors would be a source of leakage.

Obviously leakage and deception clues do not always occur. Some deceits succeed. Even when leakage and deception clues occur, they need not always be detected. Deceits that could have failed may succeed. When deceit fails it may be betrayed in words, rather than, or in addition to, face, body, and voice. The very line taken by the deceiver may, as it is elaborated, become so circuitous or improbable to suggest a lie. Or, a slip of the tongue may provide leakage of concealed information. Yet, there are times, particular types of social interaction, and particular moments within those interactions, when nonverbal behavior may be a particularly rich source of leakage and deception clues. This report will attempt to explain why and when mistakes may be most prevalent in nonverbal behavior.

While Wallace Friesen and I have been conducting experimental research on deception for the past 11 years,[1,11,12] this report is almost totally theoretical. There is a point of contact between these speculations and our quantitative research. Much of our research has been directed towards testing bits of our theory, in particular the notion that face and body differ in leakage and predictions about how deception clues are revealed in hand movements and facial expressions. The theory has developed in part to explain problems we encountered in thinking about how to study deception, and what to make of unanticipated findings.

---

† We have agreed in a previous paper (Reference 9) with Sebeok's assertion (Reference 8) that nonverbal behavior is a terrible term, but we know of no better phrase to designate facial behavior, body movement, and posture. *Motor behavior*, while technically correct, seems more appropriate to skills and abilities. *Visual behavior* refers to the sensory apparatus involved in perceiving all but the tactile events. *Kinesics* implies a particular theoretical view, first promulgated by Birdwhistell (Reference 10), that body movement can be best understood by applying methods and concepts from linguistics. *Expressive behavior* is also problematic, seeming to imply that these actions only express inner emotions or personality. By nonverbal behavior, we exclude relative distance, which is currently termed proxemics. We also exclude changes in voice tone, loudness, pitch, rate of speaking, pauses, and so on, which we consider to be *vocal behavior*. *Verbal behavior*, as we will use it here, refers to the content of spoken behavior, the words, their arrangement, but not the manner in which they are spoken, which is vocal.

Emotion Arousal as a Source
of Leakage and Deception Clues

When emotion is aroused certain changes occur in face, body and voice that can be considered *automatic*,‡ and in this way different from the changes in the content of speech. By automatic I mean that the changes occur quickly, without deliberate choice, and at least initially go unnoticed by the person showing them.

The term automatic does not mean the behavior changes are necessarily involuntary. Nor does it mean that they cannot be interrupted or inhibited. The changes in face, body, and voice during emotional arousal are *not* reflexes or fixed action patterns, which run their course until completion. Quite the contrary, the changes due to emotional arousal are susceptible to deliberate or habitually imposed control. They may be attenuated, masked, interrupted, or inhibited. The term automatic is meant to suggest that these behavior changes seem to occur without deliberate choice, very quickly. The person does not experience the changes in his behavior as something he intended to do. He may often not notice the changes in his behavior, at least at the outset. When he becomes aware of what is happening, his subjective experience is likely to be one of struggle if he tries to inhibit these changes in his own behavior.

Two separate but interrelated arguments can be made about why changes in face, body, or voice occur in any automatic fashion when emotion is aroused. One argument is based upon the proposition that certain changes in behavior are biologically programmed to occur when emotion is aroused. The other line of argument emphasizes the early development of habits linking certain behavior changes to emotion. For our purposes here it does not matter whether both or only one of these arguments is correct, although specific predictions about how leakage and deception clues may be revealed would vary with the basis that is postulated for automatic changes during emotional arousal. (Elsewhere each argument is elucidated, the evidence reviewed, and the implications for leakage detailed.[16]) All that must be granted is that on the basis of biological programming and/or learning, when emotion is aroused it is likely that certain changes may automatically occur in face, body, or voice.

Verbal behavior is not unaffected by emotion, but it is different.

‡ Mandler's use (Reference 13) of the term *automatic* is similar to what we mean. He described automatic processes as operating without requiring attentional conscious work, originating through preprogramming (innate) or habits that through some process such as overlearning become unconscious. Also see Zajonc (Reference 14). Elsewhere (Reference 15) I have described more specifically my use of the term automatic in relation to emotion.

When fear is aroused, for example, there is no pressure that impels a set of words to pop out of the mouth, tantamount to the backwards jerk of the torso, or a facial muscular contraction. What is said is deliberate, at least for that moment. The person speaks intending to transmit a message. He is aware of what he says.

Emotional arousal may cause the person to speak intemperately, saying more, or saying it more strongly than he might otherwise. Emotional arousal may interfere with the ability to speak, it may produce various speech disruptions, but we consider those phenomena as vocal behaviors. The scream that may occur when fear is aroused, or the sound "Ooaah" or something like it, may be impelled or automatic like the facial and body changes, but we consider that also as vocal not verbal behavior. And, further, the person would be aware of his scream at the moment he makes it.

The differential impact of emotional arousal on face, body, voice, or words suggests that when a person conceals an emotion he is experiencing, there should be more leakage in the face, body, or voice than in the words. Words that are more deliberate would be less likely to unintentionally leak the emotion experienced. The more automatic changes in face, body, and voice would, if not managed, leak the true feeling. Now, let us consider a second basis for leakage and deception clues—habits for monitoring and disguising behavior, which are focused more on the face and words than on the body.

Controlling Behavior and Signs of Deception

Friesen and I suggested that people learn to monitor and disguise those aspects of their own behavior for which they have been held most accountable.[1] Most people grow up receiving the greatest commentary and criticism for what they say, next most for what they show in their facial expressions, with less specific attention to most of their body movement.§ As a result of this experience, people develop the habit of monitoring carefully their own words and voice, and, to a lesser extent their facial expressions, much more than they monitor most of their body movements. They also develop skills in the management of their

---

§ It is probably no accident that people receive the most feedback and criticism about those aspects of their behavior that can provide the most information. Certainly words are a far more elaborated information transmission system than the nonverbal or vocal behaviors. And, within the nonverbal behavior, facial movement is the best sender—the quickest, most visible, most precise, and capable of assuming an enormous number of distinguishable appearances. The relationship between sending capacity and feedback from others was elaborated in our earlier theoretical article on deception (Reference 1).

behavior, learning how to withhold, simulate, and mask. These management skills are best developed for words; and management is easy for a system that does not automatically change when emotion occurs. Management skills are also developed for facial expressions. Although not as facile as word management, the skills for withholding, simulating, and masking are usually far better for the face than for the body. Many aspects of the voice are attended to, noticed by the person who speaks and by the other. The voice, therefore, should be a prime target for inhibition and simulation. But, these are very difficult skills to acquire, and few do so. It is very difficult for most people to manage their voice so that anger, fear, or distress, when experienced, is not revealed. One can of course not talk, but that tactic is not always allowable. Also, few people can convincingly simulate the sound of these emotions.

When a person explicitly sets about the business of deceiving another person it is most likely that he will attend to those aspects of his own behavior that he has learned are most scrutinized by others—his words, face, and voice. He would have the requisite awareness and skills in performance to conceal his felt emotions and simulate unfelt emotions best through his words. There should be some success, but errors as well, in facial expressions. Those changes in facial expression that occur automatically with emotional arousal will be hard to totally suppress. Yet people do learn, we believe, to quickly abort such facial expressions, interrupting, blanking, or covering them. Simulating unfelt expressions is not done with great finesse, but our evidence suggests that it is done well enough to fool most other people. Attempts will also be made to inhibit automatically occurring voice changes, and to simulate emotions with the voice, but these efforts will often not succeed. Finally, the deceiver will tend not to think of the need to manage most of his body movements. If he did decide to simulate emotions through body movements, he would not be likely to do a good job of it, not having had the practice to develop the skills for convincing body movement performances.

The discussion so far suggests that when emotion is *not* involved in the deceit, there is no reason to expect that face, body, or voice will be especially good sources of leakage and deception clues. Of course emotion can become involved even if the deceit was not undertaken for the purpose of concealing emotion.

FIVE WAYS EMOTION CAN BECOME INVOLVED IN DECEPTION

First and most simply, the concealment of affect or the substitution of an unfelt emotion for a felt emotion may be all that the deception is about. For example, a wife may wish to conceal her anger from her husband and instead have him think she is pleased. Often the deception involves the easier task of concealing how the person felt in the past or might expect to feel in the future.

The second way in which emotion can become involved in deception is when there is a feeling about what is being withheld. Suppose someone is concealing a piece of nonaffective information such as their true age. The deceiver could have strong feelings regarding his age, such as embarrassment, shame, or anxiety. The successful perpetration of his deceit involves not only concealing his true age, but also concealing his feeling about the item being concealed. In this example the central purpose of the deception was to conceal nonaffective information (age), yet there was affect about the nonaffective information. When the central purpose of the deception is to conceal emotion, there also may be affect about the affect being concealed. This secondary emotion may also need to be concealed adding to the burden of perpetrating deceit. Return to the example of the wife who is concealing her anger from her husband, trying to instead appear pleased. The wife may be ashamed of her anger, or disgusted with herself for feeling anger. These feelings about the emotion being concealed must also be concealed. If she was to look ashamed or disgusted, her husband would certainly not believe she was pleased. He would want to know why she felt that way. The emotion about the withheld emotion compounds the types and amount of affect that must be concealed.

A third way emotion becomes involved in deception is when the person fears being caught. *Detection apprehension* can be considered as a gradient, ranging from neglible to so overwhelming that it leaks, or the deceiver confesses at least in part to obtain relief from the suffering of detection apprehension. Many factors could determine the intensity of detection apprehension; the list to follow is an example of some of the possibilities. Some people may be more vulnerable to detection apprehension. Practice in deception and reported past success in perpetrating deception may attenuate detection apprehension. If the person who is being deceived has a reputation as someone who is tough to fool, detection apprehension may be greater. The greater the anticipated punishment for being caught, the greater would be the detection apprehension. The greater the reward for succeeding in deception, the greater would be the detection apprehension.

A fourth way emotion becomes involved in deception is when the person feels guilty about engaging in the process of deceit. *Deception guilt* can be distinguished from the guilt that may or may not be experienced about the item of information being withheld. Consider a student who has cheated on an exam and is concealing that fact from a suspicious teacher. The student may or may not feel guilty about cheating. If she does, that would be an affective reaction about the concealed item (cheating) and it also must be withheld. The student may or may not feel guilty about lying to the teacher. If she does, that would be deception guilt. Deception guilt also can be considered a gradient, ranging from the negligible to instances when it can become so overwhelming it leaks. Relief from its pressure may motivate a confession.

The extent of deception guilt may be due to a variety of factors, some of which may be the same as the determinants of detection apprehension. Some people may be especially vulnerable to deception guilt. These may or may not be the same personal characteristics that predispose towards detection apprehension. Practice in deception and the experience of succeeding may attenuate deception guilt, the person may become "hardened," just as practice and success attenuate detection apprehension. We also suspect that the deceiver's perception of differences in social values between himself and the person he deceives may determine his deception guilt. People may not feel very guilty about misleading those who they perceive as holding different or antagonistic social values. For example, the revolutionary may feel less deception guilt about lying to the police than would a solid non-alienated member of the middle class. Yet the revolutionary might well feel deception guilt about misleading someone with whom he shared social values. There are situations where social conventions of one kind or another encourage, sanction, or even require deception. In these instances there should be little deception guilt. The nurse concealing her feelings of disgust when cleaning up the incontinent patient, or the family member withholding the true state of affairs from the dying loved one may feel no deception guilt.

The fifth way in which emotion becomes involved in deception is in *duping delight,* which refers to the exhilaration, pleasure, glee, or satisfaction a person may experience during the process of deception. Deception can be a challenge. Like mountain climbing or chess, it may be enjoyable only if there is some risk of loss. An innocent example of duping delight occurs when "kidding" takes the form of misleading a gullible friend. The kidder has to conceal his duping delight about his achievement even though his performance may in large part be directed to others who are collusively appreciating how well the gullible person is taken in. Duping delight can also be considered a gradient, ranging from the nonexistent to the point where it becomes so great that the excitement or pleasure leaks. The person may reveal his deception in order to share his excitement in his accomplishment in having put one over.

There may be personal characteristics that distinguish those who are most likely to experience duping delight. Probably those are not the same personal characteristics that distinguish persons who are susceptible to detection apprehension and deception guilt. The gains involved for succeeding and/or the losses anticipated for being caught can enhance duping delight, making the deception more risky or challenging. If the person being deceived has a reputation as someone who is difficult to fool, this may add spice, facilitating duping delight. The presence of an audience collusively involved in the deceit also should increase the likelihood of duping delight.

Our discussion so far has emphasized the role of emotion as the source of leakage and deception clues. Let me note, before closing, another basis for deception clues, albeit one that I think is less important.

COGNITIVE CLUES TO DECEPTION IN NONVERBAL BEHAVIOR

Ekman and Friesen defined *illustrators* as movements that are intimately tied to speech rhythms, serving to illustrate what is said.[17,18] Illustrator movements can emphasize a word, trace the flow of a thought, depict the rhythm, form or action of an event or object or point to an event. Illustrators serve a number of functions including word searches, self-priming, and help in explaining certain concepts difficult to put into words. Illustrators have been found to increase when a person is involved in what they are saying, and decrease with distinterest, apathy, tiredness, or lack of concern about what is being said.

What concerns us here, however, is another observation about the conditions that influence illustrator activity. Illustrators usually will decrease, often almost entirely, when a person is focusing their efforts on exactly what it is they are in the process of saying. Careful weighing of each word and close monitoring of what is said as it is said may occur when a person is especially cautious about exact statement, is confronted with competing alternatives, has conflicting messages, only one of which is allowable, or is inventing as he proceeds and is having a tough time doing so. The drop in illustrators that will occur may also be accompanied by changes in gaze direction.¶

Care in talk, even the presence of conflicting messages, is not itself a sign of deception. In certain social interactions, when certain lines have been taken, evidence that the speaker is being careful in his talk, cautious about what he is saying or inventing with difficulty, could be a clue that deception is in progress. Such dependence upon the social context and in particular fit with the words is relevant also to the interpretation of leakage or deception clues based on emotional arousal. I do not believe there is any body movement, facial expression, or voice

¶ We have not measured gaze until very recently and have no evidence on this yet.

change that *ipso facto* is a sign of deceit. An item of behavior betrays deceit because it does not fit with the rest of the behavior.

Conclusion

This discussion of the sources of nonverbal leakage and deception clues may help to explain why some investigators have found no evidence of leakage or deception clues, while still others have obtained quite contradictory findings. They were not studying the same types of deceits. The amount and type of emotions generated in their experimental deception interactions appear to have been quite different. Before new studies are undertaken, the investigator should ask what basis there is for expecting signs of deceit in face, body, or voice.

My discussion of the sources of leakage and deception clues has implications for how such betrayals will be manifest in nonverbal and vocal behavior. The signs of deceit, the particular facial expressions and body movements that give away deception, may not be the same if there is detection apprehension but no deception guilt, or the reverse, or both, or just duping delight, and so on. Elsewhere I have described in detail the specific signs of deceit that might be expected in different types of deceit.

As I mentioned in my introduction, not all deceits fail. There may not be leakage or deception clues. The explanation that I have given of the emotional and cognitive bases of leakage and deception clues suggests when deception should be the easiest. Leakage and deception clues will be least probable when:

The central purpose of the deceit is not to withhold emotion experienced at the moment, but some nonaffective item of information is being concealed.

The person feels little affect about the nonaffective item being withheld.

There is little detection apprehension (because the person is not vulnerable to that feeling, or the deceiver is practiced and has succeeded in the past, or the object of deception has a reputation for being easy to fool, or there is little punishment or reward for either failure or success).

There is little deception guilt (because the person is not vulnerable to that feeling, or the deceiver is skilled, or the deceived and the deceiver have antagonistic social values, or there is institutional sanction for deception).

There is little duping delight (because the person is not prone to such feeling, there is little risk or challenge in perpetrating deceit, or there is no audience collusively involved in the deceit).

The deceiver has a well worked out, practiced line, and need not carefully select what he says as he says it.

Deception should be the hardest, the leakage and deception clues most probable, when the exact reverse pertains to what was just listed. These ideas form the basis for a typology of interpersonal deceits, which time does not allow me to describe. From what I have said it should be obvious, however, that leakage and deception clues will be more likely in a spousal deception about an infidelity than in conversation between Carter and Brezhnev about how many missiles each has in place.

REFERENCES

1. EKMAN, P. & W. V. FRIESEN. 1969. Nonverbal leakage and clues to deception. Psychiatry 1: 88–105.
2. FREUD, S. 1905. Fragment of an analysis of a case of hysteria. In 1959. Collected Papers. Vol. 3. Basic Books, New York, N.Y.
3. FAST, J. 1970. Body Language. Neirenberg, New York, N.Y.
4. FELDMAN, S. S. 1959. Mannerisms of Speech and Gestures. International Universities Press, New York, N.Y.
5. BEIER, E. B., E. C. VALENS & G. EVANS. 1975. People Reading. Stein & Day, New York, N.Y.
6. BLONDIS, M. N. & B. E. JACKSON. 1977. Nonverbal Communication With Patients. John Wiley, New York, N.Y.
7. KURTZ, R. & H. PRESETERA. 1976. The Body Reveals. Harper & Row, New York, N.Y.
8. SEBEOK, T. A. 1976. The semiotic web: A chronicle of prejudices. In Contributions to the Doctrine of Signs. T. A. Sebeok, Ed. Chap. 10. Peter de Ridder Press, Lisse, The Netherlands.
9. EKMAN, P. 1977. What's in a name? J. Commun. 27(1): 237–239.
10. BIRDWHISTELL, R. L. 1970. Kinesics and Context. Univ. Pennsylvania Press, Philadelphia, Pa.
11. EKMAN, P. & W. V. FRIESEN. 1974. Detecting deception from the body or face. J. Pers. Soc. Psychol. 29(3): 288–298.
12. EKMAN, P., W. V. FRIESEN & K. SCHERER. 1976. Body movement and voice pitch in deceptive interaction. Semiotica 16(1): 23–27.
13. MANDLER, G. 1975. Mind and Emotion. John Wiley, New York, N.Y.
14. ZAJONC, R. V. 1980. Feeling and thinking: Preferences need no inferences. Am. Psychol. 35(2): 151–175.
15. EKMAN, P. 1977. Biological and cultural contributions to body and facial movement. In Anthropology of the Body. J. Blacking, Ed. Academic Press, London.
16. EKMAN, P. Uncovering Deceit. Book in preparation.
17. EKMAN, P. & W. V. FRIESEN. 1969. Repertoire of nonverbal behavior: Categories, origins, usage, and coding. Semiotica 1: 49–98.
18. EKMAN, P. & W. V. FRIESEN. 1972. Hand movements. J. Commun. 22: 353–374.

Detection and Perception of Deception: The Role of Level of Analysis of Nonverbal Behavior*

VERNON L. ALLEN and
MICHAEL L. ATKINSON
Department of Psychology
University of Wisconsin
Madison, Wisconsin 53706

WITHOUT INTENDING to denigrate in the least the intellectual ability of Professor Pfungst's most famous subject, we would gently suggest that Clever Hans was not only clever but also very lucky. It was indeed clever of this horse to notice the subtle (and unwitting) nonverbal responses displayed by his human questioners. But Hans was even luckier than he was clever, since he might have concentrated his attention on many other cues that were of no importance at all (or, rather, of no importance to the humans who were interested in Hans). But through sheer good luck (we presume) Hans paid attention to those subtle nonverbal cues that happened to show a remarkable correspondence with certain critical events in the external environment; that is to say, there was a strong correlation between certain nonverbal responses displayed by the human audience and the correct answers to questions that had just been posed to Hans by his interrogators.

Human social behavior is usually much more problematic than the situation that Clever Hans faced, due in part to our ability and proclivity to "ambiguate" social reality. A multiplicity of meanings can usually be attributed to any single social response. Moreover, an objective basis is often lacking for ascertaining the validity or "correctness" (truth or falsity) of nonverbal cues. Instead, social behavior can often be characterized as falling within a shifting interval rather than at a precise point along the continuum of truth to falsity. In order to emphasize the quantitative nature of our conceptualization of deception we shall use the terms "deliberate" versus "spontaneous" to designate the endpoints of a continuum. Any ongoing stream of social behavior will exhibit fluctua-

---

* This paper was written while the first author was a Fellow at the Netherlands Institute for Advanced Study in the Humanities and Social Sciences, Wassenaar, the Netherlands.

tions in the degree to which it tends to shift toward the deceptive or authentic endpoints of this continuum. Furthermore, an individual is often uncertain about his or her own behavior in terms of the deliberate–spontaneous dimension, self-deception being a dramatic case in point. All these considerations point to the difficulty and complexity of trying to attribute any invariant meaning to the nonverbal cues occurring during human social interaction.

Many people seem to place a great deal of importance on being able to characterize ongoing social behavior in terms of its "authenticity" or "genuineness." By contrast with the present-day preoccupation, it appears that in earlier periods of history the deliberateness-spontaneity of social behavior was a much less important issue. According to Huizinga:

> To distinguish clearly the serious element from pose and playfulness, is a problem that crops up in connection with nearly all the manifestations of the mentality of the Middle Ages. We saw it arise in connection with chivalry, and with the forms of love and piety. We always have to remember that in more primitive cultural phases than ours, the line of demarcation between sincere conviction and "pretending" often seems to be wanting. What would be hypocrisy in a modern mind, is not always so in a medieval one. [Reference 1, page 217.]

Regardless of whether an observer's perception is objectively correct or not, any conclusion reached about the spontaneous or deliberate nature of another person's behavior will have serious consequences for future interaction. Hence, it is important to investigate the psychological processes that contribute to an observer's inferences about the nature of another person's behavior in terms of the deliberate-spontaneous dimension.

On occasions, an observer will become distinctly aware that another person is engaging in pretense, that the behavior is deliberate and intentional. In some cases the requirements of a particular role demand a deliberate performance or the presentation of a very positive self-image. For example, the behavior of the television actor is a fabrication not meant to be taken seriously. In his research on cues in the perception of lying, Kraut[2] suggested that an observer ignores or discounts an actor's behavior as spontaneous to the extent that it is self-serving. Many roles (e.g., defendant or job applicant) demand the presentation of a favorable self-impression and, consequently, the audience will expect deliberate and intentional behavior.[3] (It should be noted that self-serving behavior is meant to be take seriously, in contrast to the behavior of the television actor.)

In other cases, behavior may be perceived as deliberate as a result of accidental gestures. A person may inadvertently display nonverbal actions (e.g., nervous shaking, stuttering, or a guilty expression) that distract from the ongoing behavior. Observers may seriously consider these "unmeant" gestures as indicative of deliberate behavior, and erroneously infer that the individual is experiencing difficulty in presenting his or her intended performance. Ekman and Friesen[4,5] have argued that these gestures are not accidental, but rather are nonverbal actions that have escaped the censorship of the dissembling individuals. As such, "leaked" nonverbal responses may be important cues to the detection of deception. Recent research has extended this notion by identifying specific attributes of spontaneous and deliberate behavior that may be used as cues in discriminating between the two modes of encoding.[2,6-9]

In contrast to the examples given above, most social interaction involves neither accidental gestures nor obviously self-serving behavior. Nevertheless, certain types of behavior may still be perceived as deliberate, as evidenced in the following quotation from Sartre:

> Let us consider this waiter in the cafe. His movement is quick and forward, a little too precise, a little too rapid. He comes toward the patrons with a step a little too quick. . . . Finally there he returns, trying to imitate in his walk the inflexible stiffness of some kind of automaton while carrying his tray with the recklessness of a tight-rope walker by putting it in a perpetually unstable broken equilibrium which he perpetually re-establishes by a light movement of the arm and hand. All his behavior seems to us a game. [Reference 10, page 59.]

This waiter's movements were seen as mechanical, disjointed, and a little too rapid, which suggests that the key to the perception of spontaneity or deliberateness in social behavior lies in the perception of its physical structure. Research on the distinction between spontaneous (nonposed) and deliberate (posed) nonverbal behavior helps to explain Sartre's observation. Ekman, Friesen, and Ellsworth,[11] among others (e.g., Hunt[12]), have commented that the eliciting conditions for spontaneous and deliberate behavior are not identical, and, consequently, the structure of behavior encoded deliberately may differ from that of behavior encoded spontaneously. Allen and Atkinson[13] report that observers responded quite differently to spontaneous and deliberate behavior, in that observers decoded the behavior more accurately and formed more extreme impressions about the encoders when viewing deliberate as compared to spontaneous behavior. Moreover, in an objective analysis of spontaneous and deliberate behavior, Allen and Feldman[14] found quite distinctive patterns of nonverbal responses, with deliberate behavior being exaggerated, repetitious, and disjointed. In short, deliberate behavior tended to be a caricature of spontaneous behavior. Thus, the behavior of Sartre's waiter was perceived as deliberate because the observer had apparently analyzed the behavior at a molecular level—at a level where the movements were seen as disjointed and repetitious, resembling a staged, deliberate performance.

Generalizing from this example, any behavior sequence would seem to be amenable to being analyzed at either a global or fine-grained level. Newtson[15] has demonstrated that the unit of perception is variable; that is to say, a given sequence of behavior may be meaningfully viewed at either a fine or gross level. But using these different levels of analysis has the important consequence of altering a person's interpretation of the behavior in question. Thus, the kind of information that a person extracts and uses from a behavior sequence depends on the level of analysis being employed. Consider the outcome of these two levels (units) of analysis when applied to a concrete instance, for example, a man leaving a room. The global-unit observer would organize the behavior at a fairly gross level, with many component actions being combined as a single global unit. Thus, the global-unit observer might report that the man got up, picked up his coat, and left the room—or simply that the man left. The actor's behavior flows smoothly and appears to be spontaneous. In contrast, the fine-unit observer would attend to the component actions in the behavior. A description of the behavior might go as follows: The man placed his hands on the arms of the chair, pushed himself to a semi-erect position while pushing the chair backward, let go of the chair and stood up completely, then turned from the table, and so on. Soon, such close scrutiny of the behavior seems strange to the observer—the actions seem novel and unexpected. Individual movements and the behavioral minutiae may be seen as meaningful. To the observer, the behavior does not seem to flow smoothly: the action may become repetitive and novelty may make the movements appear to be exaggerated. Like Sartre's waiter, the actions resemble deliberate, intentional behavior.

We would argue that a cognitive approach based on perceived levels of analysis provides a general theoretical basis for understanding the perception and detection of spontaneity or deliberateness in ongoing behavior. According to our analysis, the perception of spontaneity or deliberateness in behavior is directly related to an observer's level of analysis. Behavior viewed at a fine-grained level does not flow smoothly, and the information extracted from the behavior sequence does not resemble spontaneous action (which tends to be organized at a more global level). It should be noted that the crucial factor in the present theory is the perceived rather than the actual (objective) structure of behavior. A given behavior sequence may appear to be either spontaneous or deliberate, depending upon the level of analysis employed by the observer. Thus, the resulting inference about the nature of behavior is related to the observer's perceptual organization rather than to events in the behavior sequence *per se*.

We have designed experiments to investigate the relation between level of perceptual analysis and the deliberate-spontaneous nature of nonverbal behavior. The first study (Experiment 1) examined the effect of cognitive set — that is, deliberate or spontaneous behavior — on a person's segmentation of ongoing behavior. Subjects unitized a standard sequence of nonverbal behavior after having been led to believe that it was produced either deliberately or spontaneously. It is hypothesized that observers in the deliberate set condition will segment the behavior at a finer level than observers in the spontaneous set condition. The rationale for this hypothesis is that an anticipatory set for deliberate behavior should elicit a preparatory shift in the observer's cognitive schema. That is, the observer adopts a fine-grained viewing strategy. A second experiment tested the converse of the relationship investigated in Experiment 1. Subjects were instructed to unitize a standard sequence of ongoing behavior at either a fine-grained or global level of analysis. Afterwards, subjects reported their perception about the nature of the behavior (deliberate or spontaneous). It is predicted that a standard sequence of behavior will be perceived as being more deliberate when subjects segment it at a fine-grained as opposed to a global level of analysis.

Experiment 1

Method and Procedure

Observers viewed one of two elementary school children who were chosen from the stimulus tape described elsewhere by Allen and Atkinson.[13] Each videotape showed a 5th-grade school student (male) watching a teacher present a lesson over a closed-circuit television system. Only the student appeared on the videotape (head-and-shoulders view). Both of the stimulus persons were role-playing; that is, they had been instructed to act in a certain manner. One of the children had been asked to pretend that he understood the lesson he was watching (when, in fact, he did not). The other child had been asked to pretend that he did not understand the lesson he was watching (when, in fact, he did). Each stimulus tape was 2.5 minutes in duration and was shown without sound. It was necessary to use role-play behavior instead of spontaneous behavior as stimuli in the experiments. Behavior segmentation requires movement, and the spontaneous behavior of children listening to a televised lesson simply did not contain much movement at all. This choice of type of behavior does not present any problem for the testing of our hypotheses, as Atkinson and Allen[16] have noted.

The stimulus tapes were presented via a closed-circuit television system that consisted of a videotape recorder (Sony model 3650) and a studio monitor (RCA Lyceum model JR345B). Observers sat at a long table divided into three booths by plywood partitions. An electronic signaling device was used to monitor the observers' behavior "unitization" (see below). This device consisted of a thumb switch (one per booth), a bank of light-emitting diodes located in the adjoining room, and a 6-volt DC power supply. The experimenter visually monitored and recorded the unitization responses of the observers.

The experimenter explained to subjects that they would see a silent videotape of a child listening to a lesson. The lesson was described as a difficult one for the child: the teacher on the videotape presented formulas for calculating wattage and resistance, the difference between series and parallel circuits, and so on. Subjects in the spontaneous instruction condition were told that the child did not understand the lesson, that a hidden camera was used to obtain the videotape, and, hence, that the child was completely unaware he was being filmed. It was stressed, then, that the child was behaving in a natural and spontaneous manner. Subjects who received the deliberate instructions were told that the child was acting, that is, that he was pretending not to understand the lesson at all in order to mislead a teacher who was watching over a closed-circuit television system. It was stated that the child was aware of being videotaped and, therefore, that he was behaving in a deliberate and intentional manner.

Following this instructional manipulation, the experimenter outlined the general concept of behavior unitization, and demonstrated how an action (closing the door) could be construed equally well as a series of small units or as one global unit. The experimenter asked the subjects to give him "a better idea of what's going on" in the videotape by segmenting the behavior. Subjects were instructed to press a thumb switch when in their judgment one meaningful unit ended and another began. It was stressed that there were no right or wrong ways to segment the behavior—that the experimenter simply wanted to know how the subjects did it.

The experiment employed a $2 \times 2 \times 2$ completely randomized factorial design. The variables were observational set (spontaneous, deliberate), sex of observer (male, female), and stimulus person (one, two). Subjects were randomly assigned to the eight conditions as they appeared for participation in the study. The data were collapsed across the stimulus persons for analysis.

Subjects were 36 undergraduate students (17 males and 19 females) recruited from an introductory psychology class. They received course credit for their participation. Subjects viewed the stimulus tapes in groups of 1 to 3 persons (in semi-isolated booths).

Results and Discussion

A postexperimental questionnaire assessed subjects' awareness of the experimental manipulation by asking them to indicate whether the child on the videotape had been filmed with a hidden camera or if he was deliberately acting. Only one subject (a female in the spontaneous condition) checked the incorrect response; she was excluded from the data analysis. Thus, the experimental manipulation was successful.

It was hypothesized that observers who believed that they were watching deliberate behavior would segment it at a finer level than subjects who believed it was spontaneous. The results of a 2 × 2 analysis of variance supported this prediction. The analysis yielded a main effect for observational set, $F(1,31) = 11.10, p < 0.005$. Observers who believed that they were viewing deliberate behavior generated approximately twice as many units for the standard behavior sequence as the spontaneous set observers (*mean* = 17.53 vs 8.61, respectively). Neither the main effect for sex of observer nor for the interaction between sex and observational set reached significance. Inspection showed that the difference between the two cognitive set conditions was, however, much greater for females than males. The sex difference was marginally significant ($p < 0.07$).

Results of Experiment 1 indicated that the observer's cognitive set (i.e., his or her belief that a sequence of behavior was deliberate or spontaneous) affects the manner in which a person actively organizes the ongoing behavior. Specifically, subjects who believed they were observing a deliberate performance unitized the ongong stream of behavior at a finer level than persons who believed it to be spontaneous. Individuals may expect that certain types of cues (e.g., small units) will occur in deliberate behavior and, therefore, actively search for them. Goffman[17] suggests that observers anticipate certain types of behaviors in "self-serving" roles. The present approach is in basic agreement, but it suggests further that observers alter their perceptual search in an effort to detect these particular features. The present results demonstrate the point clearly: Observers who anticipated deliberate behavior segmented the sequence at a finer level than persons who expected spontaneous behavior.

Trends in the data from another study conducted by the present authors[18] suggest that an individual may adopt differential preparatory cognitive sets for processing information when expecting to receive spontaneous versus deliberate information. In this study the experimenter described the scenario of a videotape that subjects expected to see later. It was stated that one person shown on the videotape knocked over a stack of papers; this action was stated as being either an accident (spontaneous) or staged deliberately by an accomplice. On the basis of this information, subjects responded to a number of relevant questions. Results showed that a larger percentage of subjects in the spontaneous than the deliberate condition indicated that when viewing the videotape they would adopt the search strategy of attempting to "get a broad, general view of the action" (60% vs 28%, respectively). Further, subjects used more words in the deliberate than the spontaneous condition when writing a description of the action (median = 64 vs 49, respectively); and subjects expected that they would see more units in the stimulus person's behavior in the deliberate (mean = 21) than in the spontaneous (mean = 17) condition.

EXPERIMENT 2

A second experiment investigated the impact of the organization of a stream of behavior on the inference made about it, i.e., whether it is spontaneous or deliberate. Observers who analyze behavior at a fine-unit level should be more likely to notice novel and repetitive actions in the behavior sequence than observers who employ a global-unit style. Since deliberate behavior does contain more novel and repetitive responses, fine-unit observers should be more likely to perceive a standard behavior sequence as being deliberate than global-unit observers.

Method and Procedure

Subjects were 52 undergraduates who viewed the same stimulus tapes used in Experiment 1. The experimenter gave the observers specific instructions for unitizing the stimulus person's behavior at either the global or fine-grained level. Subjects were told either to break the behavior into the "largest units that seem meaningful and natural to you" or into the "smallest units." After unitizing the standard sequence of behavior, subjects were asked to estimate on a 16 cm scale the probability that the stimulus child had been "acting."

Results and Discussion

A 2×2 analysis of variance was performed on the subjects' estimates of spontaneity or deliberateness in the ongoing behavior. The analysis did not produce a significant main effect for unitization instructions, $F(1,48) = 0.84$, or sex of observer, $F(1,48) = 0.02$. The interaction between the two variables was significant, however, $F(1,48) = 6.20, p <$ 0.025, which helps to account for the lack of main effects. Female subjects were more likely to see the stimulus person as acting when they analyzed the behavior at a fine-grained level (mean = 10.00) than at a global level (mean = 6.80), consistent with our hypothesis. On the other hand, male observers reacted quite differently: they attributed a higher probability of acting to the stimulus person when they were in the global rather than the fine-grained condition (mean = 8.70 vs 5.88, respectively). Tests of simple effects indicated that the difference between means was significant only for the female observers, $t(48) = 2.45$, $p < 0.025$. This sex difference is consistent with the finding in Experiment 1, which revealed that females responded more strongly than males to an experimental manipulation that was expected to influence the unitizing of a standard sequence of behavior.

According to theory, the perception of spontaneity or deliberateness in behavior is a function of the level of analysis that is actually used by the observer. In both experimental conditions of the present study, individuals were completely free to break the behavior into as many or as few meaningful units as seemed reasonable to them. And there was a considerable range among subjects in the number of units employed in each of the experimental conditions. Therefore, another analysis was performed to relate the actual unitization employed by subjects to their perception of spontaneity or deliberateness of the behavior.

The number of units generated by all 52 subjects were ranked without regard to unitization instructions or sex of observer. The resulting distribution was split at the median (two subjects whose responses fell at the median were discarded). The 25 subjects who generated fewer than 9 units for the behavior sequence were designated as "functionally global observers," and the 25 subjects who generated more than 9 units as "functionally fine-grained observers." These functionally-defined groups were then compared in terms of their perception of spontaneity or deliberateness in the behavior sequence. The results strongly supported our original prediction, $t(48) = 8.24, p < 0.001$. That is, functionally fine-grained observers were more likely to perceive the behavior as being deliberate than functionally global observers (mean = 8.79 vs 6.72, respectively, on the 16 cm scale). It should be noted that the present analysis does not simply reflect the sex effect found in the analysis of variance results; 40% of the functionally global observers and 54% of the functionally fine-grained observers were males. It is clear, then, that the utilization level actually used by subjects to organize the ongoing behavior did influence their perception of the degree of spontaneity.

It is interesting to note that subjects reported having viewed events in the behavior sequence that were consistent with the level of analysis they employed. For example, fine-unit observers reported seeing discrete nonverbal movements in the videotape (e.g., blinking of the eyes, breathing, mouth movements), whereas global-unit observers recalled more general behavior segments (e.g., periods of confusion, attention, boredom).

GENERAL DISCUSSION

Results of the present research have interesting implications for the theoretical conceptions of detecting deception formulated by Ekman and Friessen[4,5] and by Kraut.[2] Ekman and Friesen conceive of the observer as a relatively passive element in the inference process; one infers that deception has occurred by observing leakage cues and deception clues that are "revealed" by the actor. Kraut posits a similar role for the observer by suggesting that aspects of the situation (motivational cues) and slippages in the actor's self-presentation (performance cues) trigger certain "rules" that suggest that deception may be occurring. Although Kraut cautions that a direct relation between specific behaviors and attempted deception is unlikely, the theory still assumes that behavior reveals deception. According to Kraut it is from performance cues that "the audience perceives that an actor had failed to adequately control some aspect of his deceptive performance [Reference 2, page 389]." Kraut then discusses the means by which leakage and deception cues reveal different aspects of an actor's deceptive performance.

By contrast with previous theories concerning the detection of deception, the present approach construes the observer as an active organizer of experience. The perception that a stream of behavior is deliberate (deceptive) is seen as the consequence of analyzing the behavior at a fine-grained level. It should be noted that the approach of Ekman and Friesen (and, to a lesser extent, of Kraut) depends on the actual occurrence of deception in the behavior. Unless deception clues and leakage cues are available, the observer has no basis for inferring that the behavior is deliberate and intentional. The present formulation suggests that behavior will be seen as being deliberate (deceptive) if it is unitized at a fine enough level—even if the behavior is actually spontaneous (honest). How does an observer come to adopt a fine-grained level of analysis? Newtson[15] has suggested that an observer will shift to a fine level of analysis when an unexpected event occurs in the behavior stream. In a more general sense, then, perceptual organization is related to predictability in the behavior. Unexpected events (e.g., stopping in the middle of an action) serve to decrease predictability. Consistent with Heider's emphasis on predictability in social perception,[19] an observer will be motivated to increase the amount of information gleaned from the behavior sequence (accomplished through a finer level of analysis) and thereby re-establish predictability. The various cues to deception described by Ekman and Friesen and Kraut are, essentially, unexpected events (e.g., stuttering or stopping the middle of a sentence).

At a more general level, then, the basic cause for switching to a finer level of analysis can be ascribed to the existence of an unacceptable degree of uncertainty, because the fine-grained level of analysis will provide the maximum amount of potential information for the individual. An incremental shift in uncertainty can be signaled by numerous types of events: discontinuity in the behavior stream, novel or unexpected responses, or exaggerated and repetitive actions. General information about the situation or social context as well as psychological or cognitive processes within the observer can also increase uncertainty, and thus initiate a shift to a finer unit of analysis.

According to the present analysis, an observer who encounters a cue to deception shifts to a fine level of analysis. At this finer level, the observer attends closely to paralinguistic cues, bodily expressions, and so on. It is important to note that the observer does not consciously shift attention to channels that are presumed to be less under the actor's control, as implied by Kraut and Ekman and Friesen. Instead, the observer unitizes the behavior at a finer level and, consequently, behavioral channels that are not usually monitored are now included in the observer's perceptual analysis. As a result, cues such as the discrepancy between verbal and nonverbal channels may be noticed.[6] Indeed, the primary cue to deception may be the realization that one has shifted to a finer level of analysis rather than events in the behavior stream *per se*.

References

1. Huizinga, J. 1924. The Waning of the Middle Ages. St. Martin Press, London.
2. Kraut, R. E. 1978. Verbal and nonverbal cues in the perception of lying. J. Pers. Soc. Psychol. **36:** 380–391.
3. Jones, E. E., K. E. Davis & K. J. Gergen. 1961. Role playing variations and their information value for person perception. J. Abnorm. Soc. Psychol. **63:** 302–310.
4. Ekman, P. & W. Friesen. 1969. Nonverbal leakage and cues to deception. Psychiatr. **32:** 88–106.
5. Ekman, P. & W. Friesen. 1974. Detecting deception from the body or face. J. Per. Soc. Psychol. **29:** 288–298.
6. DePaulo, B. M., R. Rosenthal, R. A. Eisenstat, P. L. Rogers & S. Finkel-

STEIN. 1978. Decoding discrepant nonverbal cues. J. Pers. Soc. Psychol. **36:** 313–323.

7. EKMAN, P., W. V. FIRESEN & K. R. SCHERE. 1976. Body movement and voice pitch in deceptive interaction. Semiotica **16:** 23–27.
8. McCLINTOCK, C. C. & R. G. HUNT. 1975. Nonverbal indicators of affect and deception in an interview setting. J. Appl. Soc. Psychol. **5:** 54–67.
9. ZUCKERMAN, M., R. S. DeFRANK, J. A. HALL, D. T. LARRANCE & R. ROSENTHAL. 1979. Facial and vocal cues of deception and truth. J. Exp. Soc. Psychol. **15:** 378–396.
10. SARTRE, J. 1956. Being and Nothingness [transl. by H. E. Barnes]. Philosophical Library, New York, N.Y.
11. EKMAN, P., W. FRIESEN & P. ELLSWORTH. 1972. Emotion in the Human Face: Guidelines for Research and an Integration of Findings. Pergamon Press, New York, N.Y.
12. HUNT, W. A. 1941. Recent developments in the field of emotion. Psychol. Bull. **38:** 249–276.
13. ALLEN, V. L. & M. L. ATKINSON. 1978. The encoding of nonverbal behavior by high- and low-achieving children. J. Educ. Psychol. **70:** 298–305.
14. ALLEN, V. L. & R. S. FELDMAN. 1976. Nonverbal cues to comprehension: Encoding of nonverbal behaviors naturally and by role-play. Working Paper No. 147. Wisconsin Research and Development Center for Individualized Schooling, Madison, Wisc.
15. NEWTSON, D. 1973. Attribution and the unit of perception in ongoing behavior. J. Per. Soc. Psychol. **28:** 28–38.
16. ATKINSON, M. L. & V. L. ALLEN. 1979. Level of analysis as a determinant of the meaning of nonverbal behavior. Soc. Psychol. Quart. **42:** 270–274.
17. GOFFMAN, E. 1959. The Presentation of Self in Everyday Life. Doubleday Company, New York, N.Y.
18. ATKINSON, M. L. & V. L. ALLEN. 1980. Preparatory set and the interpretation of behavior. Unpublished manuscript, Univ. of Wisconsin, Madison, Wisc.
19. HEIDER, F. 1958. The Psychology of Interpersonal Relations. John Wiley, New York, N.Y.

Semiotics: A View from Behind the Footlights

JAMES RANDI*
The Inner Magic Circle
London, England

Committee for Scientific Investigation
of Claims of the Paranormal
Buffalo, New York

IT HAS BEEN 35 YEARS since I first donned the character of the magician. To be perfectly correct about it, that calling should be referred to, and it is in the United Kingdom, as "conjuring," rather than "magic." The term "magic" means, in my favored dictionary definition, "an attempt by Man to control the forces of nature by means of spells and incantations." I have tried both spells and incantations, and I can assure you that neither work at all well; plain skullduggery is the only method I can recommend.

Indeed, the term "conjuror" is in many ways synonymous with "liar, cheat, thief, charlatan, and fake." I relish those terms greatly—perhaps overmuch. But without those elements at work for me, I would be out somewhere extolling the virtues of Tupperware to unsophisticated housewives.

I am a student of psychology, particularly the applied kind; I also study the mathematics of probability; and I put much time and observation into optical, auditory, and other-sensory illusions. I have the great need, in my profession, to know what the spectator is apt to think or perceive under the conditions of performance. Of course, I see to it that I control those conditions, in that I direct the spectator's actions and attention—or lack thereof—and in general take control of the situation and thus the eventual denouement. To a large extent, the spectator allows me to do this, since it is, as the saying goes, "my show." But at the same time, I control much more of the activity both on and off the stage than is ever suspected. Let me demonstrate what I mean.

[The speaker chooses a member of the audience seated in the front row, and asks him to assist. The performer passes a pair of scissors along the length of a strip of newspaper slowly, back and forth, until this person says, "Stop!" At

* Address for correspondence: 51 Lennox Avenue, Rumson, N.J. 07760.

that point, the strip is cut, and the loosened section falls to the floor. It is retrieved by the spectator who is then asked to choose any word in the line chosen by the cutting process. He names it. Asked to look beneath his seat, he finds there an envelope containing a notarized letter predicting the exact word he has chosen freely.]

I presume that you are, with few exceptions, very much in the dark as to the *modus operandi* of that admittedly excellent trick. And how was it done? Rather well, I thought . . . .

Seriously, I trust that I will set you thinking when I comment that few of you will have suspected that, as I asked my assistant whether or not I had "cut through a line, or is there an intact line there," I also glanced at the floor and half-reached for the scrap I had cut away, giving the impression that I would find it easier to have the answer from him. I assure you, I was quite well aware that the cut was not through the letters of the printed line, and in your having turned your attention to the assistant's answer — as I had seemingly done — you relaxed your senses in the direction I wished. And I ask that you recall another assumption you may have made. In several ways — ways that I trust you will never solve — I set up certain expectations in your minds that permitted me to direct the discovery of the correct answer to several different locations and several different methods of arriving there. No, I did not announce in advance where the answer would be found. But what sensible person would doubt that it could only be found just where it *was* found?

I will recall for you an incident of my early life that illustrates the admitted fact that the most intelligent among us are prone to be deceived. I was deceived, at the age of fourteen. I hasten to add that I have seldom been deceived since. The occasion to which I refer involved a theatre performance of the late and very great Joseph Dunninger. He was by all odds the very paramount figure in his field, which was mentalism. This term covers the field of mental magic, an example of which I have just shown you.

At that tender age, I went to see Dunninger do his wonders, not well prepared to solve them. At one point, he asked a young lady to step forward from a group he had invited upon the stage to assist. As she did so, he said to her, "Ellen . . . ah, your name *is* Ellen, is it not?" She nodded that it was, and a murmur of astonishment passed through the audience. "But I've never met nor spoken to you before in my life, have I, Ellen?" queried Dunninger. Again, assent. And Dunninger carried on, having taken credit for a "throw-away" miracle. Much of what followed, I solved, since I had been studying this field for some time even at that age. But the "Ellen" episode bothered me. The solution, I finally was able to verify many years later with The Great Man himself, was simplicity indeed.

Upon asking for volunteer assistants to trot up onto the stage, Dunninger had asked that the orchestra favor the audience with a little marching music. As each person mounted and was assigned a seat by the performer, he had simply said "hello"—except in the case of Ellen, when he asked her name, and gave his in return, as a seemingly natural and friendly gesture. And please recall that as he asked her, a moment later, to step forward on the stage, his question seemed *to her*, to be a simple check that his recollection *was correct*, not an attempt to divine an unknown fact! To the audience, it was a minor wonder, and he allowed it to be assumed as such. Also, recall his *next* question. He asked whether he had ever met or spoken to her *before*. Before *what*? Why, before this present occasion, of course! She naturally was forced to this assumption, and answered accordingly—especially since she had no idea that the audience thought they were witnessing a miracle.

I have visited countless "spirit churches" and I have watched as the innocent and gullible were separated from their savings by unscrupulous operators giving the impression that they had direct contact with the dear departed. One of the most common of the methods used in these seances is called "billet reading," in which the faithful place questions and statements in sealed envelopes. These are presented to the "medium" who commences to show that he knows the contents of these envelopes by holding each in turn to his forehead and "reading" its contents. The method is simple: he has obtained, in advance, the knowledge of the contents of *one* of the envelopes and the information written on the outside of it. He glances at the outside of the first sealed one he picks up, apparently reads initials from it, and asks to see that person in the audience. When the victim has identified himself, the performer mentions a first name, and asks if that is correct. In a similar manner to the Dunninger method just outlined, the rest of the faithful assume that the name has been divined from the initials. In actuality, the name was written on the envelope—the secretly obtained one—not just initials. And the statement can be important, too. Often, it is, "I have the name Michael . . . ." Note that he does not mind if the named person concludes that he means by this that he has the name Michael on the envelope, and that the audience assumes he has *divined* the name Michael! If so, all well and good. If not, and he is taken up on it, which seldom if ever happens, he cannot lose, since he never claimed anything definitely except the name itself! Continuing, the performer opens the *apparently* divined envelope, as if to check the contents, thus obtaining the information needed for the *following* envelope, and so on. The method is known as, "one-ahead."

Another clever use of language in such a flim-flam—and I am not now speaking of the legitimate pursuit of entertainment by means of conjuring techniques, by these terms—is in the use of Pure Noise Questions. The performer speaks to the victim and comes up with a declaration such as, "I see a family problem," then follows it immediately with the inane question, "Can you accept that?" and nods vigorously upon receiving the inevitable assent. "Of course!" And he seems to have made another "hit." Another question is, "Can you understand?" following some statement such as, "I have here three persons in one home." *Of course* the victim "can accept" and "can understand" such simple notions. That the audience assumes that these declarations are revelations is another matter altogether.

[The performer hands out a dozen brown manila envelopes, some folded cardboard sections, and five symbol cards—the "Zener" ESP symbols—measuring 5 by 7 inches. Members of the audience are asked to shuffle all these, then seal each card between a set of cardboards and place each in any envelope. The performer handles them briefly, and arranges them in a row. He draws upon each envelope what he believes is the symbol inside. The results are 100% correct.]

You will note that I have so far been demonstrating tricks of mentalism. There is a very good reason for this, since in handling the Clever Hans phenomenon, we are also investigating the pitfalls of parapsychological research. The trick just seen involved a great number of moves of misdirection, particularly using the very heavy emphasis on security. You were assured several times that the envelopes were opaque, that any envelope or folder could be used, that they should be thoroughly mixed, and that the weight, shape, and thickness of each card was the same. All true, and all inconsequential. By suggesting what the trick was *not*, I was able to direct your attention elsewhere. And, who is to know where and at what instant the information was obtained by me? I may have—indeed, I did—have the information well before I chose to reveal it. But bear this in mind: I am allowing you to pick up all of these things without actually *telling* you. Rather, by gestures, speech, expressions, and other subtle means, I am giving you enough peripheral information to come to the conclusion I need of you. And how can you doubt the evidence of your own senses?

My personal battle with the parapsychologists in recent years has centered around their insistence on certain "rules" that the so-called "psychics" have dictated. Uncertainty is a hallmark of the true psychic, they say. Magicians can do it every time because they use trickery; psychics are expected to fail, and when they do are thus validated. Nonsense, of course. The good performer in this field can provide the observer with enough small hints and subliminal clues that he will fill in for himself a sketchy evidential picture and fall into the logical trap that has been designed for him. I am reminded of a prominent psychiatrist (he was a Freudian, so there was little hope for him in the first place, in my opinion) who exulted over the fact that one of his star psychic performers was heavily under the influence of alcohol when demonstrating. The fact that he obtained this condition from the intake of huge amounts of beer, thus necessitating frequent exits from the room for obvious purposes, was overlooked. The protocol, which was very shaky at best, never involved searching the person of this performer—when such a process would have easily revealed the *modus operandi* of his trick—and certainly such protocol was shattered by the frequent exits. But all we heard was the wonderful fact that the psychic was performing while heavily intoxicated. This particular Clever Hans was picking up from the one who was actually running the experiments—the performer—all the clues he needed to come to the wrong conclusion.

In a similar manner that observers believe the experimenters working with the "speaking" apes have been self-deceived, some of those who came in contact with the now-discredited "psychic" Uri Geller were afflicted. When tests were run under conditions preferred by Geller, they seems to produce positive results. When, at the insistence of other investigators, conditions were tightened up, Geller failed—and immediately was comforted by the believers, who told him that persons of negative feelings were present, that they noticed he did not seem up to par today, that he must be tired at this point, and that that was the kind of test he did not do well at, anyway. Geller, never one to miss an opportunity, agreed about his low state of vitality and complained of headaches and being inhibited by the conditions. Would that I could perform under those same conditions. What miracles I could produce!

I would like you, in your presently somewhat more informed state, to witness one more demonstration.

[The performer offers a 3-foot rope for examination, and his hands are tied behind his back, tightly, by two volunteers from the audience. He frees his hands instantly, and is seen to be tied again when examined. This is repeated several times.]

You will note, I hope, that *I* was in charge of the movement and actions of these two volunteers. By motions of my head, and otherwise, I directed their attention, and yours. Particularly at moments when we relaxed in laughter, I was able to move and act in ways that might not be ordinarily permitted or excused. But please take from this performance what was meant to be shown. I mean to get across the point that serious researchers who attempt to investigate so-called paranormal events and claims are considerably out of their depth when they are unprepared to examine the evidence from the point of view of the Clever Hans principle as well as other similar pitfalls.

Der Kluge Hans has long ago been laid to rest after a successful theatrical career. But when men of science assume that they cannot be persuaded by bits of language that are not spoken out loud, they should listen more carefully. For from a distance of many decades comes a sound that they should heed; it is the sound of that ubiquitous horse, and he is laughing.

Reflections on Paranormal Communication: A Zetetic's Perspective

MARCELLO TRUZZI

Department of Sociology
Eastern Michigan University
Ypsilanti, Michigan 48197

BY A "ZETETIC PERSPECTIVE" in the title of this paper, I essentially refer to a modernized version of the emphasis on neutral inquiry endorsed by the early Greek followers of the great skeptic Pyrrho of Elis (ca. 365–275 B.C.), who urged that we should suspend our judgements about facts, a position uncommon for many scientists naturally eager to make decisions about the state of the empirical world. I do not offer radical Pyrrhonic skepticism as an epistemology (for such a relativism would make science largely impossible), but I do suggest his posture as a *heuristic* approach likely to bring balance into our debates and one we should take wherever we can afford to suspend judgment. Such marginality is particularly valuable for the sociologist of science, who, after all, is less concerned with the data scientists study than with the scientists and their ways of studying as sociological data themselves. As a sociologist, I can afford tolerance towards eccentric claims made in, say, physics, more than can the physicist who must decide what priority to give to the consideration of such unusual ideas. Despite my advantage in being an outsider looking in on embroiled researchers, however, I would argue that such a zetetic posture should not be limited to the sociologist. By following Pyrrho's advice to suspend judgments by seeking "to be without beliefs, disinclined to take a stand one way or the other, and steadfast in this attitude."[1] the scientist is unlikely to close the door on further investigation. By showing patience and a reluctance to reach any but tentative decisions, rather than showing any strong interest in discrediting extraordinary claims, there is little chance of disobeying C.S. Peirce's central injunction for the scientist that he should do nothing that might block inquiry.[2] As Mario Bunge recently observed, the occasional pressure to suppress scientific dissent "in the name of the orthodoxy of the day is even more injurious to science than all the forms of pseudoscience put together."[3] Though scientists may publicly espouse such norms as disinterestedness,[4] empirical studies of them reveal their behavior is full of human factors[5] that produce vested interests of all sorts in their theories; and this is particularly true today

297

0077-8923/81/0364-0297 $01.75/0 © 1981, NYAS

when science has been gigantically institutionalized and competes for funds, public attention, and social power.[6,7]

By "paranormal communication" I mean simply those forms of communication which, if they exist, are outside the scope of our current scientific explanations. Unlike the older term "supernatural," the term paranormal implies that the phenomena in question are natural but remain theoretically anomalous.[8] Unlike merely abnormal phenomena that represent unusual or bizzare events that are atypical or statistically infrequent but still accounted for by our accepted theories, paranormal phenomena dispute our theories and are themselves in dispute as actual facts largely because they clash with our accepted explanations.[9,10] Historically, paranormal phenomena may become viewed as normal (or merely abnormal) and their existence less questioned if our theories themselves change to allow them entrance as less disputed facts.[11] It is not simply a matter of getting people to recognize the existence of an anomalous event and then getting them to alter our theories to encompass them; the existing theoretical context itself may be a major reason for our failure in accepting the evidence for the anomaly's existence as empirically real. For this reason, the context in which we discuss a paranormal communication is vastly important. Theoretical context can be contradicted by an anomalous event to a greater or lesser degree; thus an alleged event may be more or less paranormal. A communicating ape is less anomalous than a talking dog who in turn may be less anomalous than a talking frog (evidence from princesses notwithstanding). Since extrasensory perception (ESP) would fly in the theoretical face of already anticognitive behaviorism, such events would be far more anomalous (paranormal) for B.F. Skinner[12] than they would be for a modern mind–body dualist like neurophysiologist John Eccles.[13] In similar fashion, the claims of a chimpanzee using language with good grammar is far more outrageous to a follower of Noam Chomsky than they would be to those with, say, a Theosophical world view that easily allows even a horse doing difficult mathematical problems or understanding several languages to seem quite plausible.[14]

It is frequently asserted, largely following the arguments of David Hume against miracles,[15] that extraordinary claims require extraordinary proof.[16] But the point must be remembered that extraordinariness varies by degree and we may disagree about that degree. Therefore, the amount and quality of evidence we demand will significantly vary depending upon our theoretical starting points; and there are great variations in those starting points among those of us here at this conference. Since some of us here would view the plausibility of symbol-minded chimpanzees or infrequent telepathy among humans as more theoretically reasonable than we would the incredible intelligence attributed to Mr. von Osten's superhorse, even the designation of the discussion of ape signing and telepathy as topics within the same context as the Clever Hans demonstrations indicates a degree of conceptual bias, at least a modicum of guilt by association, that may or may not be warranted.

The once common simplified view of science as following fully rational and cumulative development such as that usually held among most positivists (logical and otherwise), has today been greatly undermined by contemporary philosophers and historians of science like Lakatos,[17] Feyerabend,[18] and Kuhn.[19-21] The role of the contextual Weltanschauung and the significance of paradigmatic discontinuities within science is growingly acknowledged. In a book usually not associated with his philosophy of science, Stephen Toulmin[22] has urged us to view arguments as they empirically occur rather than in terms of formal logic. Adapting his idea of a generalized jurisprudence, I would suggest that we view scientific arguments not in terms of their formal deductive logic, but as something more parallel to the adjudication procedure used in law. It might be best, then, to look upon proponents and critics of extraordinary claims as being more like lawyers in a public trial, with history and the remaining scientific community acting as a jury that must weigh the evidence in terms often involving elements like plausibility and reasonableness. Those not directly party to the actual research issues might best act as would an *amicus curiae* or friend of the court, and we should encourage the jury of our peers to thoroughly consider all the evidence without rushing the trial. Most important, we should avoid setting ourselves up as judges when we, at most, can act as foremen for the rest of the jury. It is in this spirit that I intend the rest of my remarks on paranormal communication to be taken, not as those of an advocate but as one seeking to be a friend of the court.[23]

Wittgenstein wrote, "If a lion could talk we could not understand him." David Premack quoted this aphorism in general agreement but suggested that while talking to a lion or a pigeon may be difficult, it might be substantially easier with a chimpanzee.[24] As I survey the discussions that have taken place dealing with paranormal communication, I find more and more that even though people can talk, we can hardly understand them, and they surely frequently misunderstand each other. Positions on these matters seem dominated by largely ascientific ideological positions resting on age-old differences on fundamental philosophical issues in ontology and epistemology. And unfortunately, even though both propoents and critics claim to be guided by common rules of scientific validation, there are obviously broad areas of social negotiation which seem to go on that might be better examined by sociologsits of knowledge instead of logicians.[25] I turn now to briefly consider some of these areas of miscommunication and negotiation.

Critics of extraordinary claims frequently cite Occam's razor as their reason for rejecting such claims in favor of "simpler" explanations. But the proponents of such claims are normally quite aware of the rule of parsimony and remind their critics that parsimony calls for the acceptance of the simplest *adequate* explanation, and they counter their critics by arguing that the simpler explanation being offered (in this case that results are examples of unconscious or deceptional cuing) does not adequately cover the observations reported. Some of these facts not covered may actually be spurious or malobservations, but such "loose ends" are seldom really carefully accounted for by the debunker. At the same time, the proponents may pay improper attention to their critics and misrepresent them, too. The original Clever Hans case is an excellent example. Following Pfungst's meticulous critique of the wondrous horse, his analysis was generally greeted by scientists as the final word on the affair.[26] Some defenders of Clever Hans[27] misrepresented Pfungst's argument, and many proponents of talking animals who have written in this area show little evidence that they have carefully read Pfungst.[28,29] But, alas, the same can also be said for many scientists who have accepted Pfungst's conclusions and now seek to extend them to new claims of paranormal communication. More important, the response to Pfungst by Karl Krall,[30] the later owner of Hans and the other amazing Elberfeld "talking" horses has generally been ignored by those who cite Pfungst as authority; and those who have defended Hans by citing Krall's evidence seem unaware of the rebuttal to Krall by Stefan von Maday.[31] My point here is simply to call attention to the apparent lack of importance given to the details of the evidence in the Clever Hans case. I have found this pattern consistently present in the literature on the paranormal. A good one-sided presentation debunking an extraordinary claim is seldom followed up by detailed analysis and cross-examination of the critic by the proponents of the claim within the orthodox scientific journals. Just as proponents frequently do not read their critics, the critics rarely read the rebuttals of those proponents who did take the trouble. This sort of pattern is, of course, all too frequently the case even within normal science. We need only remember B. F. Skinner's refusal for some time even to read carefully Chomsky's criticism of Skinner's work on verbal behavior, quite aside from any issue of Skinner taking the time to write a detailed rebuttal. More recently,[32] Thomas A. Sebeok dismissed an allegedly cue-free experiment with a dog by G. H. Wood and R. J. Cadoret[33] in part upon the grounds that he did not consider the *Journal of Parapsychology*, in which it appeared, to be a truly "scientific" journal. And, unfortunately, other experiments with "talking" dogs where proponents have argued cuing was precluded by the research design[34,35] were not discredited or individually considered by Sebeok's otherwise excellent review of this literature.[36]

The central point I seek to make is that reviews of the literature on paranormal claims are typically incomplete and sometimes unusually so. Too often (and I want to explicitly exempt Sebeok's from this complaint), instead of acting like scientists interested in discovering the facts wherever they might lead, researchers act more like advocates in a courtroom seeking to discredit witnesses for the opposition by improper but often effective means that will help them win their case. What should be a dialogue between peers who share a common goal of general scientific progress turns into a debate. This is essentially true for those cases where nonscientists and science journalists involve the public rather than the scientific community of experts as their jury. Too frequently scientific issues as well as criminals get tried by the papers and fail to receive due process. Such public exchanges may produce heated interchanges that make for lively reading by outsiders, but it probably does little to produce an atmosphere of mutual respect among scientists that would optimize open communication and willingness to amend past positions taken. And, as Ray Hyman has recently argued,[37] such concern with innoculating the public against what orthodox scientists prematurely view as pathology or pseudoscience may itself lead to a pathological condition within science that is ultimately dysfunctional for all concerned.

In considering paranormal communications, then, I would urge that we recognize (1) a need for us to suspend our judgments until (2) there has been a careful consideration of the intellectual context of the claims and counterclaims which determine the degree of anomaly involved in the claim; (3) a need for us to examine critically all the positions on the evidence including those that fit our preconceptions; (4) a need for us to consider the quality and weight of evidence necessary for us to accept an extraordinary claim proportionate to the degree of extraordinariness that claim involves; and (5) a need for scientists to act more like responsible lawyers, who, as officers of the court of science, should be less concerned with winning debates and should centrally encourage full cross-examination and adherence to the rules of scientific evidence, and seek to assure due process. Following these prescriptions would, I think unpack many of the problems now confronting us in evaluating claims of paranormal communications.

Our ignorance about human and animal communication remains far greater than our knowledge. Our theories are many but our observational base—as large as it has become—remains relatively small. Anomalies abound, and few serious readers of the vast parapsychological literature can emerge without recognition that anomalous scientific results exist even if extrasensory perception (psi) does not.[38] Whether the results in information transfer are due to undetermined sensory leakages or new extrasensory channels remains to be determined, but anomalous forms of communication certainly do take place. Whatever explanations are put forward, the burden of proof in science is on the claimant. This has not always been fully appreciated by critics of the paranormal though they are usually quick to demand proof when a proponent suggests an extraordinary result. Many proponents of paranormal communication have taken up the burden of proof demanded of them, but critics frequently react to such evidence by simply presenting alternative and hypothetical *post hoc* scenarios, including allegations of fraud and gross incompetence by those from whom we would expect better, without offering substantial evidence for their alternative explanations beyond vague claims of parsimony.[39,40,54] Such alternative scenarios may appear quite plausible, and they may be accurate; but where some details are not covered or where new facts are introduced, the burden of proof shifts to the debunker. Duplication through normal means of what appeared to be a paranormal effect does not logically necessitate similarity in the causes. Such duplication should lead to our suspending judgement and certainly will affect our opinions as to what actually may have happened, but it should not form the basis for our making any final conclusions. As scientists we must admit the inconclusive. Particularly important for us to recall is that any truly scientific claim must be stated in a falsifiable manner,[41] and the accusations of fraud by some critics, such as those put forward by C.E.M. Hansel[42] probably are nonfalsifiable.[40] A similar problem may exist for the critics of ape's using language. Quoting Heini Hediger, T. A. Seboek has suggested that the problem of proving leakage by completely eliminating the possibility of the Clever Hans effect may be impossible and "analogous to squaring the circle."[43] Such an indeterminism in finding a condition completely free of the Clever Hans effect would make the extreme case for such a claim a nonfalsifiable alternative claim and therefore pseudoscientific.

A latent, but perhaps the central, issue at this conference has been just what has been meant by the Clever Hans phenomenon in relation to the issues examined. For some of us, and I think this perspective is taken by Robert Rosenthal, the Clever Hans phenomenon represents a notable case of experimenter effect which constitutes a question for science. That is, we need to concern ourselves with ascertaining the mechanisms involved in producing such sources of error. The Clever Hans phenomenon, then, is a puzzle that needs to be explained and fully understood. For others, and I think this perspective is taken by

Thomas Sebeok, the Clever Hans phenomenon is seen not as a question but as an answer. Starting with the phenomenon as a given, it is used to explain away or debunk the effects claimed by those favoring the view that apes have demonstrated the use of language. Formally, Rosenthal and those on his side seem concerned with the Clever Hans phenomenon as a dependent variable (one needing explanation) while Sebeok and his allies seem concerned with it as an independent variable (offering an alternative explanation). To this degree, I would suggest, many of us have simply been talking right past one another.

Though Pfungst's Clever Hans phenomenon can be offered as an alternative explanation accounting for the extraordinary results of the ape studies, it needs to be clearly understood that the actual Clever Hans phenomenon as described by Pfungst is itself highly extraordinary.[14] In fact, if von Osten had merely claimed to train the horse to pick up such small unconscious cues of his interrogators, it seems very likely that scientists of his period would have been skeptical that it had been done. Acceptance of the explanation of the Clever Hans case given us by Pfungst occurred only because the alternative account of a humanly intelligent horse is more extraordinary. We need to remember how truly remarkable the subtle cuing of Hans is supposed to have been. I am reminded of critics of spiritualism who sought to explain away all mediumistic phenomena as merely cases of telepathy. It is of little help to explain away something by invoking causes that themselves need explanation.

Another major question in evaluating claims of paranormal communication is: What constitutes an expert witness? In this area, experimentalists are frequently critical of arm-chair theorists. Given the legal model that I have proposed for adjudicating scientific issues, the proposal by Ron Westrum[44] that science adapt a criterion for expert witnesses similar to that used by courts is especially attractive. Though there is truth in the adage that one need not be a chicken to recognize a rotten egg, there are many incompetent self-styled experts on paranormal communication, and this is true for both parapsychologists and their critics. A special problem here is the proper role for specialized experts who are not scientists. Examples would include magicians in the case of parapsychological research, animal trainers in research on extraordinary animal behavior, and polygraph specialists for studies investigating reports on communication with extraterrestrials from unidentified flying objects. Such expertise should certainly be incorporated into scientific investigations, but, in the final analysis, research judgments must be made by the scientists themselves. The useful role of such nonscientist experts should not be underestimated, for they can be important sources of information and help for the scientist. But their independent efforts should not be confused with scientific research, for such experts sometimes contradict one another, they usually know little about scientific research methods, and they can have nonscientific vested interests in the experimental outcomes, such as monetary challenges to the subjects, the desire for personal publicity, and so on. The role of the magician is particularly difficult in these matters, and there is a substantial literature on the subject.[45-47] As a specialist in deception, the magician has relevant expertise, but his public role usually associates him with debunking of the paranormal rather than with a dispassionate approach. Too frequently, the presence of a magician can turn what should have been an impartial experiment into a challenge relationship where either the magician or the alleged psychic is viewed as winning a contest. This can result in widespread publicity that is likely not only to discourage future subjects but may also discourage further impartial research. In fact, the researcher may become publicly ridiculed for having wasted his time on the research that produced the negative result.

Ray Hyman has outlined the role of "hit men" in science who are sent out or encouraged by orthodox critics to discredit—often unfairly but still quite effectively—proponents of the paranormal since they feel it necessary to innoculate the lay population against the pathological influence of the purported pseudosciences.[37] Magicians, especially since Harry Houdini,[48] have frequently played this role and are often viewed as heroes by critics but blackguards by those defending the paranormal.

The self-fulfilling[49,50] or self-altering[51] prophecy is particularly interesting in this context. Critics who contend that paranormal communications by both animals and humans actually result from unconscious experimenter effects,[26,52] seem to accept such an approach—even without knowing the exact mechanisms that produce the effects—as essentially debunking. But when parapsychologists talk about sheep–goat (believer-disbeliever) effects in psi experiments,[53] they have been ridiculed for introducing such a notion. Without falsifiable specification of the mechanism involved, the claim of unconscious experimenter effects may itself border on the paranormal, if not the pseudoscientific, and raises many questions. For example, Pfungst[14] says he found he was unable to keep himself from cuing Clever Hans even when he sought to control himself.[14] Yet, how could he then be certain he had in fact cued the horse at all? This seems especially unlikely in light of the supposed failures by Hans with those who are skeptical of his abilities and who Pfungst concluded did not send any cues. Clearly, the hypothesis of the self-fulfilling prophecy is a neutral one since the mechanisms through which it might operate could be either normal or paranormal. Just as proponents have been accused of having a "will to believe," critics have been accused of having a "will to disbelieve."

In the case of magicians, this situation has an interesting additional problem. Many alleged psychics, e.g., Uri Geller, have refused to participate in any experiment with a magician present, on the grounds that such disbelievers have a negative effect on the psychic's power. Critics see this simply as a ruse to avoid being caught in trickery by a fellow professional deceptionist who is a disbeliever. This stereotype of the magician as disbeliever may be true of some magicians, but a survey I recently conducted among the members of the Psychic Entertainers Association, the leading organization of professional pseudopsychics and conjurors most specialized and expert in this form of deception, indicates that the vast majority of these mentalists actually believe in the reality of the paranormal communication (ESP) they daily simulate. This suggests that the highly visible magician debunkers are not necessarily representative of those specializing in mentalism (and most of whom hold the debunking magicians in great disdain). More important, this survey indicates that it should readily be possible to conduct experiments with alleged psychics in collaboration with expert magicians who can help construct controls in research designs but who also believe in ESP. The work done with Clever Hans indicates that the expert animal trainers that were brought in by the scientific commissions failed to discern the cuing that Pfungst later claimed had been consistently present. Though no survey has yet been conducted on the attitudes of animal trainers, I suspect that they, like the mentalists I have been studying, are not so negative on the possibilities of paranormal communication by animals as are the scientific critics. This is another good example of the need for empirically checking our preconceptions, for most of my fellow skeptics had led me to expect negative opinions on ESP from those magicians most skilled in simulating it.

A final problem relating to magicians and paranormal communication concerns the magicians' ability to duplicate paranormal effects through natural means. Critics have stressed the need for replicability in psi research. On the surface, this seems a simple and reasonable demand, but careful examination reveals a complex problem. The work of Harry Collins[55,56] has demonstrated that replicability is itself a largely socially negotiated agreement on when some replication is truly a replication. Parapsychologists have insisted that there is a relatively high level of replication in psi research and point out that complete reproducibility of a variable upon demand is an accepted impossibility in many other science areas.[57] Critics of psi research demand strict criteria for replication from the parapsychologists, but critics may have overlooked the fact that a magician's duplication of a psi effect also must be assessed as a replication. Yet the critics allow very weak criteria for replication in this area since the conditions of the magician are seldom very similar to those claimed for the allegedly psychic event being duplicated. In this sense, critics may be using a "rubber ruler" just as they have been accused of using one when comparing children and apes.[58]

These, then, have been some of my reflections on paranormal communication. Like all zetetics, I have produced more new questions than answers. Though I began my investigations completely convinced by Pfungst's arguments, I am now wary of extending them to new areas and find myself—like Charles Richet[59] who wanted to dispel the wonders of Clever Hans for very different reasons—forced to suspend judgment even about Hans. I offer you doubts rather than conclusions. I can only echo the words of the great agnostic T.H. Huxley who has been quoted as saying, "God give me the strength to face a fact, though it slay me."[27] I may personally still disbelieve in talking dogs, mindreading humans, and real conversation with gorillas, but such beliefs must be distinguished from those things about which I have scientific knowledge. I hope my remarks elicit your zetetic spirit as well.

REFERENCES

1. STOUGH, C. L. 1969. Greek Skepticism: A Study in Epistemology. p. 26. Footnote 21. Univ. of California Press, Berkeley, Calif.
2. HARTSHORNE, C. & P. WEISS, Eds. 1965. Collected Papers of Charles Sanders Peirce. Vol. 1: 50. Harvard University Press, Cambridge, Mass.
3. BUNGE, M. 1980. Comments on Ray Hyman's paper on pathological science. Zetetic Scholar (6): 45-46.
4. MERTON, R. K. 1973. In The Sociology of Science: Theoretical and Empirical Investigations. N. Storer, Ed. Papers 12-14. Univ. of Chicago Press, Chicago, Ill.
5. MAHONEY, M. J. 1979. Psychology of the scientist: An evaluative review. Soc. Studies Sci. 9: 349-375.
6. FEYERABEND, P. 1978. Science in a Free Society. NLB/Schocken Books, New York, N.Y.
7. NIEBURG, H. L. 1970. In the Name of Science. Quadrangle Books, Chicago, Ill.
8. TRUZZI, M. 1978. Editorial: A word on terminology. Zetetic Scholar (2): 64-65.
9. WESTRUM, R. & M. TRUZZI. 1978. Anomalies: A Bibliographic introduction with some cautionary remarks. Zetetic Scholar (2): 69-78.
10. HUMPHREYS, W. C. 1968. Anomalies and Scientific Theories. Freeman, Cooper and Co, San Francisco, Cal.
11. TRUZZI, M. 1979. Discussion: On the reception of unconventional scientific

claims. *In* The Reception of Unconventional Science. S.H. Mauskopf, Ed. pp. 125–137. Westview Press, Boulder, Col.

12. SKINNER, B. F. 1937. Is sense necessary? Saturday Rev. Lit. **26** (Oct. 9): 5–6.
13. POPPER, K. R. & J. C. ECCLES. 1977. The Self and Its Brain: An Argument for Interactionism. Springer-International, New York, N.Y.
14. PFUNGST, O. 1911. Clever Hans: The Horse of Mr. von Osten. Henry Holt and Co., New York, N.Y.
15. HUME, D. 1902 [orig. 1748]. *In* An Enquiry Concerning Human Understanding. L.A. Selby-Bigge, Ed. Section 10. Oxford Univ. Press, London.
16. TRUZZI, M. 1978. On the extraordinary: An attempt at clarification. Zetetic Scholar (1): 11–19.
17. LAKATOS, I. 1978. The Methodology of Scientific Research Programmes; Philosophical Papers. J. Worral & G. Currie, Eds. Vol. 1. Cambridge Univ. Press, New York, N.Y.
18. FEYERABEND, P. 1975. Against Method. New Left Books, London.
19. KUHN, T. S. 1970. The Structure of Scientific Revolutions. 2nd edit. Univ. of Chicago Press, Chicago, Ill.
20. SUPPE, F., Ed. 1977. The Structure of Scientific Theories. 2nd edit. Univ. of Illinois Press, Urbana, Ill.
21. BROWN, H. I. 1979. Perception, Theory and Commitment: The New Philosophy of Science. Univ. of Chicago Press, Chicago, Ill.
22. TOULMIN, S. 1960. Uses of Argument. Cambridge Univ. Press, New York, N.Y.
23. TRUZZI, M. 1978. Editorial. Zetetic Scholar (1): 2, 34.
24. PREMACK, D. 1978. On the abstractness of human concepts: Why it would be difficult to talk to a pigeon. *In* Cognitive Processes in Animal Behavior. S.H. Hulse, H. Fowler & W.K. Honig, Eds. pp. 423–451. Larence Erlbaum Associates, Hillsdale, N.J.
25. WALLIS, R., Ed. 1979. On the Margins of Science: The Social Construction of Rejected Knowledge. Sociological Review Monograph 27. Univ. of Keele, Keele, Stratfordshire, England.
26. ROSENTHAL, R. 1965. Introduction. *In* Clever Hans (The Horse of Mr. von Osten) by Oskar Pfungst. R. Rosenthal, Ed. pp. ix–xlii. Holt, Rinehart and Winston, New York, N.Y.
27. ANASPACHER, L. K. 1947. Challenge of the Unknown: Exploring the Psychic World. p. 49. Hill and Wang, New York, N.Y.
28. MAETERLINCK, M. 1914. The Unknown Guest. Dodd, Mead and Co., New York, N.Y.
29. CARRINGTON, H. 1919. Modern Psychical Phenomena: Recent Researches and Speculations. pp. 232–249. Dodd, Mead and Co., New York, N.Y.
30. KRALL, K. 1912. Denkende Tiere. Beitrage zur Tierseelenkunde auf Grundeigener Versuche. Friedrich Engelmann, Leipzig. Germany.
31. VON MADAY, S. 1914. Gibt es denkende Tiere? Eine Entgegnung auf Krall's "Denkende Tiere." Wilhelm Engelmann, Leipzig, Germany.
32. SEBEOK, T. A. 1979. Response to letter from James W. Davis. Zetetic Scholar (5) 2–3.

33. Wood, G. H. & R. J. Cadoret 1958. Tests of Clairvoyance in a Man-Dog Relationship. J. Parapsychol. **22:** 29–39.
34. Bechterev, W. 1949. "Direct influence" of a person upon the behavior of animals. J. Parapsychol. **13:** 166–176.
35. White, R. A. 1964. The investigation of behavior suggestive of ESP in dogs. J. Am. Soc. Psychical Res. **58:** 250–279.
36. Sebeok, T. A. 1979. Close encounters with canid communication of the third kind. Zetetic Scholar (3/4): 3–20.
37. Hyman, R. 1980. Pathological science: Towards a proper diagnosis and remedy. Zetetic Scholar (6): 31–41.
38. Wolman, B. B. 1977. Handbook of Parapsychology. Van Nostrand Reinhold Co., New York, N.Y.
39. Palmer, J. 1978, Extrasensory perception: Research findings. *In* Adv. Parapsychol. Res. 2(Extrasensory Perception): 59–243 (especially 64–73).
40. Martin, M. 1979. The problem of experimenter fraud: A re-evaluation of Hansel's critique of ESP experiments. J. Parapsychol. **43:** 129–139.
41. Popper, K. 1959. The Logic of Scientific Discovery. Hutchinson, London.
42. Hansel, C. E. M. 1980. ESP and Parapsychology: A Critical Re-evaluation. Prometheus Books, Buffalo, N.Y.
43. Sebeok, T. A. & J. Umiker-Sebeok 1979. Performing animals: Secrets of the trade. Psychology Today **13** (Nov.): 91.
44. Westrum, R. 1976. Scientists as experts: Observations on "Objections to Astrology." The Zetetic **1**(1): 34–46.
45. Dingwall, E. J. 1921. Magic and mediumship. Psychic Res. Quarterly **1**(3): 206–219.
46. Webb, J., Ed. 1976. The Mediums and the Conjurors. Arno Press, New York, N.Y.
47. Bayless, R. 1972. Experiences of a Psychical Researcher. pp. 208–228. University Books, New Hyde Park, N.Y.
48. Houdini, H. 1924. A Magician Among the Spirits. Harper & Bros., New York, N.Y.
49. Merton, R. K. 1957. Social Theory and Social Structure. Revised edit. p. 423. The Free Press, Glencoe, Ill.
50. Jones, R. A. 1977. Self-Fulfilling Prophecies: Social, Psychological, and Physiological Effects of Expectancies. Lawrence Erlbaum Associates, Hillsdale, N.J.
51. Henshel, R. L. 1976. On the Future of Social Prediction. p. 12. Bobbs-Merrill, Indianapolis, Ind.
52. Rosenthal, R. 1976. Experimenter Effects in Behavioral Research. Enlarged edit. Irvington (Halsted), New York, N.Y.
53. Schmeidler, G. & R. McConnell. ESP and Personality Patterns. Yale Univ. Press, New Haven, Conn.
54. Pinch, T. J. 1979. Normal explanations of the paranormal: The demarcation problem and fraud in parapsychology. Soc. Studies Sci. **9:** 329–348.
55. Collins, H. M. 1978. Science and the rule of replicability: A sociological study of scientific method. Paper delivered at the Ann. Meet. of the Am. Assoc. for the Advancement of Sci., Washington, D.C.

56. COLLINS, H. M. 1976. Upon the replication of scientific findings: A discussion illuminated by the experiences of researchers into parapsychology. Paper read at the Society for Social Studies of Science — International Sociological Association Conference, Cornell University.
57. HONORTON, C. 1975. Has science developed the competence to confront claims of the paranormal? Presidential address given to the Parapsychological Assoc.
58. BAZAR, J. 1988. Catching up with the ape language debate. Am. Psychol. Assoc. Monitor (Jan.): 4–5, 47.
59. RICHET, C. 1923. Thirty Years of Psychical Research. pp. 240–244. Macmillan, New York, N.Y.

Index of Contributors